Integration of Public Health with Adaptation to Climate Change

Lessons learned and new directions

The long lifetime of the anthropogenic greenhouse gases in the atmosphere and the inherent inertia in the climate system imply that global climate change will continue for decades after effective implementation of mitigation measures. Overall, the negative health impacts of global climate change are anticipated to significantly outweigh positive impacts. One challenge is to find effective ways to reduce the potential future impacts of climate change in the context of other pressing needs. This process can be informed by an examination of the history of public health interventions, as well as by evaluation of the current capacity to cope with the adverse consequences of climate variability and change. Public health has more than 150 years of experience in implementing policies and measures to increase human well-being. Case studies in public health are explored to identify what modification to public health systems will be necessary to enhance their capacity to adapt to climate variability and change, and what public health has learned about the process of developing and implementing effective intervention policies and measures. These lessons are of interest to climate adaptation specialists in developed and developing countries.

Integration of Public Health with Adaptation to Climate Change

Lessons learned and new directions

Edited by
Kristie L. Ebi, Joel B. Smith, and
Ian Burton

Taylor & Francis
Taylor & Francis Group
LEIDEN / LONDON / NEW YORK / PHILADELPHIA / SINGAPORE

Typesetting: Charon Tec Pvt. Ltd, Chennai, India
www.charontec.com
Printed by: Antony Rowe Ltd, Chippenham, Wiltshire, UK

Library of Congress Cataloging-in-Publication Data
A catalog record for this book has been requested.

Published by: Taylor & Francis, Leiden, The Netherlands,
a member of Taylor & Francis Group plc.
www.tandf.co.uk

ISBN 90 5809 686 6

We dedicate this book to John Weyant. John's vision, expertise, creativity, and enthusiasm in running the Energy Modeling Forum, particularly the annual meetings in Snowmass, Colorado, created the opportunity for the multidisciplinary discussions that led to this volume.

Contents

List of tables

List of figures

List of contributors

Martha Anker formerly with the Department of Communicable Disease Surveillance and Response, World Health Organization, CH-1211 Geneva 27, Switzerland. e-mail: anker@geneva-link.ch.

Ian Burton Scientist Emeritus, Meteorological Service of Canada, University of Toronto, 72 Coolmine Road, Toronto M6J 3E9, Canada. e-mail: ian.burton@ec.gc.ca.

Kristie L. Ebi Exponent, 1800 Diagonal Road, Suite 300, Alexandria, VA 22314, USA. e-mail: kebi@exponent.com.

Andrew K. Githeko Kenya Medical Research Institute, Centre for Vector Biology and Control Research, Climate and Human Health Research Unit, PO Box 1758, Kisumu, Kenya. e-mail: agitheko@kisian.mimcom.net; andrew@kenya.cdc.gov.

Duane J. Gubler Asia-Pacific Institute of Tropical Medicine and Infectious Diseases, Leahi Hospital, 3675 Kilauea Avenue, Honolulu, HI 96816, USA. e-mail: dgubler@hawaii.edu. Formerly with the National Center of Infectious Diseases, CDC, Fort Collins, CO, USA.

Samuel Hitz formerly with Stratus Consulting Inc., 3620 Helms Avenue, Culver City, CA 90232, USA. e-mail: samhitz@comcast.net.

Heather Jones-Otazo Institute for Environmental Studies, University of Toronto, 45 St. George St., Room 209, Toronto, Ontario M5S 1A1, Canada. e-mail: heather.jones@utoronto.ca.

Christina Koppe Deutscher Wetterdienst, Business Unit Human Biometeorology, Stefan-Meier Strasse, 4, D-79104 Freiburg, Germany. e-mail: christina.koppe@dwd.de.

R. Sari Kovats Centre on Global Change and Health, London School of Hygiene and Tropical Medicine, Keppel Street, London WC1E, United Kingdom. e-mail: Sari.kovats@lshtm.ac.uk.

Stephen H. Linder The University of Texas, Health Science Center at Houston, School of Public Health, PO Box 20186, Houston, TX 77225, USA, e-mail: stephen.h.linder@uth.tmc.edu.

Robyn M. Lucas National Centre for Epidemiology and Population Health, The Australian National University, Canberra 0200, Australia. e-mail: robyn.lucas@anu.edu.au.

Abdel Maarouf Adaptations and Impacts Research Group, Environment Canada, 4905 Dufferin St., Downsview, Ontario M3H 5T4, Canada. e-mail: abdel.maarouf@ec.gc.ca

A.J. McMichael National Centre for Epidemiology and Population Health, The Australian National University, Canberra 0200, Australia. e-mail: director.nceph@anu.edu.au.

David Mills Stratus Consulting Inc., PO Box 4059, Boulder, CO 80306-4059, USA. e-mail: dmills@stratusconsulting.com.

Joel Scheraga U.S. Environmental Protection Agency, Ariel Rios Building, 1200 Pennsylvania Avenue, NW, MC 8601N, Washington, DC 20460, USA. e-mail: scheraga.joel@epa.gov.

Michael Sharpe Senior Advisor to the Assistant Deputy Minister, Health Environments and Consumer Safety Branch, Health Canada, 5th Floor, Holland Cross, Tower B, 1600 Scott Street, Ottawa, Ontario K1A 0K9, Canada. e-mail: msharpe@hc-sc.gc.ca.

Clive Shiff Johns Hopkins University, School of Hygiene & Public Health, 615 N. Wolfe Street, Baltimore, MD 21205, USA. e-mail: cshiff@jhsph.edu.

Joel B. Smith Stratus Consulting Inc., PO Box 4059, Boulder, CO 80306-4059, USA. e-mail: jsmith@stratusconsulting.com.

Roger Street Adaptations and Impacts Research Group, Environment Canada, 4905 Dufferin St., Downsview, Ontario M3H 5T4, Canada. e-mail: roger.street@ec.gc.ca.

Philip Weinstein School of Population Health, University of Western Australia, M431, 35 Stirling Highway, Crawley WA 6009, Australia. e-mail: philip.weinstein@uwa.edu.au.

Mark L. Wilson Department of Epidemiology and Global Health Program, School of Public Health, The University of Michigan, Ann Arbor, MI 48109-2029, USA. e-mail: wilsonml@umich.edu.

Rosalie E. Woodruff National Centre for Epidemiology and Population Health, The Australian National University, Canberra 0200, Australia. e-mail: rosalie.woodruff@anu.edu.au.

Alistair Woodward School of Population Health, University of Auckland, Private Bag 92019, Auckland, New Zealand. e-mail: a.woodward@auckland.ac.nz.

Gary Yohe John E. Andrus Professor of Economics, Wesleyan University, Middletown, CT 06459, USA. e-mail: gyohe@wesleyan.edu.

Acknowledgements

Producing a book requires more than the scientific expertise of the authors. In particular, producing a readable and coherent book requires expert technical editing. We are fortunate to have had the outstanding editing skills of Christina Thomas, formerly of Stratus Consulting. Christina helped structure the book, edited the entire book, and guided us throughout the process. She also worked closely with each chapter author. Sam Hitz, formerly of Stratus Consulting, now an independent consultant, helped in structuring the book, managing the process, and organizing the meeting of authors in Boulder, Colorado in November, 2002. Shiela DeMars, formerly of Stratus Consulting, also deserves our thanks for helping to manage the process of producing the book and making arrangements for the authors' meeting. Finally, we wish to thank Diane Blagusz and Erin Miles of Stratus Consulting for pulling the manuscript together.

Without the financial support of the Electric Power Research Institute and the cofunding of the U.S. Environmental Protection Agency's Office of Research and Development, specifically its Global Change Research Program and the National Center for Environmental Assessment, this book would not have been possible. We'd also like to thank the Energy Modeling Form for the unparalleled opportunities provided.

Preface

The health of a population is an integrating measure of all factors affecting society. Globally, the challenges to maintaining population health are daunting. Billions of people, particularly in developing countries, lack adequate nutrition, access to clean water and a viable public health system. Many of these countries face the additional challenges of diseases such as HIV/AIDS, tuberculosis and malaria, which consume economic and human capital resources. The worldwide burden of disease is large, with significant negative impacts on the quality of life, including economic productivity. It will be a continuing task to overcome these obstacles. Climate change is an additional challenge that could work for or against efforts to control climate-sensitive diseases.

In one sense, adaptation to climate change is not new—climate has always been changing. The record of prehistoric human life is a record of adaptation to a more slowly changing climate via accommodation or migration. Partly as a result of this evolution, human beings have a wide biological tolerance for climate and have considerable capacity to "acclimatize". There is a need to adapt more quickly to climate change because anthropogenic forcing is leading to rapid change. The new challenge is to develop and deploy a capacity for more rapid social and technological adaptation[1], and to ensure that this capacity is universally and equitably distributed in the face of global environmental change. The history of adaptation offers limited lessons for the challenges we face because some of the adaptations used by primarily hunter/gather societies are not available to today's societies with higher population density and fixed infrastructure. In this sense, adaptation to climate change is new in the conscious, planned, anticipatory approach being proposed by the climate community.

The promotion of adaptation is important to both the public health and climate change communities. The global community of public health scholars and practitioners is being challenged to take into account the added threats from climate change. At the same time, the climate change community is attempting to come to terms with the fact that anthropogenic forcing of climate change cannot be sufficiently reduced at a rate fast enough to prevent significant impacts in the near and medium term. How severe the impacts will be depends to a significant degree on the capacity to adapt and its effective deployment. The box describes assumptions about adaptation that underlie the themes in this book.

1 Although the term "adaptation" is not in common use in public health, many public health protection and promotion activities are analogous to activities that are labeled as adaptation by the climate change and development communities.

Assumptions about adaptation

1 A primary assumption is that adaptation matters, that informed responses to rapid climate change will increase in importance to policymakers and the public, as evidence regarding unacceptable impacts continues to accumulate. Adaptation, in this context, signals the possibility of desirable, timely courses of action in the face of such impacts.
2 Adaptation should focus on pragmatic strategies, policies and measures that aim to prevent possible adverse impacts and to take advantage of opportunities that arise. This shifts the question from whether impacts from climate change will occur in the near term, and whether some portion will be unacceptable, to the hows of achieving some control over the more dire consequences expected.
3 Strategies, policies and measures are needed to adapt to the increased uncertainty that climate change is bringing. The climate change issue is characterized by a high level of confidence that climate is going to change (indeed, is already changing), but high uncertainty about the rate of change and the actual changes that will occur in specific places. In other words, there is good global sense but a poor local sense. Adaptation to uncertainty will be needed as much as to climate change itself. Individuals, communities and nations will need to decide how much risk they are willing to accept under particular conditions of uncertainty.

The principal purpose of this book is to contribute to the evolving dialogue between the climate change and public health communities over adaptation. Accordingly, two lines of questioning were set out for discussion in the following chapters.

1 How can public health practitioners and policy makers take climate variability and change into account in strategies, policies and measures to reduce the potential added burden of disease? Strategies for adapting to climate change might necessitate a diverse range of modifications to public health systems. These changes can be thought of in three categories.
   - Lessons learned but forgotten or not applied. Lessons learned in one part of the world may be applicable to other public health systems, or historical lessons may be informative for current and future issues. For instance, the re-emergence of a number of diseases has resulted from no longer aggressively pursuing once-successful vector control programs, demonstrating the importance of maintaining institutional capacity and vigilance. Knowledge gleaned from experience must be coupled with political and institutional will if change is to be widespread or lasting.
   - Win/win or no-regrets strategies. These might include improvements partially motivated by climate that could enhance efficiency or advance sustainable development goals, and thus render public health systems more capable of confronting challenges.
   - Changes that confront new risks posed by climate change. Public health systems in the future could face novel threats, such as a change in the spatial range of a disease or the introduction of a new disease, which would require deliberate and planned adaptation. In some instances, climate change might demand innovative and/or large-scale modifications to existing approaches.

2 What lessons can be drawn from the long history of public health managing external environmental and other threats that can be applied to adaptation to climate variability and change? Public health has more than 150 years of experience in trying to increase human well-being in the face of significant challenges. Some of this experience may be of interest in the context of adaptation to climate change. Public health has been successful in increasing human well-being by lowering the burden of particular diseases, such as the eradication of smallpox, vaccinations for many major infectious diseases, etc. The long history of public health allows the exploration of how interventions have played out over time.

These questions form the theme that runs throughout the book. The first chapters introduce the basic concepts. The subsequent chapters focus on case studies in public health that may have relevance to adaptation to climate change. The final chapters turn to the policy implications for adaptation to climate change.

Foreword

I welcome this timely analysis of adaptive response to climate change in the human health sector. Without adaptation, a wide range of health impacts can be expected with the projected changes in temperature and precipitation, including deaths, diseases, and injuries caused by changes in the distribution of disease vectors and possible increases in extreme weather events such as droughts and cyclones. These impacts will be in addition to the acute and growing health problems faced by a large part of the human population, especially among the poor. According to the United Nations Development Programme's Human Development Report 2003, 10 million children die every year of preventable illnesses. Around the world 42 million people are afflicted with HIV/AIDS, the overwhelming majority of whom (39 million) live in developing countries. Tuberculosis kills up to 2 million people annually, and some 1 million children die of malaria every year. More than 1 billion people in developing countries lack access to safe water and 2.4 billion people lack access to improved sanitation. The current situation is intolerable and is likely to become worse as several regions of the world suffer from the growing impacts of climate change. For example, increased drought and floods would only increase the extent and incidence of diseases from inadequate and unsafe drinking water as well as lack of proper sanitation. Climate change is, therefore, a major reason for concern in the context of attendant adverse health impacts for a large part of the world's population.

The long lifetime of the anthropogenic greenhouse gases currently in the atmosphere and the inherent inertia in the climate system mean that global climate will continue to change for decades or even centuries to come, even with effective implementation of mitigation measures. Thus, adaptation is crucial for reducing the potential negative impacts of climate change. Countries, regions, communities, and individuals will have to adapt if there is to be any hope of reducing the global burden of disease.

If public policy and action are to successfully ameliorate adverse health conditions imposed by climate change, then we have to not only create the necessary infrastructure for successful efforts, but also directly involve communities who have considerable experience in adaptation to past natural changes. There needs to be regular and intensive interaction between policy-makers and communities at the grassroots level.

There is little literature on the sharing of lessons between the public health and climate adaptation communities. This book contributes to that dialogue by facilitating an exchange of lessons learned and potentially inspiring new approaches for confronting the impacts of climate change. Furthermore, this discourse points a way toward a stronger collaborative engagement among all the sectors concerned with adaptation to climate change.

R.K. Pachauri, Ph.D.
Chair, Intergovernmental Panel on Climate Change (IPCC)
Director, The Energy and Resources Institute (TERI)

1 Adaptation to climate variability and change from a public health perspective

Kristie L. Ebi, Joel B. Smith, Ian Burton, and Samuel Hitz

ABSTRACT: Lessons learned from more than 150 years of public health research and intervention can provide insights to guide public health professionals and institutions as they design specific strategies, policies, and measures to reduce potential negative impacts and exploit potential opportunities of climate variability and change. In many cases, continuing and augmenting established interventions may be the most effective approach to reducing vulnerability and increasing adaptive capacity. Interventions often will be specific to particular health outcomes. For some cases, new strategies, policies, and measures will be needed. This chapter provides an overview of some basic principles of public health, and then discusses some lessons learned in the practice of public health that may have relevance for other sectors as they design approaches to reduce vulnerability to climate change.

1. INTRODUCTION

The Earth's climate has been warming since the middle of the 19th century. Although the warming before the middle of the 20th century was most likely not the result of human causes, the Intergovernmental Panel on Climate Change (IPCC) recently concluded that most of the warming since 1950 is the result of anthropogenic emissions of greenhouse gases [1]. Furthermore, the IPCC concluded that global average temperatures are likely to rise by 1.4 to 5.8°C between 1990 and 2100 [1]. This projected rate of warming is much larger than the changes observed during the 20th century and, if it occurs, will be without precedent during at least the last 10,000 years. Global climate change will not be spatially uniform, and is expected to include changes in the hydrologic cycle and changes in extreme events.

One of the main concerns about climate variability and change is how they may alter human risks. Figure 1.1 summarizes some possible health consequences of climate change [2]. While there clearly will be deleterious effects, there may also be some beneficial impacts. For example, although higher average temperatures will most likely result in more extreme high temperatures, which can increase heat stress, warmer winters may reduce deaths associated with cold temperatures. In addition, there is the possibility that climate change will enable new health risks to be introduced, most likely in the form of emerging or re-emerging diseases.

The vulnerability of a particular population to climate change-induced impacts will depend on the degree to which individuals and systems are susceptible to, or unable

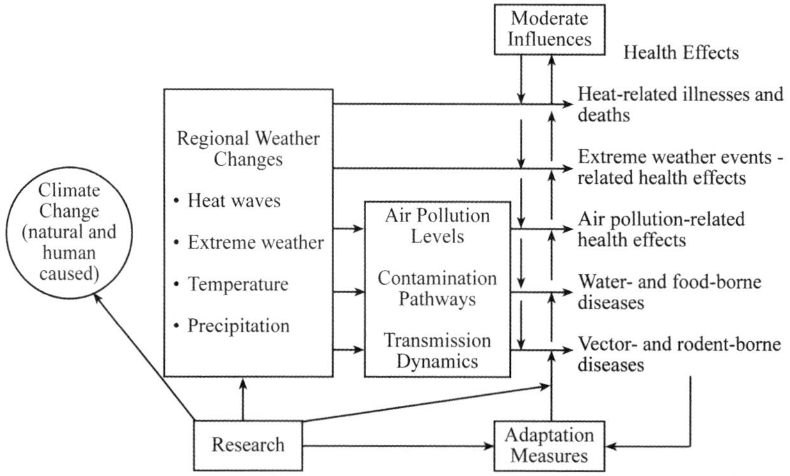

Figure 1.1 Potential health effects of climate variability and change (adapted from Ref. 2)

to cope with, the adverse effects of climate change as well as variability and extremes [3]. Vulnerability is a function of the following:

- The exposure to the weather or climate-related hazard (includes the character, magnitude, and rate of climate variation, e.g., increase in extreme heat).
- The extent to which health, or the natural or social systems on which health outcomes depend, is sensitive to changes in weather and climate – the exposure-response relationship.
- The adaptation measures and actions in place to reduce the burden of a particular adverse health outcome (e.g., the penetration of air conditioning within an urban area will help to reduce the impacts of heat extremes). The effectiveness of these interventions partially determines the exposure-response relationship.

Populations, subgroups, and systems that cannot or will not adapt are more vulnerable, as are those who are more susceptible to weather and climate changes. Understanding a population's capacity to adapt to new climate conditions is crucial to the realistic assessment of the potential health and other impacts of climate change [4]. In general, the vulnerability of a population to a health risk depends on the level of material resources, effectiveness of governance and civil institutions, quality of public health infrastructure, access to relevant local information on extreme weather threats, and existing burden of disease [5]. These factors are not uniform across a region or nation; rather, there are geographic, demographic, and socioeconomic differences.

Working Group II of the IPCC [6] concluded in their Third Assessment Report that:

> advances have been made since previous IPCC assessments in the detection of change in biotic and physical systems, and steps have been taken to improve the understanding of adaptive capacity, vulnerability to climate extremes, and other critical impact-related issues. These advances indicate a need for initiatives to begin designing adaptation strategies and building adaptive capacities.

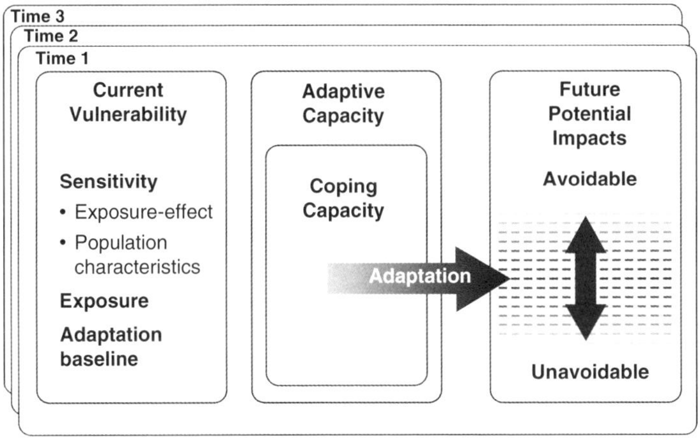

Figure 1.2 Public health framework for vulnerability and adaptation

Public health professionals and institutions should be involved in designing and implementing these initiatives, because the underlying goal of the strategies proposed is to protect and enhance human well-being.

Adaptive capacity is a theoretical construct that describes the general ability of institutions, systems, and individuals to adjust to potential damages, to take advantage of opportunities, or to cope with the consequences [3]. Adaptive capacity encompasses coping capacity (what could be implemented now to minimize potential damages due to climate variability and change) and the strategies, policies, and measures that have the potential to expand future coping capacity. Specific adaptation actions arise from the coping capacity of a community, nation, or region. The relationships between vulnerability and adaptation are shown in Fig. 1.2. The primary goal of building adaptive capacity is to reduce future vulnerability to climate variability and change. Similarly, the goals of public health are to reduce population vulnerability to adverse conditions and exposures. Specifically, public health aims to reduce the amount of current and future avoidable disease, premature death, and disease-related discomfort and disability in a population.

Lessons learned from more than 150 years of public health research and intervention can provide insights to guide the design and implementation of effective and efficient strategies, policies, and measures to reduce the potential negative impacts and exploit the opportunities of climate variability and change. In addition, lessons from public health may be applicable to other sectors that will be affected by climate change. This chapter first provides an overview of public health, and then discusses some lessons learned in the public health sector that may have relevance within public health and for other sectors as they design approaches to reduce vulnerability to climate variability and change.

2. OVERVIEW OF PUBLIC HEALTH

Public health is the science and art of preventing disease, prolonging life, and promoting health through the organized efforts of society [7]. "It is the combination of sciences, skills, and beliefs that is directed to the maintenance and improvement of the health of all people through collective or social actions. The programs, services, and institutions involved emphasize the prevention of disease and the health needs of the population as a whole. Public health activities change with changing technology and social values, but the goals remain the same: to reduce the amount of disease, premature death, and disease-produced discomfort and disability in the population. Public health is thus a social institution, a discipline, and a practice" [8]. This definition highlights the different conceptualizations of public health: one is the institutions involved, such as the World Health Organization and the US Centers for Disease Control and Prevention; and the other includes the professionals who study causes of disease and interventions to improve population health.

2.1 Guiding principles

The practice of public health is guided by four ethical principles [9]: respect for autonomy, nonmaleficence, justice, and beneficence. Respect for autonomy suggests a concern for human dignity and freedom, and the rights of individuals to make choices and decisions for themselves rather than having others decide for them. Nonmaleficence is the principle of not intentionally harming, derived from the ancient medical maxim: first, do no harm. Justice in this sense means social justice – fairness, equity, and impartiality. Beneficence is the principle of doing good, and is the dominant ethical principle of public health. Public health practitioners and institutions believe in providing benefits, and they have an impressive record of doing so.

Public health is frequently faced with situations where it is not possible to uphold these principles simultaneously. One example is quarantine to control the spread of specific communicable diseases (respect for autonomy and justice versus beneficence). In general, the need to protect society is a higher imperative in public health than are the rights of an individual. Public health practitioners believe that as long as the community perspective and the pursuit of common good remain in force, actions can occasionally be taken that limit individual autonomy or infringe on justice.

2.2 Prevention is a multitiered approach

Measures to reduce disease and save lives are categorized as primary, secondary, or tertiary prevention. Primary prevention is the "protection of health by personal and community wide efforts" [8]. It aims to prevent the onset of disease in an otherwise unaffected population (e.g., supply of bed nets to all members of a population at risk of exposure to malaria). When primary prevention is not feasible, secondary prevention includes "measures available to individuals and populations for the early detection and prompt and effective intervention to correct departures from good health" [8]. It focuses on preventive actions taken in response to early evidence of health impacts (e.g., strengthening disease surveillance and responding adequately to disease outbreaks such as the West Nile virus outbreak in the United States). As discussed in Chapter 10, surveillance is necessary to gather early evidence of adverse health impacts. Finally, tertiary prevention "consists of the measures available to reduce or eliminate long-term impairments and disabilities, minimize suffering caused by existing departures from good health, and to promote the patient's adjustment to irremediable conditions" [8]. These measures include health care actions taken to lessen the morbidity or mortality caused by the disease (e.g., improved diagnosis and treatment of cases of malaria).

For many adverse health outcomes, secondary and tertiary prevention is generally less effective and more expensive than primary prevention. The same may be true for some interventions to reduce the impacts of climate variability and change. Encouraging primary prevention actions is appropriate when specific impacts can be anticipated with a comfortable degree of uncertainty. For example, measures to limit new construction in a susceptible coastal zone might be undertaken when projections suggest that future sea level rise could pose a problem. If these measures are not undertaken in advance of sea level rise, then secondary and tertiary prevention measures are likely to have more adverse consequences for human well-being. Secondary prevention could include the selective abandonment of infrastructure and retreat from the coastal zone after adverse impacts of sea level rise are experienced. Tertiary prevention would be responses to reduce injuries and deaths in the event of a disaster.

2.3 Individual- versus population-based approaches

There are individual- and population-based approaches to prevention. In the first approach, public health entities seek to identify high-risk individuals and to offer them some individual protection (e.g., identify individuals more likely to suffer from heat stroke and give them access to air conditioning). The purpose of identifying susceptible individuals is to detect and treat those who are at risk, some of whom may not be aware that they are at risk. Basically, the aim of this approach is to truncate the risk distribution. Advantages of this approach include that the intervention is appropriate to the individual, there is greater motivation, it is a cost-effective use of resources, and there is a favorable ratio of benefit to risk [10]. Disadvantages include the difficulties and costs of screening, that treatment does not try to alter the underlying causes of the disease and so is palliative and temporary, and the limited long-term potential for individuals and populations. This approach has limited potential for individuals because most individuals with risk factors will remain well, at least for some years.

The population strategy seeks to control the determinants of incidence in the population (e.g., understanding why some populations experience more heat stroke than others to plan effective interventions). It aims to lower the mean level of risk factors and to shift the whole distribution of exposure in a favorable direction [10]. Advantages include the large potential for the population and that the measures undertaken are behaviorally appropriate. For example, lowering the blood pressure distribution by 10% would correspond to about a 30% reduction in total attributable mortality [11]. Disadvantages include the small benefit for each individual (many of whom will not go on to develop disease), poor motivation, and a poor ratio of benefit to risk. With a mass intervention, each individual has only a small expectation of benefit, and this small benefit can easily be outweighed by a small risk.

The choice of whether to take an individual or a population approach depends on the situation. If information is available on the individuals who are vulnerable and a cost-effective control measure is available, then the individual risk approach may be most effective.

Public health recognizes that health benefits to populations differ from health benefits to individuals. Rose identified the preventive paradox as a preventive measure that brings much benefit to the population but offers little to each participating individual [10], but because populations are made up of individuals, how can one benefit a population without benefiting the individuals who comprise it? The answer is that public health measures reduce the incidence and mortality rates of many diseases within populations. These savings are of real lives and real disease, but the savings cannot be linked to specific persons. Improvements in the health of populations often result in disappointingly small improvements to the health of any given individual. In most instances, the health of an individual next year is not likely to be much better or worse if they accept or reject prevention advice.

Anticipatory adaptations to climate change may have similar tensions between individual and population level approaches, with similar advantages and disadvantages. An anticipatory measure such as spending public funds to subsidize air-conditioning units in public buildings in a city may impose costs on those who ultimately do not benefit individually (through increased taxes or reduced availability of public funding for other needed measures), but the expectation is that society will benefit through reduced morbidity and mortality during periods of extreme temperatures.

2.4 Prerequisites for public health action

To solve any public health problem, the following must be present [12]:

- an awareness that a problem exists,
- an understanding of the causes,
- a sense that the problem matters,
- the capability to influence, and
- the political will to deal with the problem.

These prerequisites for action are similar to the determinants of adaptive capacity (discussed in Chapter 2) [13]. Much of the literature on adaptive capacity focuses on the capability to control. Effective control requires the commitment to deploy economic resources, engage institutions, develop technology, and develop information and skills.

3. LESSONS LEARNED FROM PUBLIC HEALTH

Climate change will have impacts over a variety of spatial and temporal scales. Public health professionals have a wealth of experience in responding to both gradual long-term changes (disease rates changing over decades) and stimuli with severe consequences across short time scales (disease outbreaks). In addition, some intervention programs have been in place for decades. As such, public health offers a wealth of case studies that illustrate how a particular system responded to specific stress, whether on a small or large scale. Numerous examples exist of the efficacy and effectiveness of adaptation (prevention) measures, as do examples that demonstrate limits to the resilience of public health systems. We chose the following 10 lessons because they are relevant when considering approaches to decrease the vulnerability and increase the adaptive capacity of individuals and communities to climate variability and change. Understanding how these lessons were learned and their consequences can provide guidance when designing and implementing strategies, policies, and measures to address the potential impacts of climate variability and change.

3.1 Populations are greater than the sum of individuals

Vaccination is a clear example of a situation in which populations are greater than the sum of individuals. In the course of learning why epidemics often end before all susceptible individuals become diseased and the role that vaccination could play, researchers discovered that the risk for one individual was not independent of the risk for other individuals, it was contingent on the risk of others. Before this understanding, it was not understood that all individuals in a population do not need to be immunized to protect everyone in the population. This concept, referred to as "herd immunity", is the basis for immunization programs, which have been one of the core successes of public health over the last century. As long as a sufficient proportion of susceptible individuals (generally children) are immunized against a disease, then protection against the disease will be conferred to everyone in the population. The percentage of the population that must be immunized depends on the agent, its transmission characteristics, the vulnerability of the population, and environmental factors [14]. For example, to confer 100% protection, a greater proportion of children need to be vaccinated against measles, which is highly infectious, than against mumps, which is less so. A higher proportion of children in a crowded urban area will need to be vaccinated than in a rural area. There also are differences between countries. On average, to achieve 100% protection against measles, vaccination coverage needs to be about 95% in the United Kingdom versus about 99% in India [15]. Knowledge of such differences is needed for each vaccine-preventable disease to determine the institutional infrastructure requirements to administer vaccination programs.

The aim of vaccination programs is to stop disease from spreading through a community. The reproduction rate (R) is the average number of new infections generated by each case of disease; it is a function of the biologic mechanism of transmission and the rate of contact or interaction between members of the host population. If R is less than 1, then the disease will eventually die out, because there will not be enough individuals susceptible to the disease for the disease to spread. Table 1.1 summarizes the average number of new infections generated by each case of disease (R) and the minimum proportion of the population to be immunized for elimination of infection. An R of 12–18 for measles means that each new measles case is expected to generate about 12–18 additional cases in a susceptible population. The minimum proportion of the population to be immunized represents a form of threshold that, if exceeded,

Table 1.1 Approximate basic reproduction rates in developed countries and implied herd immunity thresholds for common diseases potentially preventable by vaccination

Infection	*Reproduction rate*	*Herd immunity threshold[a] (%)*
Diphtheria	6–7	85
Measles	12–18	83–94
Mumps	4–7	75–86
Pertussis	12–17	92–94
Polio[b]	5–7	80–86
Rubella	6–7	83–85
Smallpox	5–7	80–85

Source: Data from Ref. 14.

Notes

a Herd immunity threshold is defined as the minimum proportion to be immunized in a population for elimination of infections.

b The distinct properties of different polio vaccines need to be considered in interpreting the herd immunity thresholds.

will lead to an explosive increase in the incidence of an introduced infection. Public health measures aim to take advantage of herd immunity by reducing both the proportion of susceptibles and the potential sources of infection below the epidemic threshold, e.g., to reduce R to less than 1.

A current concern in public health is the degree of vaccine coverage in both developed and developing countries. Over time, as the proportion of children who are immunized increases, the number of new cases should drop. However, if children are not vaccinated, more cases will appear, putting the entire community at risk. For example, vaccine coverage is often limited in areas of armed conflict, putting large numbers of children at risk. For example, WHO estimated that 35,000 Afghani children die of measles every year; measles is the largest killer of children in a country where 25% of children do not live to celebrate their fifth birthday [16,17]. In developed countries, as parents decide not to have their children vaccinated because of perceived risk (or other reasons), more cases will start to appear. In the United Kingdom, the number of children being immunized for measles, mumps, and rubella has been reduced by 5% [18]. This reduction is sufficient to allow the reproduction rate to rise, with the result that a measles epidemic has been predicted [15].

These issues are relevant to climate change because vaccination is apt to be one adaptation tool to combat the potential spread of vector-borne diseases. For example, knowledge of the vaccination coverage rates by disease and target population will be needed to determine resource requirements of new immunization programs. Improved infrastructure and methods need to be developed to ensure that vaccine coverage is maintained in populations over time. Also, new questions may need to be addressed if the distribution of diseases is influenced by climate change. The more general lesson for adaptation to climate change is that proven tools and methods will continue to play a role in adaptation, although they may have to be modified or applied in new ways.

3.2 Wealth is not the sole determinant of health

Table 1.2 reports the results of an attempt to quantify the relative importance of key determinants of mortality reductions from 1960 to 1990 for 115 countries [19]. It is based on a statistical assessment of how the relationship between various health

Table 1.2 Sources of mortality reduction, 1960–1990[a]

Reduction	*Percentage contribution of gains in*		
	Income	*Educational level of adult females*	*Generations and utilization of new knowledge*
Under-5 mortality rate	17	38	45
Female adult mortality rate	20	41	39
Male adult mortality rate	25	27	49
Female life expectancy at birth	19	32	49
Male life expectancy at birth	20	30	50
Total fertility rate	12	58	29

Note
a Results are based on analysis of data from 115 low and middle income countries in Ref. 19.

indicators and both income levels and average education levels of adult females changed over time. The study looked at improvements in health by the following three factors: increases in average income levels, improvements in average educational levels, and a favorable shift over time in the underlying curve. This favorable shift was ascribed to the generation and application of new knowledge. All three factors can affect health through a variety of mechanisms. Better understanding is needed of the roles these factors play, and how these factors can be modified to reduce vulnerability to climate variability and change.

In addition, a growing body of literature suggests that populations living in areas with unequal income distributions have higher mortality rates than do populations in more homogeneous areas, irrespective of the income level [20–22]. In the United States, mortality is more strongly correlated with various measures of how income is shared within a state than it is with mean or median state income [21]. US metropolitan areas with greater income inequality have significantly higher mortality rates than do metropolitan areas with more equal income distributions, independent of the median income of the metropolitan area [20]. These associations are age-specific and are strongest for working age populations [22]. Recent analyses found no significant association between income inequality and mortality in Canada at the provincial or metropolitan area level [22]. This absence of an effect in Canada suggested two possibilities: that the association between income inequality and mortality is nonlinear, such that at higher levels of equality there is a diminishing effect on health, or that the association is not universal but instead depends on social and political characteristics specific to place. The first possibility suggests that reducing income inequality would be beneficial for population health. The second suggests that specific policies could be implemented to buffer the health effects of income inequality. Further understanding of when income inequality has health consequences and when it does not provides opportunities to examine the role of various economic and social policies in modifying individual experiences in such a way as to influence patterns of health and disease.

Additional research on the neighborhood social environment has shown that the characteristics of where a person lives, such as living in a disadvantaged area, are health risk factors independent of lifestyle and socioeconomic differences [23,24]. These results suggest that disease prevention strategies may need to combine person-centered approaches with approaches aimed at changing residential environments.

Altogether, these results suggest that assuming that increasing wealth alone will reduce vulnerability to climate variability and change is overly simple. Similarly, it is not reasonable to assume that wealth is the only determinant of adaptive capacity. Adger showed that income gaps might increase vulnerability to climate change [25]. Regardless of their income level, individuals may be unable to diversify livelihood strategies, and consequently reduce vulnerability to natural hazards, if the productive resources of a society are concentrated in the hands of a few. Further research is needed to clarify in what instances this relationship holds and how other determinants interact with wealth.

3.3 High risk groups may not bear most of the risk

When considering the burden of exposure to a risk factor, epidemiologists consider not only the relative risk associated with the factor but also the population attributable risk, defined as the incidence of a disease in a population that is associated with (attributable to) exposure to the risk factor [8]. The population attributable risk depends on the number of persons exposed, the total size of the population, and the incidence rates in the exposed, unexposed, and total population. It is often expressed as a percentage. A large relative risk between an exposure and disease does not always translate into a large burden of disease. For example, a woman with a mother and sister who both had breast cancer is at a 3.6 to 14-fold increased risk of developing breast cancer herself, but this "exposure" is rare with a population prevalence of 0.063% [26]. This can be compared with a relative risk of breast cancer of 1.7 for having a first child over the age of 30; this "exposure" has a prevalence of 23%. The latter results in more cases of breast cancer.

The relevance for climate change is in ensuring that information is available on both those groups with the highest relative risk and those groups from which most of the cases of disease arise. For instance, concern about heat-related mortality focuses on cities in more temperate latitudes because their populations are not physiologically adapted to high temperatures and presumably have the highest relative risks. Indeed, the elderly in inner cities comprise two-thirds of excess mortality in US cities during heat waves [27]. Little information is available on heat-related mortality in megacities of developing countries; although their mortality rates are expected to be lower, the number of people exposed may mean that more deaths occur.

3.4 Human behavior is difficult to predict; changing behaviors can be difficult

Human behavior is difficult to predict – otherwise, economic projections, intervention programs, advertising campaigns, and many other programs would be more successful. There is a rich literature on trying to understand and predict human health behavior, with considerable opportunity for improved understanding.

The limited capability to predict how individuals will react is one of the difficulties in trying to project future vulnerability because vulnerability is partially a function of autonomous adaptation. If affected individuals understand that climate change is occurring and react appropriately, adverse impacts may be substantially reduced [28]. It is difficult to predict which behaviors people will readily change in response to a changing climate and which may prove to be intractable.

3.5 It is important to develop quantitative indices of health

Composite indices of health status aid decision-makers in the allocation of resources among competing health outcomes. Such indices need to weigh different types of health outcomes such as mortality reduction or disability prevention. If the weighting is left to the political or bureaucratic process, then there is a high probability that similar health outcomes may be weighted inconsistently [29]; this could lead to inappropriate allocations of resources. To address this issue, and to provide a comparable measure of output for intervention, program, and sector evaluation and planning, an indicator of the burden of disease was developed, called the disability-adjusted life year (DALYs) [30]. As in other sectors, quantitative indicators of the burden of disease attributable to climate change can be used to facilitate the allocation of resources and to guide policy development.

3.6 When to intervene requires careful consideration

Assuming that the benefits of intervention outweigh the risks, when to intervene hinges on the question of what is the least amount of information that is needed for intervention. It is relatively straightforward to reduce exposures to a chemical in the workplace that is shown to be carcinogenic when readily available alternative chemicals or technologies exist. The answer to when to intervene is far less clear when risks are lower, when there may be other factors that could account for a reported association between exposure and disease, and when clear alternatives are not available. This may be further complicated when risks change or new risks are introduced. Evidence of the change in risk may be limited, and delaying intervention until more certainty is reached may result in many additional cases. Conversely, formulating a response based on incomplete information may lead to errors being made.

An example that demonstrates these issues is the ongoing question of the potential adverse health impacts of power-frequency electromagnetic fields (EMF). EMF can be viewed as an analogue for climate change in that it is a relatively new risk that, even after years of study, is still not well understood. Over 20 years of research have not resolved scientific questions about the possible adverse health effects of exposure, and evaluations of exposure assessment and epidemiologic studies have been hampered because of the lack of knowledge of what, if any, is the biologically relevant exposure and the lack of a biological mechanism [31]. Faced with a variety of uncertainties regarding a potential causal association between exposure to magnetic fields and adverse health outcomes, a number of researchers have suggested that "prudent avoidance" of EMF exposure may be justified [32,33]. This sort of no regrets approach, which is inexpensive, safe, and easy to implement, may be indicated in adaptation to climate change when information is lacking. In the near future, it may be difficult to differentiate natural climate variability from longer term changes in the climate system. Therefore, the adoption of no regrets policies aimed at adapting to or coping with climate variability is the appropriate course of action. In addition to adaptation strategies, policies, and measures, mitigation of greenhouse gases is needed to address longer term climate change concerns.

3.7 Intervention is a process, with success varying spatially and temporally; vigilance is required

There is increasing concern over emerging and re-emerging diseases. Climate change may facilitate the introduction of new diseases and increase or decrease the potential for outbreaks of diseases that the public health system has previously controlled. Prevention and control programs for most major vector-borne diseases were developed in the first few decades of the 1900s. These programs were based on breaking the transmission cycle at its weakest link by using mosquito control; this was aided by the development and use of residual insecticides. By 1970, many of these diseases, which had been major causes of disease and death for centuries, were effectively controlled [34]. Malaria had been eliminated from North America and Europe, effectively controlled in Asia, the Pacific, and Central and South America, and progress had been made in controlling it in Africa. Urban yellow fever had been effectively controlled in both Africa and the Americas, as had dengue in the Americas and the Pacific [35]. Plague was no longer a major public health problem, and antibiotics and other drugs effectively controlled most of the important infectious diseases. This led the US Surgeon General and Sir MacFarlane Burnett, a prominent infectious disease scientist in Australia, to declare that the war on infectious disease had been won [34]. These triumphs and the policy decisions that followed resulted in a 30-year period of inadequate funding for field research in tropical medicine. Policy makers could see no reason to continue to fund research and control programs for diseases that were no longer public health problems.

Failures began to occur when global demographic and societal changes, combined with decaying public health infrastructures and de-emphasis on prevention, opened the door to the dramatic global resurgence of infectious diseases [36]. Public health had become complacent after successes were achieved and relied too much on the magic bullet approach to disease control. The unprecedented population growth since World War II, the unplanned and uncontrolled urbanization, the changing land use and agricultural practices, the emergence of pesticide resistance, and the rapid movement of people, animals, and commodities via modern transportation have contributed to the dramatic resurgence in malaria, dengue, and cholera. The reinfestation of many South American tropical cities with the mosquito vector for yellow fever, combined with low vaccination rates, provides the potential for explosive urban outbreaks [37,38]. This potential was recently demonstrated by the massive epidemic of dengue fever in Brazil. The mosquito vector for dengue is the same as for yellow fever. During 2001/2002, there were approximately 700,000 cases of dengue in Brazil, with nearly 700 deaths from dengue hemorrhagic fever [39].

Gubler identified lessons learned from previous triumphs and failures in vector-borne disease control [34]. Several of these lessons are appropriate to the process of adapting to climate change. Effective programs maintained a research component to further understanding of disease ecology and of technologies to improve control. Without this research component, programs did not have the resilience to deal with changes in exposure or response. Long-term surveillance and research will be important if climate change facilitates a change in the geographic distribution of vector-borne diseases, or facilitates changes in transmission dynamics. Similar issues are found in other sectors. For example, in the agriculture sector, climate change is unlikely to result in a single, simple shift in crops grown in a particular region. Ongoing monitoring of changes in weather variables and other factors as they relate to crop decisions is needed.

Another lesson was that success often breeds failure. Effective control programs must not become complacent after achieving success, but rather must maintain constant vigilance. In the past, once a disease was successfully controlled, the resources devoted to that program were redirected to other, competing priorities. The past 30 years have taught that the consequence of stopping control activities is that many infectious diseases will return with a vengeance; intervention is a process that must be supported.

3.8 Interventions have multiple effects, including unintended consequences

Public health interventions are carefully designed to increase population health and well-being. However, interventions can have unintended adverse consequences not anticipated at the time of implementation. An extreme example is the contamination of drinking water by arsenic in Bangladesh (see Chapter 5); this has been called the largest mass poisoning of a population in history [40]. In summary, tubewells were installed beginning in the 1970s in an effort to provide a "safe" source of drinking water to a population that was experiencing high morbidity and mortality from water-related diarrheal diseases (e.g., cholera, dysentery, and other intestinal diseases). In many regions of Bangladesh, the groundwater that was accessed has naturally occurring high concentrations of arsenic, a known carcinogen. WHO recommends a maximum concentration of arsenic in drinking water of 10 μg/L. The maximum level permitted in Bangladesh is 50 μg/L. Health consequences of exposure above these levels range from skin lesions to a variety of cancers. A British Geological Survey estimated that the number of people exposed to arsenic concentrations above 50 μg/L to be about 21 million [41]. The prevalence of skin lesions in villages drinking from contaminated tube wells is 26–30% [42]. Because skin lesions have a typical latency of 10 years, and since the majority of wells were installed within the past 20 years, the expectation is that both skin lesions and arsenic-caused cancers, which have a typical latency of 20 years, will increase dramatically in coming years. The actual extent of the contamination and the number of people with skin diseases caused by arsenic are highly uncertain.

One obvious lesson learned is that even with careful planning, interventions can have deleterious and previously unimaginable, and unintended, consequences. Potential interventions should be evaluated as thoroughly as possible, and should be monitored to ensure that the intervention is safe and effective. Another lesson is the importance of developing general institutional capacity to deal with unexpected events. A crisis is not the time to implement or reorganize systems and institutions. Rather, it is the time to take advantage of existing governmental and nongovernmental organizations that have contacts in the field and that can respond quickly. Such capacities should be developed before they are needed.

It is expected that climate change, in particular sea level rise, may have significant impacts on Bangladesh [43]. The organizations currently working to address the problem of arsenic in drinking water should be taking the projected changes into account when developing new sources of clean drinking water.

3.9 Don't be over-confident

There is a wealth of examples illustrating the tendency for scientists and policy makers to be over-confident of the success of research and intervention programs. Let it suffice to cite a passage published in *Scientific American* in January 1901 [44]:

> It should not be surprising to make this prediction for the next century: Insect screens will be unnecessary. Mosquitoes will be practically exterminated. Boards of health will have destroyed all the mosquito haunts and breeding grounds, drained all stagnant pools, filled in all swamp lands and chemically treated all still-water streams.

Predictions of disaster and triumph tend to be overstated, to say the least. This point is relevant for climate change because there is a tendency in the literature to be overly pessimistic or optimistic about the consequences of climate change. For example, studies about climate change impacts that do not take into consideration changes in adaptation, such as improvements in public health over time, may overestimate potential future impacts.

In contrast, other studies assume perfectly efficient adaptation, which tends to minimize estimates of impacts (e.g., [45]). For instance, Tol and Dowlatabadi [46] estimated that malaria risks might be reduced to virtually zero through increased per capita income, a trend that presumably results in improved public health systems. However, malaria in tropical countries is more severe than malaria in countries with temperate climates, and has proved more difficult to control despite the efforts of the WHO and others.

The real impacts of climate change are likely to lie in between these extremes. Many developing countries are likely to have higher incomes and improved public health systems over the course of the 21st century. It is uncertain how many will be able to control outbreaks of infectious diseases. Similarly, adaptations in other sectors will be made in response to changes in climate related risks – the success of these adaptations remains to be seen.

3.10 Surprises do happen

In the early 1900s, as today, the United States and Europe had state-of-the-art public health systems. Despite the presence of established institutions with economic resources and professional staff, the 1918 influenza pandemic resulted in approximately 20 million deaths worldwide, either directly or indirectly from secondary bacterial pneumonia [47]. The pandemic appeared with little warning, strained resources to the limit, then ended for unknown reasons [48]. The effect on the world is hard to overestimate. For example, in 1917 life expectancy in the United States was about 51 years. In 1918, life expectancy decreased 12 years to 39 years. About 25% of the general US population became ill and about 40% of the armed forces were ill [49]. Approximately 500,000 Americans died during the pandemic. The fatality rate averaged about 2.5% and ranged up to almost 10%. Philadelphia was particularly hard hit. Influenza struck the city toward the end of September 1918. In the week ending 5 October, about 2,600 died. During the following week, more than 4,500 died. In all, within one month about 11,000 died and hundreds of thousands were ill. At the peak of the pandemic, 759 died on one day alone. The disease was so severe that death could occur within 24 hours; children were orphaned between the start and end of school. The city services were barely able to cope.

One function of public health is to be on the alert for new trends, to avoid future pandemics of influenza and other diseases. However, surprises continue to happen; climate change may facilitate or hinder these surprises. There is always a risk of new diseases emerging, from Ebola fever to AIDS to SARS, or of diseases changing their ranges. Maintaining strong public health institutions and surveillance programs, with sufficient capacity, will continue to be a high priority under all climate change scenarios.

4. DISCUSSION

Public health has been a key factor in the tremendous improvements in human health and well-being over the last century. The societal changes it has facilitated have improved the health of billions of people. To continue that trend, in particular to successfully exploit the benefits and limit the potential adverse impacts of climate variability and change, public health infrastructure must be strengthened and maintained. In addition, specific adaptation strategies, policies, and measures that address specific health risks need to be identified. Because the health issues identified as possible impacts of climate variability and change are issues that public health grapples with today, in one form or another, there is 150+ years of history from which to draw information concerning the effectiveness of various interventions. From this history and from ongoing research, public health needs to determine in which situations it should continue and augment present policies and measures, and when and where new policies and measures need to be implemented to address specific risks. In either case, the dimensions of the potential impacts of climate change will necessitate even more cross-sector collaboration and coordination in the development of strategies, policies, and measures; i.e., effective water resource management should consider the impacts of changes in infrastructure on human health, agriculture, etc. If the lessons learned are forgotten or not applied, then they will need to be learned again in another context – wasting time and resources in the process. These lessons also may have value to other sectors developing adaptation strategies to climate variability and change.

ACKNOWLEDGMENT

The authors wish to thank Drs. John Last and Jonathan Patz for their useful and informative comments.

REFERENCES

1. Albritton, D.L. and Meira Filho, L.G.: Coordinating lead authors: *Technical Summary. Working Group 1, Intergovernmental Panel on Climate Change.* Cambridge University Press, Cambridge, UK, 2001.
2. National Assessment Synthesis Team: Health Sector. In *Overview of Climate Change Impacts on the United States: The Potential Consequences of Climate Variability and Change.* U.S. Global Change Research Program. Cambridge University Press, Cambridge, UK, 2000.
3. Smit, B. and Pilifasova, O.: Adaptation to Climate Change in the Context of Sustainable Development and Equity. In *Intergovernmental Panel on Climate Change: Climate Change 2001: Impacts, Adaptation and Vulnerability. Contribution of Working Group II to the Third Assessment Report of the Intergovernmental Panel on Climate Change.* Cambridge University Press, Cambridge, UK, 2001, pp.877–912.
4. Smithers, J. and Smit, B.: Human Adaptation to Climatic Variability and Change. *Global Environ. Change* 7 (1997), pp.129–146.
5. Woodward, A., Hales, S., and Weinstein, P.: Climate Change and Human Health in the Asia Pacific Region: Who Will Be the Most Vulnerable? *Climate Res.* 11 (1998), pp.31–38.
6. IPCC: *Summary for Policymakers. Climate Change 2001: Impacts, Adaptation and Vulnerability. Contribution of Working Group II to the Third Assessment Report of the Intergovernmental Panel on Climate Change.* Cambridge University Press, Cambridge, UK, 2001, pp.1–17.
7. Committee of Inquiry into the Future Development of the Public Health Function: *Public Health in England.* Cmnd 289. HMSO, London, 1988.
8. Last, J.M.: *A Dictionary of Epidemiology, 4th Edition.* Oxford University Press, Oxford, UK, 2001.
9. Beauchamp, T.L. and Childress, J.F.: *Principles of Biomedical Ethics, 5th Edition.* Oxford University Press, Oxford, UK, 2001.

10. Rose, G.: *The Strategy of Preventive Medicine.* Oxford Medical Publications, Oxford University Press, Oxford, UK, 1992.
11. Rose, G.: Sick Individuals and Sick Populations. *Int. J. Epidemiol.* 14 (1985), pp.32–38.
12. Last, J.M.: *Public Health and Human Ecology, 2nd Edition.* Prentice Hall International, London, UK, 1998.
13. Yohe, G. and Tol, R.S.J.: Indicators for Social and Economic Coping Capacity – Moving Toward a Working Definition of Adaptive Capacity. *Global Environ. Change* 12 (2002), pp.25–40.
14. Fine, P.E.M.: Herd Immunity: History, Theory, Practice. *Epidemiol. Rev.* 15 (1993), pp.265–302.
15. Berger, A.: How Does Herd Immunity Work? Science Commentary. *BMJ* 319 (1999), pp.1466–1467.
16. World Health Organization: Afghanistan Crisis Health Update, 31 December 2001. Available from www.who.int/disasters.repo/7487.html.
17. World Health Organization. Press Statement 22 January. Tokyo Donors Conference on Afghanistan, 2002.
18. CDR: Fall in MMR Vaccine Coverage Reported as Further Evidence of Vaccine Safety is Published. *CDR Weekly* 9 (1999), pp.227–230.
19. Wang, J., Jamison, D.T., Bos, E., Preker, A., and Peabody, J.: *Measuring Country Performance on Health: Selected Indicators for 115 Countries.* Human Development Network. Health, Nutrition and Population Series. The World Bank, Washington, DC, 1999.
20. Lynch, J.W., Smith, G.D., Kaplan, G.A., and House, J.S.: Income Inequality and Mortality: Importance to Health of Individual Income, Psychosocial Environment, or Material Conditions. *BMJ* 320 (2000), pp.1200–1204.
21. Kaplan G.A., Pamuk, E.R., Lynch, J.W., Cohen, R.D., and Balfour, J.L.: Inequality in Income and Mortality in the United States: Analysis of Mortality and Potential Pathways. *BMJ* 312 (1966), pp.999–1003.
22. Ross, N.A., Wolfson, M.C., Dunn, J.R., Berthelot, J.-M., Kaplan, G.A., and Lynch, J.A.: 2000. Relation between Income Inequality and Mortality in Canada and in the United States: Cross Sectional Assessment Using Census Data and Vital Statistics. *BMJ* 320 (2000), pp.898–902.
23. Diez Roux, A.V., Stein Merkin, S., Arnett, D., Chambless, L., Massing, M., Neito, F.J., Sorlie, P., et al.: Neighborhood of Residence and Incidence of Coronary Heart Disease. *N. Engl. J. Med.* 345 (2001), pp.99–106.
24. Yen, I.H. and Kaplan, G.A.: Neighborhood Social Environment and Risk of Death: Multilevel Evidence from the Alameda County Study. *Am. J. Epidemiol.* 149 (1999), pp.898–907.
25. Adger, W.N.: Social Vulnerability to Climate Change and Extremes in Coastal Vietnam. *World Dev.* 27 (1999), pp.249–269.
26. Goodman, M., Kelsh, M., Ebi, K., Iannuzzi, J., and Langholz, B.: Evaluation of Potential Confounders in Planning a Study of Occupational Magnetic Field Exposure and Female Breast Cancer. *Epidemiology* 13 (2002), pp.50–58.
27. Kalkstein, L.S. and Greene, J.S.: An Evaluation of Climate/Mortality Relationships in Large U.S. Cities and the Possible Impacts of a Climate Change. *Environ. Health Persp.* 105 (1997), pp.2–11.
28. Tol, R.S.J., Fankhauser, S., and Smith, J.B.: The Scope for Adaptation to Climate Change: What Can We Learn from the Impact Literature? *Global Environ. Change* 8 (1998), pp.109–123.
29. Murray, C.J.L.: Quantifying the Burden of Disease: The Technical Basis for Disability-Adjusted Life Years. *Bull. WHO* 72 (1994), pp.429–445.
30. Murray, C.J.L. and Lopez, A.D.: Quantifying the Burden of Disease and Injury Attributable to Ten Major Risk Factors. In: C.J.L. Murray and A.D. Lopez (eds): *The*

Global Burden of Disease. A Comprehensive Assessment of Mortality and Disability from Diseases, Injuries, and Risk Factors in 1990 and Projected to 2020. Harvard University Press, Cambridge, MA, USA, 1996, pp.295–324.
31. Ebi K.L., von Ehrenstein, O., and Radon, K.: Electromagnetic Fields. In: G. Tamburlini, O. v. Ehrenstein, and R. Bertollini (eds): *Children's Health and Environment: A Review of Evidence.* A Joint Report from the European Environment Agency and the WHO Regional Office for Europe. Environmental Issue Report 29. EEA, Copenhagen, 2002, pp.172–187.
32. Johnsson, A. and Mild, K.H.: Electromagnetic Fields – A Threat to Children's Health? In: T. Christensen and S. Stephens (eds): *Children and Radiation. Selected Topics Raised at an International Conference.* Norwegian Centre for Child Research, Trondheim, Norway, 2000, pp.53–86.
33. Kheifets, L.I., Hester, G.L., and Banerjee, G.L.: The Precautionary Principle and EMF: Implementation and Evaluation. *J Risk Res.* 4 (2001), pp.113–125.
34. Gubler, D.J.: Prevention and Control of Tropical Diseases in the 21st Century: Back to the Field. The President's Address. *Am. J. Trop. Med. Hyg.* 65 (2001), pp.v–xi.
35. Halstead, S.B.: Emergence Mechanisms in Yellow Fever and Dengue. In: W.M. Scheld, W.A. Craig, and J.M. Hughes (eds): *Emerg. Infect. 2.* ASM Press, Washington, DC, 1998, pp.65–79.
36. Gubler, D.J.: *Aedes aegypti* and *Aedes aegypti*-Borne Disease Control in the 1990s: Top Down or Bottom Up. Charles Franklin Craig Lecture. *Am. J. Trop. Med. Hyg.* 40 (1989), pp.571–578.
37. IOM: *Emerging Infections: Microbial Threats to Health in the United States.* Institute of Medicine, Washington, DC, 1992.
38. Monath, T.P., Giesberg, J.A., and Fierros, E.G.: Does Restricted Distribution Limit Access and Coverage of Yellow Fever Vaccine in the United States? *Emerg. Infect. Dis.* 4 (1998), pp.698–702.
39. SIGFUNASA-DOENCAS. Sistema de Informacoes Geranciais da FUNASA. Relatorios Geranciais-Dengue-Casos. 2001. http://sis.funasa.gov.br/dw/dm01/menu_p/0201.htm.
40. Smith, A.H., Lingas, E.O., and Rahmanm, M.: Contamination of Drinking-Water by Arsenic in Bangladesh: A Public Health Emergency. *Bull. WHO* 78 (2000), pp.1093–1103.
41. British Geological Survey: Groundwater Studies for Arsenic Contamination in Bangladesh. Final Report. Mott MacDonald Ltd., London, UK, 1999.
42. Tondel, M., Rahman, M., Magnuson, A., Chowdhury, I.-A., Faruquee, M.-H., and Ahmad, S.-A.: The Relationship of Arsenic Levels in Drinking Water and the Prevalence Rate of Skin Lesions in Bangladesh. *Environ. Health Persp.* 107 (1999), pp.727–729.
43. Lal, M., Harasawa, H., and Murdiyarso, D.: Asia. In: J.J. McCarthy, O.F. Canziani, N.A. Leary, D.J. Dokken, and K.S. White (eds): *Climate Change 2001: Impacts, Adaptation and Vulnerability. Contribution of Working Group II to the Third Assessment Report of the Intergovernmental Panel on Climate Change.* Cambridge University Press, Cambridge, UK, 2001, pp.533–590.
44. Scientific American. 50, 100 and 150 Years Ago. *Sci. Am.* January 2001, pp.16–17.
45. Mendelsohn, R., Nordhaus, W., and Shaw, D.: The Impact of Global Warming on Agriculture: A Ricardian Analysis. *Am. Econ. Rev.* 84 (1994), pp.753–771.
46. Tol, R.S.J. and Dowlatabadi, H.: Vector-Borne Diseases, Development, and Climate Change. *Integr. Environ. Assess.* 2 (2002), pp.173–181.
47. Pyle, G.F.: *The Diffusion of Influenza: Patterns and Paradigms.* Rowman & Littlefield, Totowa, NJ, USA, 1986.
48. Wain, H.: *A History of Preventive Medicine.* Charles C. Thomas Publisher, Springfield, IL, USA, 1970, pp.382–385.
49. Crosby, AW.: *America's Forgotten Pandemic.* Cambridge University Press, Cambridge, UK, 1989.

2 Approaching adaptation: parallels and contrasts between the climate and health communities

Gary Yohe and Kristie L. Ebi

ABSTRACT: There has been limited cross-fertilization between public health and the climate change adaptation community, in part because of underlying differences in philosophy and conceptualization of key concepts. Natural hazards research on discrete and irregular events informed the climate community's conceptualization of vulnerability. Public health views vulnerability in the context of the current distribution of exposure and response in a population. These different views thus affect each discipline's understanding and use of thresholds and coping ranges. The communities share similar conceptualizations of what is required for a system to reduce its vulnerability, the climate community through the determinants of adaptive capacity and public health through the prerequisites for prevention. Both approaches result in a broad range of potential strategies for adaptation. Public health experience in the components of adaptive capacity can enrich the climate community's understanding of adaptation. Added value goes in both directions, with concepts formulated by the climate community facilitating progress in the understanding of how to examine and prepare for the potential health impacts of climate change. Discussion of these similarities and differences provides insights that increase the ability of both communities to better reach their shared goal of increasing the ability of individuals, communities, and nations to effectively manage the challenges and changes likely to arise with climate change.

1. INTRODUCTION

Public health prevention and climate change adaptation share the goal of increasing the ability of nations, communities, and individuals to effectively and efficiently cope with challenges and changes. Indeed, that is what is meant by adaptation to an external stress. The two communities have approached this goal from different viewpoints: public health from the perspective of protecting and enhancing the health and well-being of individuals and communities, and adaptation from a perspective that can trace its roots to the natural hazards community. The activities of these two communities have been running in parallel, with few opportunities for crossover learning.

This chapter is an attempt to illuminate the similarities and differences between the health and adaptation perspectives. Its intent is to enrich the approaches used by each. Public health has more than 150 years of experience in developing prevention strategies, policies, and measures, but it is only beginning to apply that experience in the context of global climate change. The insights of the adaptation community may prove instructive for how to plan, prepare for, facilitate, and implement measures that moderate harm or exploit beneficial opportunities arising from climate change. In addition, understanding a population's capacity to respond (i.e., to adapt) to new climate conditions is crucial to the realistic assessment of the potential health impacts of climate variability and change [1].

Public health is "the combination of sciences, skills, and beliefs that is directed to the maintenance and improvement of the health of all people through collective or social actions. The programs, services, and institutions involved emphasize the prevention of disease and the health needs of the population as a whole. Public health activities change with changing technology and social values, but the goals remain the same: to reduce the amount of disease, premature deaths, and disease-produced discomfort and disability in the population" [2]. Health is viewed as a fundamental human right; it is a state of complete physical, mental, and social well-being and not merely the absence of disease or infirmity [3].

Public health aims to achieve its goals through prevention (its version of adaptation). Strategies and measures to reduce disease and save lives are categorized as primary, secondary, and tertiary prevention. Primary prevention aims to prevent the onset of disease in an otherwise unaffected population (e.g., supplying bed nets to all members of a population at risk of exposure to malaria). Where primary prevention is not feasible, secondary prevention focuses on preventive actions taken in response to early evidence of health impacts (e.g., strengthening disease surveillance and responding adequately to disease outbreaks such as the West Nile virus outbreak in the United States). As discussed in Chapter 10, surveillance is necessary to gather early evidence of adverse health impacts. Finally, tertiary prevention includes health care actions taken to lessen the morbidity or mortality caused by a disease (e.g., improved diagnosis and treatment of cases of malaria).

The analogs between these public health strategies on the one hand and autonomous and planned adaptation to climate change on the other may appear obvious, but there are differences in language and underlying philosophy that must be understood if we are to make progress in cross-fertilizing both fields. This chapter explores these differences even as it identifies similarities. We begin in Section 2 with a general comparison of approaches before turning to some of the specifics; we find that the two communities mean two different things when they speak of "vulnerability". Section 3 focuses on the determinants of adaptive capacity and compares them with the prerequisites for prevention. Section 4 then turns to the "coping range". It is one of the fundamental tools with which researchers approach adaptation to climate variability and change, and we find comparable concepts in the public health approach. Subsequent sections explore the robustness of this comparability first in the context of a hypothetical climate example and an alternative portrait of the coping range, and then in the context of two applied examples drawn from our experience in public health. Concluding remarks ultimately put these similarities and contrasts into a context that will, we hope, facilitate progress in the understanding of exactly how to examine the potential health impacts of climate change. Working definitions of some of the critical terminology are provided as the chapter proceeds.

2. COMPARING PERSPECTIVES

In the Report of Working Group II to the Third Assessment Report (TAR) of the Intergovernmental Panel on Climate Change (IPCC), Chapter 18 considered adaptation to climate change in the context of sustainable development and equity [4]. The authors of that chapter focused on the role of adaptation in assessing the economic, social, and perhaps political implications of climate variability and change. Although the chapter primarily took a natural hazards approach to climate-related stresses, the authors expected that their approach would be applicable in other sectors. In particular, it was hoped that the results would be relevant for those who contemplate the potential for adaptation to diminish the costs (or to enlarge the benefits) of change in the mean or variability in any variable that defines a system's environment. The IPCC authors also recognized, however, that adaptation has been a way of life in many sectors of modern societies and that the climate researchers should be able to learn from the experiences, processes, and insights of other disciplines. This chapter responds to this recognition by offering a first step in bringing centuries of learning from the health community to bear on the validity of what is, for the most part, a collection of working hypotheses offered in 2001 by the climate community.

2.1 Chapter 18 perspectives

The IPCC viewed the vulnerability of any system to an external stress or collection of stresses (the degree to which a system is affected by external stress) as a function of exposure (the degree to which a system might experience manifestations of the stress), sensitivity (the degree to which those manifestations influence the system), and adaptive capacity (the ability of the system to cope with those influences by reducing its exposure, its sensitivity, or both). Human and natural systems tend to adapt autonomously to gradual change and to change in variability. In addition, human systems can plan and implement adaptation strategies to further reduce potential vulnerability or exploit emerging opportunities. The cost of vulnerability from an external stress can be viewed as the incremental cost of adaptation plus any residual damages that cannot be avoided. The IPCC also emphasized that the ability to cope with an external stress, termed adaptive capacity, varies significantly from system to system, from sector to sector, and from region to region. Climate researchers have come to recognize that it is crucial to identify clearly who might be adapting to what and why. Of course, it is equally critical, in assessing the potential for future adaptation to a changing environment, to determine the baseline of current experience and current adaptations.

2.2 Public health perspectives

Public health uses the concept of vulnerability in two different senses. One acknowledges that advances in public health are not permanent and that deterioration of the public health infrastructure could permit the return of adverse health outcomes that are currently under control. Examples include the rapid increase in morbidity and mortality, with the attendant decline in life expectancy, from a variety of different diseases after the break-up of the former Soviet Union. Another example is the current concern over the possible use of smallpox, a disease that has been eradicated, by bioterrorists. In this sense, all societies are potentially vulnerable; the degree of vulnerability depends on maintaining and improving public health systems.

The second sense relates to specific health outcomes. The classic approach to evaluating environmental health risks is the four-step risk assessment paradigm:

1 hazard identification,
2 dose (exposure)-response assessment,

3 exposure assessment, and
4 risk characterization.

Under this paradigm, the evaluation of information on the hazardous properties of environmental agents and on the extent to which sensitive receptors (i.e., humans, animals, and ecosystems) are exposed to them results in a quantitative or qualitative statement about the probability and degree of harm to the exposed populations. Thus, vulnerability is a function of exposure to an agent and the exposure-response relationship between that exposure and a particular health outcome. It is important to identify what, if any, preventive measures were implemented before an exposure-response relationship is determined. Vulnerability from a health perspective includes adaptation only to the extent that measures currently in place influence the exposure-response relationship (e.g., estimates of the impact of a heat wave on the elderly will be influenced by the number of the elderly who used air conditioning or who spent adequate time in air conditioned environments). The combination of the current exposure-response relationship, the extent of exposure, and the preventive measures in place creates a vulnerability baseline against which the effectiveness of future policies and measures can be measured by changes in the burden of disease (e.g., the impact on lung cancer risk of convincing smokers to quit smoking). Vulnerability changes over time with the implementation of appropriate and effective interventions. That is, within the health sector, vulnerability is viewed as the current or baseline state, not as the residual state after adaptive capacity has been considered.

One notable project to take future adaptations into account is the Global Burden of Disease study sponsored by the World Health Organization (WHO). The 1990 study used DALYs (disability adjusted life years) to compare disease risk factors (malnutrition; poor quality of water supply, sanitation, and personal and domestic hygiene; unsafe sex; tobacco use; alcohol use; occupational exposure to hazards; hypertension; physical activity; illicit drug use; and air pollution) by world region [5] for 1990 and 2020. One goal of these projections is to show how health can be improved if countries undertook various interventions (e.g., determine how many DALYs would be gained if countries decreased the percentage of smokers by particular amounts). The utility of these estimates has been demonstrated by their use to underpin health policy in a number of areas. The 2000 Global Burden of Disease study includes 18 risk factors, including climate change, and will focus on the time period from 2000 to 2030.

2.3 Discussion

The different perspectives of the climate change and public health communities are a possible source of confusion in any conversation between health and climate researchers. Neither is conceptualizing vulnerability incorrectly, rather they are addressing two different concepts of vulnerability. Perhaps one of the most important lessons is that both communities must be made aware that the same word can mean fundamentally different things to different people. Panel A of Table 2.1 offers a tabular summary of the differences between the two perspectives. Panel B depicts these differences functionally in the context of a hazards perspective. The climate community would see each alternative listed as a possible response to climate change. The goal of public health is to improve the burden of disease; accepting a risk is to accept the

Table 2.1 Contrasting the climate and health perspectives

Climate	*Hazards*	*Health*
Panel A – A descriptive comparison		
Focuses on exposures		Focuses on burden of disease
Accepts a long time horizon		Works on a shorter time scale
Can consider all species		Focuses on human beings
Social buffering is less important		Social buffering is critical
Panel B – A practical comparison		
A possible response	Accept a risk (bear or share)	Not preferred Bear = status quo Share via WHO, etc.
A possible response Reduce exposure Reduce sensitivity	Reduce a risk (modify events) (prevent events)	Primary focus Reduce exposure Reduce sensitivity
Possible in the extreme	Effect radical change	Possible in the extreme

status quo. It aims, instead, to create interventions to reduce the risk and, perhaps in extreme cases, work to effect dramatic change.

It should be noted that the two communities often work on different scales. The current approach of the climate community originated in natural hazard research; the focus is on discrete and irregular events such as floods and droughts. The health community focuses not on one event (such as a particular death from malaria) but on the distribution of exposure and response in a population (such as the 1 million children who die each year from malaria). These distributions are often continuous. This difference in scale is one source of the different conceptualizations of vulnerability and of thresholds. At the extreme, the climate community can think of thresholds in the sense of whether or not a dam will hold. The health community often deals with distributions where thresholds can be somewhat arbitrary cut-points (e.g., the risk of heart disease for individuals with serum cholesterol concentrations of 195 is not much different from that for individuals with serum cholesterol concentrations of 205).

3. ADAPTIVE CAPACITY

3.1 Chapter 18 perspectives

To explain the observed diversity in the ability of systems to adapt (primarily to natural hazards), the IPCC offered the hypothesis that adaptive capacity is a function of a series of determinants:

1 the range of available technological options for adaptation;
2 the availability of resources and their distribution across the population;
3 the structure of critical institutions, the derivative allocation of decision-making authority, and the decision criteria that would be employed;
4 the stock of human capital, including education and personal security;
5 the stock of social capital, including the definition of property rights;

6 the system's access to risk-spreading processes;
7 the ability of decision-makers to manage information, the processes by which these decision-makers determine which information is credible, and the credibility of the decision-makers themselves; and
8 the public's perceived attribution of the source of stress and the significance of exposure to its local manifestations.

Moreover, the IPCC emphasized that exposure to climate variability and to extreme events is an important source of vulnerability. In fact, Chapter 18 noted that systems typically respond to climate variability and extreme events before they respond to gradual changes in the mean.

3.2 Public health perspective

The public health sector also recognizes that the ability to influence a public health problem (i.e., to adapt to a perceived level of vulnerability) depends on a number of factors that are equally path dependent and site specific. It is not possible to influence a public health problem when any of the following are missing:

1 an awareness that a problem exists;
2 a sense that the problem matters;
3 understanding of what causes the problem;
4 capability to influence; and
5 political will to influence the problem [6].

3.3 Discussion

In terms of climate change, the awareness of a problem (and a consensus that the problem exists) is required before implementing specific adaptation policies and measures. One consequence of impact research has been to raise this awareness. One goal of the IPCC assessments has been to facilitate consensus that climate change matters; e.g., What might be the potential consequences of sea level rise? To plan and implement effective responses, there needs to be an understanding of what causes the problem. For example, to formulate adaptation options to a climate stress, it is necessary to understand how the stress translates into exposure and sensitivity. Is the exposure gradual or sudden? Is it a manifestation of mean conditions or the extremes of variability? Are sensitivity and exposure linear, or are there thresholds of exposure beyond which sensitivity is exaggerated? Once sufficient understanding of what causes the problem exists, then capacity to deal with the problem is needed. This is the focus of much of the research on adaptive capacity, including the technologies required to estimate the costs of policies and measures, and who will bear those costs. Finally, even when all the other factors are present, policies and measures will not be instituted without the political will to address the problem. A sufficient example is an informal comparison of the approaches of different countries with similar resources to adapt to climate change.

In terms of public health, these requirements for prevention do not proceed in a linear manner, as illustrated by the HIV/AIDS epidemic. By and large, there was little awareness that a problem existed or that it mattered until men in developed countries began to die from AIDS. The recognition of the problem led to large research programs to understand both the causes of the disease and measures to alter risk factors. The results led to behavioral education programs to help prevent the spread of the disease and to drugs to help influence the course of the disease – at least for developed countries. The manner in which the epidemic was brought to the attention of the public created the political will to fund the research and to implement the measures that slowed the epidemic in developed countries. The recognition of the size and impact of the HIV/AIDS problem in Africa and India took longer to develop; the capacity and the political will to deal with the epidemics in these countries are still being developed.

An important issue implicit in these prerequisites for prevention is that they are imbedded in social values. Society plays a critical role in determining when a problem matters. An example is the change in attitude in the United States over the past 50 years about the risks of smoking tobacco or driving after consuming alcoholic beverages. The social acceptability of the burden of various health outcomes changes over time. Along with differences within a society, there are differences between societies. These differences may arise for a range of reasons, from social preferences to economic realities.

It is not difficult to see that these prerequisites provide evidence that the hypotheses offered by the climate community have some validity. Table 2.2 makes this point on a relatively abstract level by mapping the list of public health prerequisites to the determinants of adaptive capacity as articulated in Chapter 18. The only IPCC determinant that is not immediately matched to a public health prerequisite is access to risk spreading; but even this factor is functionally part of the "capability to influence". In part, this is because the scale at which risk can be spread varies by health outcome and by disease determinant (e.g., the risk of a heat-related event during a heat wave cannot be shared amongst individuals, while not vaccinating a particular child for measles increases the risks of that child contracting measles and of the community experiencing a measles epidemic). Public health institutions and medical insurance programs (formal in developed countries and sometimes informal in developing countries) can and do contribute to spreading risk via prevention. Experience in the public health context offers evidence that the IPCC hypotheses are valid, especially with regard to public infrastructure (governance, social capital); human capital (education and behavior); and the ability to manage information. It also suggests some problems in claiming universal validity for the IPCC hypotheses.

Tables 2.3 and 2.4 illustrate how the differential expression of adaptive capacity, through different choices in the provision of health services, can influence population

Table 2.2 Mapping of determinants of adaptive capacity and prerequisites for prevention

Determinants of adaptive capacity[a]	*Prerequisites for prevention*[b]
Availability of options (1)	Capability to influence (4)
Resources (2)	Capability to influence (4)
Governance (3)	Political will (5)
Human (4) and social capital (5)	Understanding of causes (3); political will (5)
Access to risk-spreading mechanism (6)	Capability to influence (4)
Managing information (7)	Understanding of causes (3); problem matters (2)
Public perception (8)	Awareness (1); problem matters (2)

Notes

a Numbers in parentheses are determinant numbers discussed in Sections 3 and 8.

b Numbers in parentheses are prerequisite numbers discussed in Section 3.

health. Table 2.3 compares a number of health indicators for Costa Rica and the former USSR [7], and Table 2.4 compares socioeconomic, health, and finance indicators for Costa Rica and the Russian Federation [8]. Costa Rica, with lower per capita income and educational attainment, has much higher life expectancy for males and

Table 2.3 Trends in selected health indicators and their determinants in Costa Rica and the former USSR, 1960–1990

	Costa Rica			*Former USSR*		
	1960	*1990*	*% due to technical progress*	*1960*	*1990*	*% due to technical progress*
Health indicators						
Under-5 mortality rate	124	14	55	39	27	40
Female adult mortality rate	203	73	48			
Male adult mortality rate	246	122	59			
Female life expectancy	64.5	78.6	59	72.0	74.0	43
Male life expectancy	61.6	74.0	60	65.0	63.0	46
Total fertility rate	7.0	3.3	38	2.7	2.2	25
Determinants						
Income per capita	2,001	3,381		2,397	7,453	
Female education (years)	4.0	5.6		7.6	10.3	
Male education (years)	4.1	5.5		8.5	10.8	

Source: Data from Ref. 7.

Notes
Under-5 mortality: the probability of dying between birth and age five, expressed per 1,000 live births.
Adult mortality rate: The probability of dying between age 15 and 60, expressed per 1,000 persons reaching age 15.
Life expectancy at birth: The number of years a newborn infant would live if the prevailing age-specific levels of mortality at the time of its birth were to stay the same throughout its life.
Total fertility rate: The number of children that would be born to a woman if she were to live to the end of her childbearing years and bear children at each age in accordance with prevailing age-specific fertility rates.
Income per capita is measured in 1985 constant international dollars (dollars adjusted for purchasing power).
Education: the average years of education in the female and male populations 15–60 years of age.

Table 2.4 Socioeconomic and health services and finance indicators for Costa Rica and the Russian Federation, 1960–1990

	Costa Rica	*Russian federation*
Socioeconomic indicator: Malnutrition stunting among children under age 5 (%) around 1995		
Males	6	12
Females	7	13
Health services and finance indicators		
Children immunized against measles (%)	99[a]	92[b]
Health expenditures around 1995		
Total (% of GDP)	8.5	4.8
Public sector (% of GDP)	6.3	4.1
Public sector (% of total)	74	87

Source: Data from Ref. 8.

Notes
a 1997.
b 1987 (data not available for 1997).

females. Note the decrease in male life expectancy in the former USSR from 1960 to 1990; this is partly due to the well-documented consequences of the break-up of the USSR. Part of the differences between Costa Rica and the Russian Federation can be attributed to higher health expenditures (in 1995, 8.5% of GDP in Costa Rica versus 4.8% in the Russian Federation). Another factor is technical progress. The Wang et al. study [6] looked at improvements in health by increases in average income levels, improvements in average educational levels, and a favorable shift in the underlying curve that was ascribed to the generation and application of new knowledge (technical progress). Overall, technical progress accounted for the biggest percentage of health improvement across the 115 countries evaluated in the Wang et al. study; it contributed an average of 42% to the improvement in female life expectancy and 43% in male life expectancy. Technical progress can be viewed as an indicator of adaptive capacity.

From the public health perspective, the hypotheses from the climate community suggest ways to conceptualize approaches to realizing adaptive capacity by developing intervention strategies, policies, and measures. Some of the determinants of adaptive capacity are explicitly incorporated, such as the availability of resources, the range of available options, institutional restraints, and the public perception of the issue. Others are incorporated, but in a slightly different manner than conceptualized by the climate community; e.g., critical institutions include both national and international institutions. Many successful public health interventions have been conducted in a structured, top-down fashion. In those situations, the human and social capital required may be shared between a country and an international partner such as the WHO or UNICEF. Public health has an impressive record of improving population health. As a consequence, there is confidence in the ability of decision-makers to manage information, in the processes by which these decision-makers determine which information is credible, and in the credibility of the decision-makers themselves.

Finally, the IPCC viewed autonomous and planned adaptation as reducing vulnerability by reducing exposure, sensitivity, or both; but the authors emphasized that there would most likely be residual damage in most if not all cases. The cost attributed to change was therefore the sum of the cost of adaptations and these residuals. The public health perspective also recognizes the residual burden of disease after preventative measures are employed, but those measures are not often costed. Indeed, many measures (such as reducing tobacco smoking) simultaneously reduce the burden of many diseases, so calculating incremental costs attributed to one disease can be difficult. The public health perspective leads to descriptions of the global burden of disease and explores how this burden may change over time with changes in individuals' behaviors, public health initiatives, or both.

4. COPING RANGE

4.1 Chapter 18 perspectives

The climate literature on vulnerability and adaptation recognizes explicitly that systems' environments are inherently variable from day to day, month to month, year to year, decade to decade, and so on [9,10]. It follows that changes in the conditions that define those environments can be experienced most noticeably through changes in the nature or frequency of variable conditions that materialize across short time scales, and that adaptation necessarily involves reaction to this sort of variability. This is the fundamental point in Hewitt and Burton [11], Kane et al. [12], Yohe et al. [13], Downing [14], Schimmelpfennig and Yohe [15], and Yohe and Schlesinger [16]. Some researchers, like Smithers and Smit [1], Smit et al. [17], and Downing et al. [18], use the concept of "hazard" to capture these sorts of stimuli, and claim that adaptation is warranted whenever either changes in mean conditions or changes in variability have significant consequences. For most systems, though, changes in mean conditions over short periods of time fall within a "coping range" – a range of circumstances within which, by virtue of the underlying resilience of the system, significant consequences are not observed [18,19]. There are, however, limits to resilience for even the most robust of systems.

Taking these insights as a point of departure, it is important to understand the boundaries of systems' coping ranges – thresholds beyond which the consequences of experienced conditions become significant. A growing literature is showing the utility of this approach. Parry et al. [20] may provide the most extensive use; impact thresholds were provided for a wide range of crops over a range of locations in Europe in response to a variety of climate-induced stress.

For our purposes, it is equally important to conceptualize how adaptation might alter those boundaries and their significance. Several authors, including de Vries [21], de Freitas [22], and Smit et al. [23], linked this insight to adaptive capacity. They make it clear that adaptive capacity depends critically on both defining a coping range and understanding how the efficacy of any coping strategy might be enhanced (or diminished in maladaptation) by implementing new or modified adaptations. Adaptation could, for example, alter exposure by moving the time series trajectory of stress up or down, by shrinking variability, by truncating peaks or filling in valleys, and so on. Adaptation could also affect sensitivity and thereby work to define or to redefine the thresholds. Finally, contingent or responsive adaptations might be portrayed as changes in strategy that are triggered by certain events or frequency of events. In summary, adaptation baselines of the sort prepared by Parry and his collaborators reflect initial thresholds against which current and future climate might operate, but planned or autonomous adaptations to future conditions could make them less binding (or, in maladaptation, more binding).

4.2 Public health perspectives

The climate literature has primarily focused on variability of environments, whereas public health generally focuses on variability in both exposures and responses. Public health seeks to identify and reduce both the background level of disease and any epidemics or outbreaks.

Public health does not use the terminology or the concept of a coping range. As defined above, use of the term suggests a range within which significant consequences are not observed. It is difficult to say that developing countries are "coping" given the current burden of preventable diseases, from diarrheal diseases to vaccine-preventable diseases. Or that developed countries are coping with the number of premature deaths from smoking-related diseases, etc. The idea that adaptation is warranted only when either changes in mean conditions or changes in variability have significant consequences appears inappropriate when current conditions themselves are unacceptable. Adaptation policies and measures are needed now to address current conditions.

Public health has recognized thresholds for centuries. However, it is difficult to generalize approaches to thresholds because each is specific to a particular exposure-response relationship. Categories of exposure-response relationships include the following:

- Exposures that exhibit J-shaped relationships with health outcomes, where either too little or too much is detrimental to health (e.g., ambient temperature and oxygen).
- Exposures that have threshold relationships with health outcomes, where low doses are not associated with increased morbidity and mortality (e.g., arsenic, and dose required to develop a case of cholera). A subset of this category is "exposures" that are necessary for health at low concentrations, but can be toxic at high concentrations (e.g., some vitamins and minerals).
- Exposures that have linear relationships with health outcomes, where any amount of exposure may convey some risk (e.g., asbestos).

Establishing thresholds when exposure and sensitivity produce a continuous distribution can be problematic. The choice of a threshold is often a social construct, based on social acceptance of that particular risk, availability of technological solutions to prevent exposure, etc. Setting a threshold above zero can mean that it is acceptable that x% of the population will contract a particular disease and perhaps die. For example, various countries are developing smallpox vaccination strategies that accept that some individuals will have severe adverse vaccine reactions, including death. An additional issue is that setting positive thresholds for exposure to certain risks that vary from country to country explicitly raises the issue of different values for human life. The threshold approach can, nonetheless, depict these differences by accepting different norms; and it can illuminate the cost of setting a higher threshold. Moreover, setting health thresholds at zero can also be accommodated, and the approach can then reveal the value of expending the extra effort involved. However, there are opportunity costs of not expending that effort elsewhere, and these costs can be enumerated in human lives or other metrics of the burden of disease.

In health, there are situations where populations are so far above a threshold value that interventions can be problematic. For example, in some parts of sub-Saharan Africa, individuals can receive 2,000 or more malaria-infected mosquito bites per year. Reducing exposure by 95% means that individuals would still receive 100 infective bites per year. Different approaches may be needed for particular populations, depending on how close exposure levels are to a threshold.

Besides the coping range, adaptive capacity also depends on understanding how the efficacy of a coping strategy might be enhanced or diminished by implementing new or modified adaptations. This is central to public health – the constant search for interventions that will reduce suffering and disease. These interventions can focus on the exposure, the response, or both. Reduction of the disease burden in a population enhances a population's adaptive capacity by increasing health, productivity, etc. [24]. Good population health is critical for adaptive capacity.

5. ILLUSTRATIVE EXAMPLE OF THE CHAPTER 18 PERSPECTIVE

Figure 2.1 illustrates how coping ranges can be employed to portray adaptation baselines graphically. Both panels work with hypothetical river flows at some artificial location; the figure is redrawn from Yohe and Tol [25]. Figure 2.1a establishes how an initial environment with a historical trend in mean flow might be advanced into the future with a historical variance in flow over prescribed periods of time; it was actually produced from flows along the Nile River in Egypt over the past 50 years. Figure 2.1a also portrays upper and lower thresholds that define a current coping capacity for our location. The upper threshold was set at 120% of the current mean flow to reflect, for example, circumstances when inhabited parts of the flood plain would be flooded. The lower threshold, meanwhile set at 80% of current flow, might indicate the flow level below which operation of an existing local power plant or subsistence fishing would be infeasible. As drawn, this location would see 5 periods of flooding and 0 periods of power interruption over the 50 period series. Figure 2.1b portrays the same

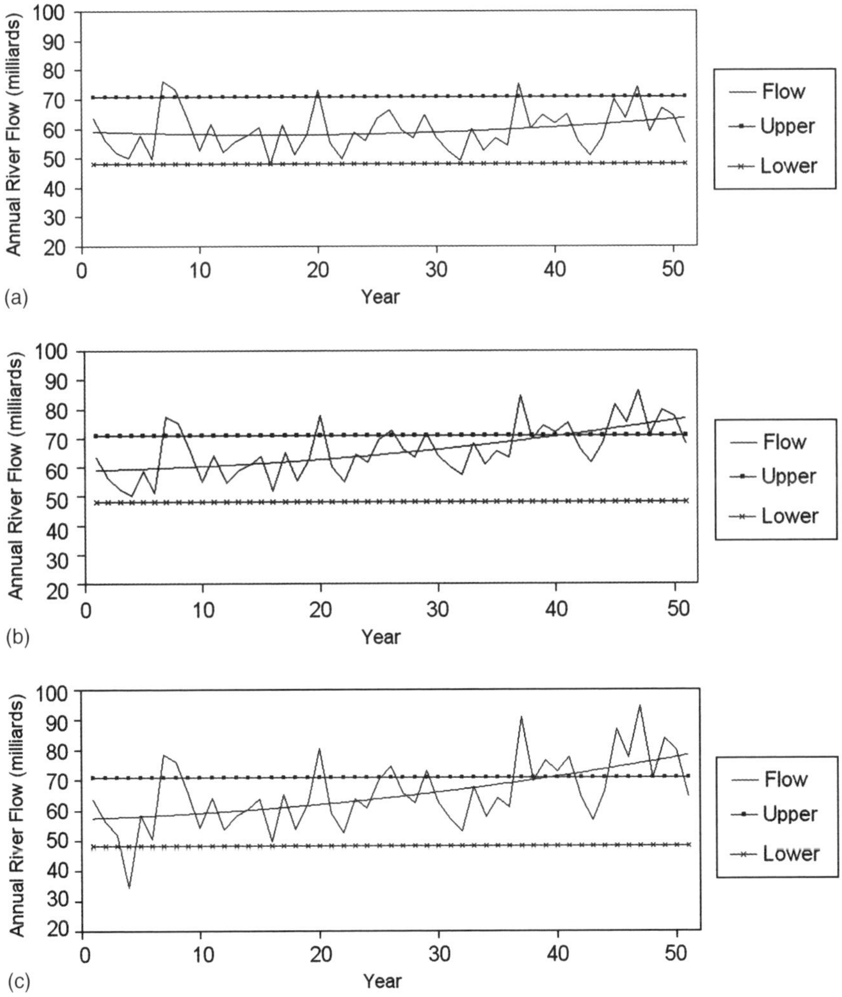

Figure 2.1 (*a*) Historical context – adaptation baseline. (*b*) Amplifying the historical trend in the mean adaptation baseline revisited. (*c*) Superimposing an enlarged variance

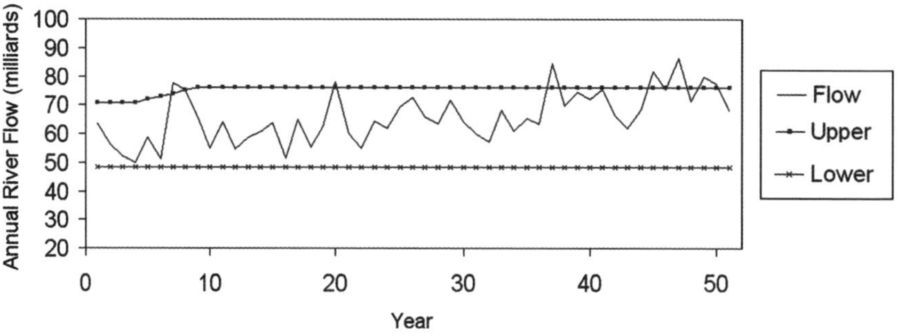

Figure 2.2 Building a levee in the fifth period

series with a gradually increasing mean – the result, perhaps, of climate change or some alteration in upstream land use. Even without any change in variation around the mean, the frequency of flooding climbs to 15 of 50 periods. When variability expands, as in Fig. 2.1c, the frequency of flooding need not increase as significantly (up to 16 of 50 periods), but the severity of the flooding can expand noticeably.

Exploring three alternatives can now illustrate how future adaptation might be represented in the same context. A first option would involve constructing a series of protection levees, and Fig. 2.2 portrays the effect of building levees by expanding the upper threshold of the coping range. If the levees were constructed high enough, flooding could be eliminated over the 50-year period for the historical trend; but those levees would eliminate only 7 of the 15 floods that could be anticipated along the amplified trend. Because most of the remaining floods would occur late in the series, levees built on the basis of historical trends and variability alone could eventually be overwhelmed if mean flow (or variability, for that matter) rose over time. Could the levees be enlarged later on as needed? Perhaps. That would be an engineering question whose cost implications might be mitigated if that possibility were included in near-term planning (most likely at a higher cost). In the meantime, construction and maintenance costs would be incurred, and these could be local environmental effects, local amenity costs, and increased flooding downstream.

Building a dam upstream could also be contemplated, and Fig. 2.3 portrays its potential effect by reducing the variability in observed river flow at our location. This option would, in particular, allow managers to release water from the dam during low flow periods and hold water back during high flow periods so that the actual flow below the dam would be a moving average of current conditions. Figure 2.3, in fact, depicts observed flow as a four-period moving average. Exposure could, with a dam of sufficient capacity, limit variability in actual flow to the size of the original coping range of Fig. 2.1 along the historical trend, so this option holds the potential of eliminating vulnerability to crossing either the high or low threshold. Along the accelerated trajectory, however, the dam could also fail to accommodate future change and actually produce periods of long-lived flooding. A bigger dam might work, but that would probably involve a much larger short-term expense. In any case, construction and maintenance costs would again be incurred, as would significant environmental impacts upstream (although energy and recreational benefits could be also be anticipated).

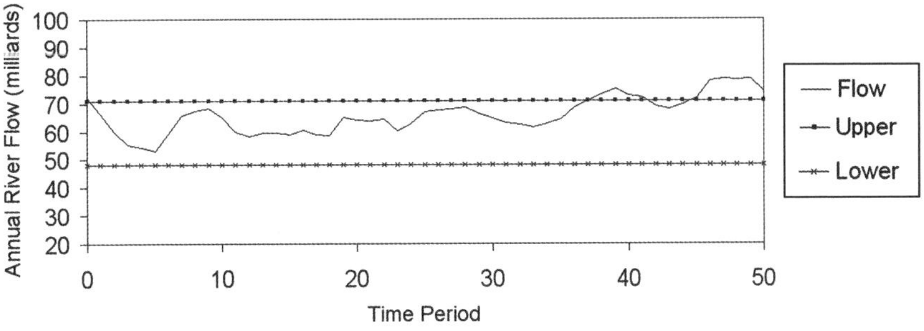

Figure 2.3 Smoothing variation with an upstream dam

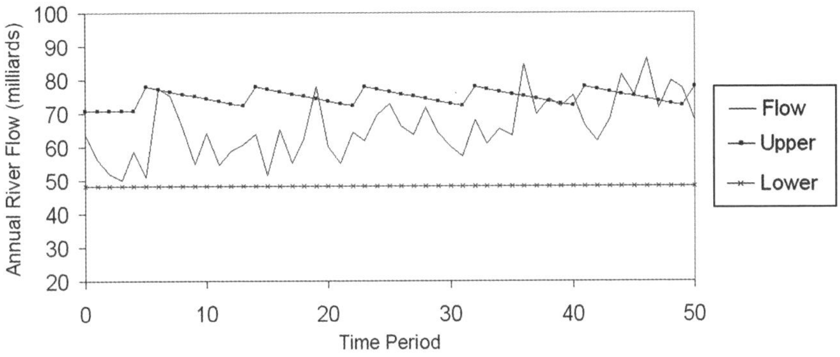

Figure 2.4 Reducing flood threat by dredging

Periodically dredging the river could also be contemplated. Figure 2.4 portrays the hypothetical effects of periodic dredging as a saw-toothed pattern for the upper threshold. Dredging would allow the river to accommodate more water and thereby increase the upper threshold, but this benefit would depreciate over time as silt re-deposits on the riverbed. To maintain long-term benefits, therefore, dredging would have to be repeated on a regular basis. This option also holds the potential of eliminating exposure to flooding along the historical trajectory, but only for short periods of time. On the other hand, opting for dredging regime could allow managers to accommodate unexpected changes in the underlying mean or variance by increasing or decreasing the frequency of dredging operations. The recurring cost of dredging plus some environmental damage could be expected, as could increased flooding risk downstream.

6. AN ALTERNATIVE PORTRAIT OF THE COPING RANGE

Coping ranges have been used to construct unitless indices of vulnerability (or the inverse of extreme vulnerability – sustainability) that can, in principle, be used to judge the relative severity of exposures to change across diverse sectors and regions [14]. To see how, consider a distribution of variables that depicts current short-term variability for a particular system around its current "average" condition. Region A_0 in Fig. 2.5a depicts such a distribution for two sources of stress; the subscript denotes time zero, and the region itself might reflect something like a 95% confidence region for short-term variability. It should be possible to superimpose a comparable distribution of combinations of the same variables that are "tolerable" for the current mix of sensitivities and adaptations. Region B_0 in Fig. 2.5a might, for example, apply as an alternative representation. The intersection of A_0 and B_0 can therefore be interpreted as a region of feasible sustainability at time 0; and probabilistically weighting the area of that intersection can be interpreted as an index of sustainability (also equal to 1 minus a similarly unitless index of extreme vulnerability). More formally, the index would simply be the probability that actual experience at time 0 would fall within the bounds of sustainability. Vulnerability could be indexed equivalently by subtracting this probability from 1 to reflect the likelihood of falling outside the bounds of tolerable experience.

As the future unfolds, both regions might move. The distribution of critical variables could move to something like A_{50} in Fig. 2.5b, where the means of both variables have changed and the second order moments have expanded. Along an adaptation baseline, then, the index of vulnerability would be the intersection of A_{50} and B_0 – a smaller area because the adaptations embodied in B_0 were designed to accommodate (as well as possible) the range of experiences reflected in A_0. Of course, the boundaries of "tolerable" experience could also change over time as adaptive strategies were adjusted to the new experiences. As a result, something like B_{50} in Fig. 2.5b might be a more accurate depiction of future conditions.

Figure 2.5c displays the content of Fig. 2.4 as a time series of sustainability indices that could be created by replicating the adjustments depicted in Fig. 2.5b over time. Note that the adaptation baseline trajectory (denoted Locus I) falls precipitously; Fig. 2.5b shows experience moving away from the area that could be accommodated by current adaptations, so Locus I must indicate that the frequency of intolerable combinations of stressing variables has increased. Locus II would depict a corresponding trajectory for the case where adaptation to change is undertaken. It is a little more optimistic, but the index must show some erosion in sustainability because Fig. 2.5b showed a case where adaptation was not able to keep up entirely with the changing environment. Locus III shows the trajectory for a case in which intertemporal adjustments in adaptive capacity kept pace with changing conditions.

Figures 2.6a–d portray the trajectories for a sustainability index derived for the three adaptations described in the flooding example of the previous subsection under the assumption that the planners and citizens are concerned about the frequency of flooding (measured as the proportion of years over the past decade that have seen flooding episodes). Note that the baseline holds around 0.8 (two floods per decade) through year 39, but then falls dramatically. The levee holds until almost year 50, but then starts down; and it accomplishes a perfect score for a few time periods earlier in the time series. The dam, meanwhile, achieves perfect scores through year 35, but then falls as severely and as permanently as the adaptation baseline. And over the very long term, dredging seems to do best for longer. This is, of course, not an endorsement of dredging; it is, instead, an illustration suggesting that aggregate

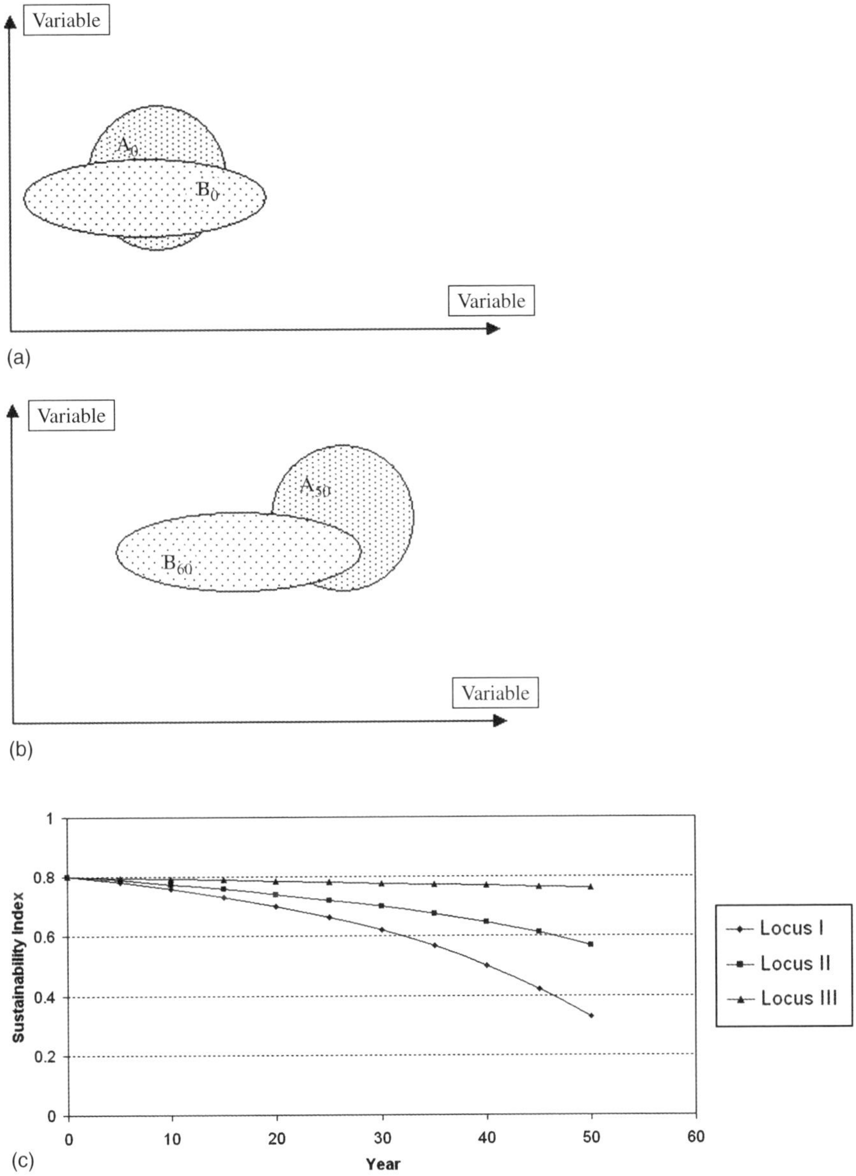

Figure 2.5 (*a*) Initial conditions. (*b*) Conditions in 50 years. (*c*) Sample trajectories

indices of sustainability can depict dramatically different time profiles for alternative adaptations (or for the same adaptations applied in different sectors or regions). Figure 2.7 combines the trajectories for the baseline and three adaptations on one graph.

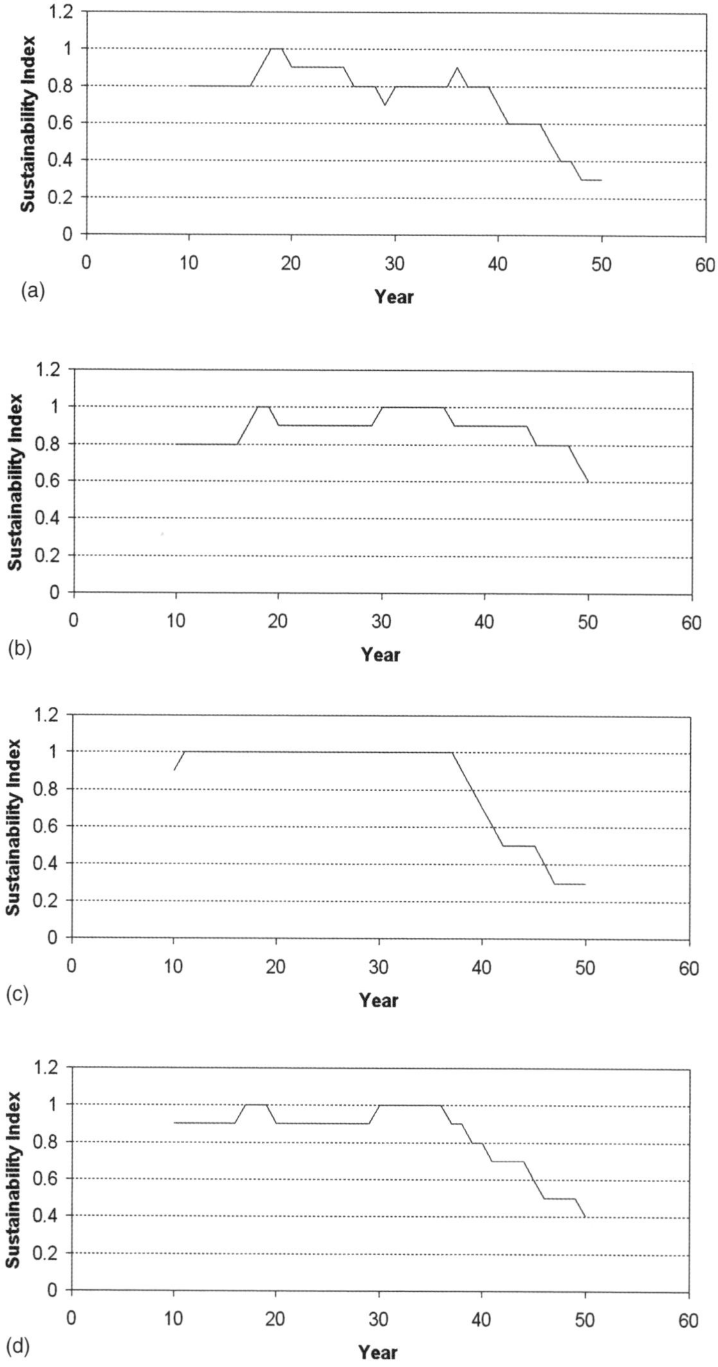

Figure 2.6 (*a*) Baseline from fig. 2.1b. (*b*) Building a levee from fig. 2.2. (*c*) Building an upstream dam from fig. 2.3. (*d*) Periodic dredging from fig. 2.4

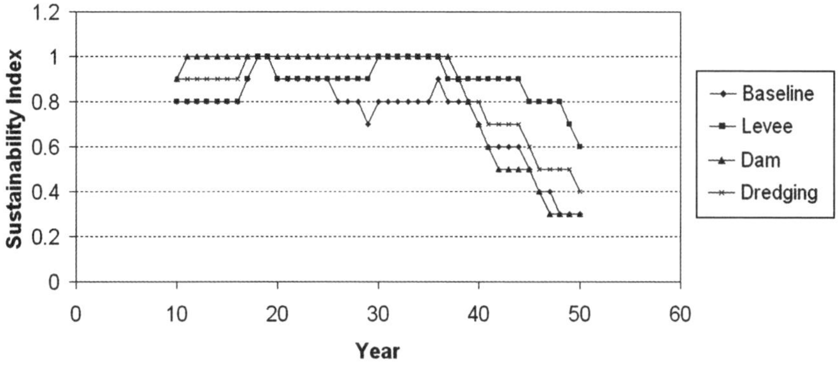

Figure 2.7 Sustainability indices for the river example

7. EXAMPLES FROM THE HEALTH CONTEXT

7.1 Example of a zero threshold – eradication of smallpox

On 8 May 1980, the 155 member states of the WHO, represented by their delegates to the 33rd World Health Assembly, unanimously accepted the conclusions of the Global Commission for the Certification of Smallpox Eradication that smallpox had been eradicated and that there was no evidence that smallpox would return as an endemic disease [26]. Smallpox is a highly infectious viral disease whose repeated epidemics have decimated populations and altered the course of history. It is spread from person to person, with a case fatality rate of up to about 25% [26]. There is no known effective treatment.

Eradication was successful because the smallpox virus (variola) produces an acute illness that is readily recognizable, there are no carrier stage or nonhuman reservoirs, recurrent infectivity does not occur, and there is an effective, heat-stable vaccine that protects in a single dose [27]. These factors, combined with international surveillance, public education, and research, led to success. However, eradication was achieved by only the narrowest of margins [28]. The progress of the eradication campaign was not straightforward; it wavered between success and disaster over space and time. Success was achieved by quixotic circumstances and extraordinary performances of field staff. Nor was support generous: a number of endemic countries were persuaded only with difficulty to participate; the industrialized countries initially were reluctant contributors; and UNICEF declined to participate [28]. For example, cash donations to WHO during the first seven years of the campaign (1967–1973) totaled $79,500.

The institutional requirements for the eradication campaign were significant. The total WHO budget from 1967 until completion of the eradication campaign in 1980 was $81 million, $43 million of which came from extrabudgetary sources [26]. In addition, WHO estimated that there was $32 million worth of assistance on a bilateral basis, and, accurate only in terms of orders of magnitude, individual governments spent about $200 million. Most workers were at the national level, supplemented by volunteer workers and 687 international personnel from the 73 countries who worked on the campaign.

Not only was the smallpox eradication campaign successful, but it also established programs and institutional structures that are being relied on today. When the campaign began in 1967, smallpox vaccine was one of only two vaccines widely used throughout the developing world [28]. Few countries had organized national vaccination programs. Among those that did, the programs seldom extended beyond the larger towns and cities. Substandard or poorly preserved vaccines were commonly used, often with little to no effective supervision. The Global Commission for the Certification of Smallpox Eradication envisaged a campaign that included the development of a management structure within the existing health service structure, with mechanisms to ensure that fully potent and stable vaccine was used to vaccinate at least 80% of inhabitants [28]. The campaign required that a national surveillance system be established and that planning reach established goals within a given period; most national health ministries had never attempted this type of effort. With a campaign budget of $2.4 million per year, it was not possible to underwrite more than a small proportion of the personnel and program costs. The campaign functioned within existing health service structures and took advantage of available resources.

Another key to success was concurrent research, which was conducted despite the opposition of senior WHO leadership, who insisted that only better management was required [28]. According to Henderson [28], it is highly unlikely that eradication would have succeeded without the following research initiatives: development of new vaccine devices; field studies that revealed the epidemiology of the disease to be different than that described in textbooks, which led to modification of basic operations; discovery that the duration of vaccine efficacy was longer than stated, making revaccination less important; operational research into more efficient vaccine delivery and case detection; and studies that demonstrated there is no animal reservoir.

7.2 Example of a threshold greater than zero – arsenic

Arsenic is a metalloid abundant in the earth's crust; it occurs in trace quantities in rock, soil, water, and air [29]. It is widely distributed in surface fresh water and in groundwater, with concentrations in rivers and lakes generally below 10 μg/L. Arsenic concentrations in groundwater average about 1–2 μg/L; under certain conditions, they can range up to 3 mg/L [29]. Elevated groundwater concentrations have been found in a wide range of locations, including Taiwan; West Bengal (India); Bangladesh; Chile; northern Mexico; several areas of Argentina; parts of China, parts of the United States (California, Utah, Nevada, Washington, and Alaska); and Finland [29].

Environmental exposure to arsenic is primarily through the ingestion of food and water, with food the principal contributor to the average daily intake of 20–300 μg [29]. The human body has evolved mechanisms for elimination of minerals, including arsenic. Adverse health effects from arsenic exposure begin once an individual's threshold body burden is exceeded; there is variability in the individual body burden at which symptoms begin to occur. Early symptoms from long-term exposure to arsenic in drinking water can range from development of dark spots on the skin to a hardening of the skin into nodules, often on the palms of the hands and the soles of the feet. Chronic arsenicosis can affect multiple organ systems, including lungs, gastrointestinal system, liver, spleen, genitourinary system, hemopoietic system, eyes, nervous system, and cardiovascular system. Arsenic exposure also is causally related to increased risks of cancers in the skin, lungs, bladder, and kidney [29].

Morales et al. [30] concluded that although the shape of the exposure-response curve is uncertain at low levels of arsenic exposure, over a lifetime, one out of every 100–300 people who consume drinking water containing ⩾50 μg/L arsenic may suffer an arsenic-related cancer death (lung, bladder, or liver cancer) [30].

7.3 Discussion

Comparing the three examples – a hypothetical one from the climate arena and two actual experiences from the health sector – clearly makes at least one point. Researchers and practitioners who confront exposure to climate variability and change focus attention on measures designed to reduce exposure, sensitivity (exposure-response), or both. Their counterparts in the health sector focus much of their attention on reducing exposure – the eradication of a disease or the isolation of particularly harmful substances. Health professionals can, however, also turn their attention to reducing the response to an exposure, for example, through the invention and distribution of vaccines and other strategies that diminish the likelihood that exposure to a particular threat would actually do some harm. Despite using different language, therefore, both perspectives look across the same broad spectrum of potential strategies.

8. EXPANDED INSIGHTS

The major lesson to be drawn from the flooding example is that implementing alternative adaptations to an anticipated stress could easily produce different effects on either exposure to that stress or an effective coping range of sensitivity. We already noted, though, that diversity is a fact of life in adaptation; we simply added more complexity to the issue. A major lesson from the two health examples is that the public health community can contribute to our ability to understand this diversity, particularly with respect to the limitations of a coping capacity approach.

8.1 The determinants of adaptive capacity and the prerequisites for prevention

If the sources of adaptive capacity are so diverse, how are we to make progress in discovering fundamental insights that can be applied widely across multiple sectors and multiple locations? How, in other words, can a collection of case studies provide more general understanding? Fortunately, some of the determinants of adaptive capacity listed in Table 2.2 should operate on macro-scales in which national or regional factors play the most significant role even while other determinants function on more micro-scales that are truly location specific. But is that enough? Both the climate and the health perspectives offer some encouraging signs.

8.2 Adaptive capacity, prerequisites for prevention, and scale

As suggested by the illustrative examples, the set of available, applicable, and appropriate technological options (the first determinant in Table 2.2) for a given exposure at a particular location should, for example, be defined on a micro-scale, even though the complete set of possible remedies might have macro-scale roots. Flood influence options of the sort portrayed above are, for example, determined by the local conditions of the river bed and available engineering knowledge; and this knowledge may be restricted to indigenous knowledge on the one hand or informed by worldwide consultants on the other. Similar examples are available in public health, such as vector control activities where general approaches need to be modified in response to local conditions.

Determinants 2 through 6 should also have large, macro-scale components to them, but their micro-scale manifestations could vary from location to location or even from adaptation option to adaptation option. Resources (determinant 2) are often distributed differently across specific locations; adaptive capacity may be more sensitive to larger scale distributional issues across different locations. The essential questions here focus on whether sufficient funds are available to pay for adaptation and whether the people who influence those funds are prepared to spend them on adaptation (political will). Yohe and Tol [25] report some suggestive empirical results from international comparisons; they show that these questions can be critical at the most fundamental level. They found, for example, that poorer people are more likely to fall victim to natural violence than are richer people. The relationship is highly significant. For every 1% increase in economic growth, vulnerability falls by 1%. They also found that more densely populated areas are more vulnerable. This is as expected, because the same disaster affects more people. Moreover, they found a positive relationship between income inequality and vulnerability; i.e., people in more egalitarian societies seem to be less likely to fall victim of natural violence than are people in a society with a highly skewed income distribution. Similar conclusions have been reached for overall health [31–33]. Both sets of results were expected and both are consistent with the negative correlation between income and vulnerability. Measures designed to highlight a skewed distribution would confirm the notion that the poorest communities within a country would most likely face resource deficiencies when it comes to protecting themselves.

Macro-scale and international institutions (determinant 3) could certainly matter even at a micro-scale level, especially in determining how decisions among various adaptation options might be made and who has access to the decision-making process. For example, the World Bank follows certain procedures in its investment decisions, and adaptation projects in countries seeking World Bank support must satisfy Bank criteria before being considered. The European Union also has a framework (on procedures as well as consequences) into which all water management projects must fit, so macro-scale influences can be felt even in developed countries. On the other hand, adaptation projects in other places can be decided and implemented completely according to local custom alone.

The stock of human capital (determinant 4) could be a local characteristic as well, but its local manifestation would most likely be driven in large measure by macro-scale forces such as national education programs. The stock of social capital (determinant 5) and efficacy of risk-spreading processes (determinant 6) should be largely functions of macro-scale structures and rules; but they could again take different forms from location to location and option to option. Risk can be spread through national markets for commercial insurance and the international reinsurance markets, but some companies refuse to sell flood insurance. Risk can also be spread through mutual obligations in the extended family, the strength of which varies between cultures and city and countryside. By way of contrast, determinants 7 (information management) and 8 (attribution of signals of change) may have some general macro-scale foundations, but their primary import would be felt on a micro-scale. Indeed, decision rules and public perceptions could take on forms that would be quite particular to the set of available options.

9. DISCUSSION AND A CAVEAT

Taken in their most general context, a careful review of the hypothesized determinants of adaptive capacity from the climate literature offers several derivative insights that are supported by experience in public health. First, the local implications of macro-scale determinants of adaptive capacity are their most critical characteristics. Second, most if not all of the determinants of adaptive capacity can be seen working through specific adaptation options. Finally, the potential of most if not all adaptation options to affect sensitivity or exposure might not be too difficult to assess. As a result, careful review of multiple sectors and sites can identify macro-scale determinants or prerequisites that are consistently reflected across applications even if specific manifestations are still (development) path dependent.

Looking at different health thresholds for the same health risk in different countries can, however, highlight a critical caveat in either approach. Public health can be a natural laboratory for exploring the implications of working in the "second-best" arena – taking underlying structures and constructs as they are in specific locations rather than assuming that "best" available practices are employed. In the climate arena, these constraints are frequently derived from distortions that lie outside the specific study focus and can therefore be difficult to portray. In either case, though, the question of what to do with the distortions that define the "second best" must be confronted directly. Should the adaptation baseline adopted by a researcher interested in future effects in either climate or health take the implications of those distortions as given, and thereby limit either the range of adaptation options or their respective efficacies? Or should the baseline contemplate a world in which those impediments are dismantled (perhaps for other reasons)?

Proper vulnerability and adaptation analyses must confront these issues directly by comparing results from a series of runs into the future. One might, for example, look at the future with a given adaptation baseline (with existing distortions and impediments) and no climate change. A second set of runs into the future might also assume no climate change but include adjustments in adaptation that could be anticipated to reduce exposure or sensitivity to present variability in the specified stress or present variability from some other related stress. A third set of runs could then impose climate change on the adaptation baseline (the first set) to see how they might work; and a fourth collection could repeat the climate analysis with anticipated adjustments (the adjusted baseline for the second set of runs). In every case, however, it is critical that the analysis presumes neither "dumb agents" (like the "dumb farmers" of the first vulnerability assessments) who will not respond to any changes in environment nor clairvoyant agents who know everything from the very beginning. The future will be fraught with uncertainty, just like the present, and any considerations of adaptation must recognize this fact. Indeed, a complete vulnerability and adaptation analysis of a particular region or sector would contemplate a range of "not-implausible" futures and perhaps contemplate the value of information at various points in time. To do so could uncover an entire list of new adaptations directed primarily at providing *useful information.*

A qualitative description of a coastal zone example might make these distinctions more clearly. Consider, for the sake of argument, that a coastal community is facing the added threat of sea level rise; it already faces the threat of coastal storms. One might consider two adaptations and two thresholds. In the first case, decisions to protect or not protect coastal property would be made on the basis of a sea level rise threshold that had been determined by current practices. In the second, the same decisions to protect or not would be made on the basis of a flooding threshold that had been expanded by the storm-damage setback zone. A setback zone defines boundaries within which property owners whose properties have been severely damaged by a storm may not be able to rebuild. In this case, an analyst would run storm scenarios (no climate effect on frequency or intensity) without sea level rise but without and with expanded setback zones. Call these analyses S1 and S2, respectively. The next steps would repeat the storm scenarios with sea level rise; call these S3 and S4. Comparisons of S3 and S1 would indicate costs attributable to rising seas along the "no-adjustment" baseline. Comparisons of S2 and S4 would produce comparable estimates in a modified future. Comparisons of S1 and S2 would demonstrate the value of those modifications without climate-induced sea level rise, while comparisons of S3 and S4 would estimate their value in a changing environment.

10. SYNTHESIZING CONCLUSIONS

We have noted that public health and climate approaches to adaptation and prevention are comparable even though vulnerability means different things to the two communities. The critical link seems to be the analogous concepts of the determinants of adaptive capacity and prerequisites for prevention. It must be noted, though, that the more practical approach of the health community helps illuminate some of the definitional problems for the climate community. For example, the IPCC defines vulnerability as a function of adaptive capacity, but the definition poses the question, What is the difference between adaptation, coping, and response? Indeed, the climate community is not of one mind when it comes to differentiating between these concepts. To some, the IPCC definition is incomplete – If a system copes with an external stress in the same way it has coped with similar stresses in the past, is it adapting? More to the point, the term "adaptation" suggests *changes* in behavior. If a community were to employ a commonly used short-term change in behavior to cope with a transient stress and then revert to its original behavior after the stress had passed, would this be adaptation? Some would say "yes", but others would say "no".

There are also competing views of vulnerability. Should vulnerability be viewed as a function of hazard and geography (as in the IPCC definition), or is it an intrinsic property of a system that is not defined by a particular hazard? For human systems, the latter view, consistent with the approach in the natural hazards literature, presents vulnerability as a social construct arising from a set of social, economic, and political factors rather than the frequency, severity, and duration of climate stresses (hazards). It is the interaction of hazard and social vulnerability that produces risk, which in turn determines the potential impacts of a given climate stress or climatic event.

The public health perspective on the components of adaptive capacity can shed some light on the issue of adaptation for the climate community. Value added does not, however, move only in one direction. While it is correct to emphasize that adaptive capacity, like vulnerability, is highly path dependent and site specific, the term "political will" is potentially problematic for the climate community. The concern expressed by some researchers is that use of the term consigns some very important and complex factors to a "black box" – a process that discourages further discussion and analysis. Do we measure political will and then simply state its influence on a society's adaptive capacity, or do we break down the political will into a number of other factors, much in the way we assess adaptive capacity itself? We could replace the concept of political will with that of governance, and attempt to assess the implications of different modes of governance and governance structures for adaptation. However, from the health perspective, governance does not capture all that political will implies. For example, Mothers Against Drunk Driving provided the political will to make significant changes in laws about driving under the influence of alcohol.

There are certainly problems in trying to develop a universal, cross-sectoral theory of vulnerability and adaptation – problems that are as likely to arise from differences of scale and process within the climate (or health) arena as from systematic differences between the nature of climate change impacts and health risks. Nonetheless, even determinants like risk spreading (via insurance against climate-related damage) may have immediate analogues in the health sector if we think more broadly. In both climate and health, insurance generally pays for remedial action once a health or climate event has taken place, rather than preventative action. As well as minimizing undesirable health outcomes, public institutions can reduce vulnerability to climate change through education and programs to promote and financially assist adaptive measures such as strengthening housing against flooding or storm damage. An analogy can indeed be drawn between the climate and health cases, even though the factors that lead to vulnerability are quite different – but then, that is exactly the point.

REFERENCES

1. Smithers, J. and Smit, B.: Human Adaptation to Climatic Variability and Change. *Global Environ. Change* 7 (1997), pp.129–146.
2. Last, J.M.: *A Dictionary of Epidemiology. Fourth Edition.* Oxford University Press. Oxford, UK, 2001.
3. World Health Organization: *Preamble to the Constitution of the World Health Organization* (as adopted by the International Health Conference, New York, 19–22 June 1946 and signed on 22 July 1946 by the representatives of 61 States and entered into force on 7 April 1948; the Definition has not been amended since 1948), Official Records of World Health Organization, no. 2, p.100. World Health Organization, Geneva, 1948.
4. Intergovernmental Panel on Climate Change: *Climate Change 2001: Impacts, Adaptation, and Vulnerability – The Contribution of Working Group II to the Third Scientific Assessment of the Intergovernmental Panel on Climate Change.* Cambridge University Press, Cambridge, UK, 2001.
5. Murray, C.J.L. and Lopez, A.D.: Quantifying the Burden of Disease and Injury Attributable to Ten Major Risk Factors. In: C.J.L. Murray and A.D. Lopez (eds): *The Global Burden of Disease. A Comprehensive Assessment of Mortality and Disability from*

Diseases, Injuries, and Risk Factors in 1990 and Projected to 2020. Harvard University Press, Cambridge, MA, USA, 1996, pp.295–324.

6. Last J.M.: *Public Health and Human Ecology*, 2nd Edition. Prentice Hall International, London, UK, 1998.
7. Wang, J., Jamison, D.T., Bos, E., Preker, A., and Peabody, J.: *Measuring Country Performance on Health: Selected Indicators for 115 Countries.* Human Development Network. Health, Nutrition and Population Series. The World Bank, Washington, DC, 1999.
8. World Health Organization: *The World Health Report 1999: Making a Difference.* World Health Organization, Geneva, 1999.
9. Mearns, L.O., Rosenzweig, C., and Goldberg, R.: Mean and Variance Change in Climate Scenarios: Methods, Agricultural Applications and Measures of Uncertainty. *Climatic Change* 34 (1997), pp.367–396.
10. Karl, T.R. and Knight, R.W.: Secular Trends of Precipitation Amount, Frequency and Intensity in the United States. *B. Am. Meteorol. Soc.* 79 (1998), pp.231–241.
11. Hewitt, J. and Burton, I.: *The Hazardousness of a Place: A Regional Ecology of Damaging Events.* University of Toronto, 1971.
12. Kane, S.J., Reilly, J., and Tobey, J.: An Empirical Study of the Economic Effects of Climate Change on World Agriculture. *Climatic Change* 21 (1992), pp.17–35.
13. Yohe, G., Neumann, J., Marshall, P. and Amaden, H.: The Economic Cost of Greenhouse-Induced Sea-Level Rise for Developed Property in the United States. *Climatic Change* 32 (1996), pp.387–410.
14. Downing, T.E. (ed.): *Climate Change and World Food Security.* Springer, Berlin, 1996.
15. Schimmelpfennig, D. and Yohe, G.: Vulnerability of Crops to Climate Change: A Practical Method of Indexing. In: G. Frisvold and B. Kuhn (eds): *Global Environmental Change and Agriculture: Assessing the Impacts.* Edward Elgar Publishing, Northampton, MA, USA, 1999.
16. Yohe, G. and Schlesinger, M.: Sea-Level Change: The Expected Economic Cost of Protection or Abandonment in the United States. *Climatic Change* 38 (1998), p.437–472.
17. Smit, B., Burton, I., Klein, R.J.T., and Street, R.: The Science of Adaptation: A Framework for Assessment. *Mitig. Adapt. Strat. Global Change* 4 (1999), pp.199–213.
18. Downing, T.E., Ringius, L, Hulme, M., and Waughray, D.: Adapting to Climate Change in Africa. *Mitig. Adapt. Strat. Global Change* 2 (1997), pp.19–44.
19. Pittock, B. and Jones, R.N.: Adaptation to What and Why? *Environ. Monit. Assess.* 61 (2000), pp.9–35.
20. Parry, M. Park, S., Dockerty, T., Harrison, P., Jones, P., Harrington, R., Osborn, T., Rounsevell, M., Shao, J., Viner, D., Wheeler, T., Arnell, N., Butterfield, R., Park, J., and Rehman, T.: Investigation of Thresholds of Impact of Climate Change on Agriculture in England and Wales. Research Report Number 4. Jackson Environment Institute, University of East Anglia, Norwich, UK, 2002.
21. De Vries, J.: Analysis of Historical Climate-Society Interaction. In: R.W. Kates, J.H. Ausubel, and M. Berberian (eds): *Climate Impact Assessment.* John Wily and Sons, New York, 1985, pp.273–291.
22. De Freitas, C.R.: The Hazard Potential of Drought for the Population of the Sahel. In: J.I. Clarke, P. Curson, S.L. Kayastha, and P. Nag (eds): *Population and Disaster.* Blackwell, Oxford, UK, 1989, pp.98–113.
23. Smit, B., Burton, I., Klein, R.J.T., and Wandel, J.: An Anatomy of Adaptation to Climate Change and Variability. *Climatic Change* 45 (2000), pp.223–251.
24. Sachs, J.D.: *Macroeconomics and Health: Investing in Health for Economic Development.* Report of the Commission on Macroeconomics and Health. World Health Organization, Geneva, 2001.
25. Yohe, G. and Tol, R.: Indicators for Social and Economic Coping Capacity – Moving Toward a Working Definition of Adaptive Capacity. *Global Environ. Change* 12 (2002), pp.25–40.

26. World Health Organization: *The Global Eradication of Smallpox – Final Report of the Global Commission for the Certification of Smallpox Eradication.* World Health Organization, Geneva, 1980.
27. Nelson, A.M.: The Cost of Disease Eradication – Smallpox and Bovine Tuberculosis. *Ann. NY Acad. Sci.* 894 (1999), pp.83–91.
28. Henderson D.A.: Eradication: Lessons from the Past. *MMWR* 48 (1999), pp.16–22.
29. Gomez-Caminero, A., Howe, P., Hughes, M., Kenyon, E., Lewis, D.R., Moore, M., Ng, J., Aitio, A., and Becking, G.: *Environmental Health Criteria 224: Arsenic and Arsenic Compounds.* Second Edition. International Programme on Chemical Safety, World Health Organization, Geneva, 2001.
30. Morales, K.H., Ryan, L., Kuo, T-L., Wu, M-M., and Chen, C-J.: Risk of Internal Cancers from Arsenic in Drinking Water. *Environ. Health Persp.* 108 (2000), pp.655–661.
31. Kaplan, G.A., Pamuk, E.R., Lynch, J.W., Cohen, R.D., and Balfour, J.L.: Inequality in Income and Mortality in the United States: Analysis of Mortality and Potential Pathways. *BMJ* 312 (1996), pp.999–1003.
32. Lynch, J.W., Smith, G.D., Kaplan, G.A., and House, J.S.: Income Inequality and Mortality: Importance to Health of Individual Income, Psychosocial Environment, or Material Conditions. *BMJ* 320 (2000), pp.1200–1204.
33. Ross, N.A., Wolfson, M.C., Dunn, J.R., Berthelot, J-M., Kaplan, G.A., and Lynch, J.A.: Relation between Income Inequality and Mortality in Canada and in the United States: Cross Sectional Assessment using Census Data and Vital Statistics. *BMJ* 320 (2000), pp.898–902.

3 The global resurgence of vector-borne diseases: lessons learned from successful and failed adaptation

Duane J. Gubler and Mark L. Wilson

ABSTRACT: Vector-borne diseases have historically been among the most important public health problems facing humans. In the first half of the 20th century, effective disease prevention and control programs were successfully designed and implemented; these primarily controlled arthropod vectors to interrupt pathogen transmission. By 1970, malaria, yellow fever, dengue fever, plague, and other regionally important diseases had effectively been brought under control, relegating the once great plagues of humans to relatively unimportant regional public health problems. Success resulted in complacency, which in turn led to policy changes that resulted in dramatically decreased resources and deterioration in the public health infrastructure needed to deal with vector-borne diseases. To make matters worse, support for research on new and more effective vector control strategies and approaches was greatly reduced. These public health changes, combined with population growth, modern transportation, increased movement of people, animals, and commodities, and other demographic and societal changes over the past 50 years, resulted in a global resurgence of many of the old diseases that were once effectively controlled (malaria, dengue fever, yellow fever, and plague), as well as other vector-borne diseases, in the waning years of the 20th century. We review the history of successful interventions against three such diseases (yellow fever, dengue fever, and malaria) and identify lessons learned from the successes and failures of the past for application to developing effective prevention strategies to reverse the trend of resurgent vector-borne diseases, and to mitigate any adverse effects of future climate change.

1. INTRODUCTION

During the past two decades, vector-borne diseases such as yellow fever, dengue fever, and malaria have become increasingly prevalent and more widespread [1,2]. These re-emerging or resurging diseases now cause even more morbidity and mortality than occurred during the early 1900s, when successful prevention and control programs were initiated. As many as 100 million people each year are afflicted with dengue fever, with approximately 500,000 experiencing the severe form of disease, dengue hemorrhagic fever, which has an average 5% case fatality rate [3]. An estimated 200,000 people are infected annually with yellow fever, with up to 20 to 30,000 deaths, but this is believed to be an underestimate due to inadequate surveillance [4]. Estimates for malaria range as high as 200 to 300 million cases and 1.1 million deaths per year [5]. As staggering as these figures are, it is the rate at which the incidence of these diseases has increased in last two decades that is of concern. This is particularly problematic because these diseases were successfully controlled in the 1950s and 1960s [1,2].

What has led to this dramatic resurgence of disease? This chapter reviews the experience of yellow fever, dengue, and malaria, with attention to insights that can be gleaned to help direct public health policy and intervention aimed at reversing this alarming trend, lessons learned from the past, and what kind of renewed action is required to reduce the present burden of these diseases and lessen the chances that climate-related risks will exacerbate the current situation.

2. YELLOW FEVER AND DENGUE/DENGUE HEMORRHAGIC FEVER

Yellow fever and dengue/dengue hemorrhagic fever are both caused by mosquito-borne viruses belonging to the family Flaviviridae, genus Flavivirus [6]. Yellow fever, the prototype of this group, is caused by a single virus; dengue and dengue hemorrhagic fever are caused by four closely related viruses designated DENV-1, DENV-2, DENV-3, and DENV-4 [6]. Both yellow fever and dengue viruses are maintained in primitive cycles involving lower primates and canopy dwelling mosquitoes in the rainforests of Africa, and in the Americas (yellow fever) [4] and Africa and Asia (dengue) [3] (Fig. 3.1). Both diseases can be transmitted in an urban cycle by the highly domesticated *Aedes aegypti* mosquito, which is a highly efficient epidemic vector because of its close association with humans and its blood-feeding behavior (it will often bite several people before feeding to repletion) [3].

Both dengue and yellow fever virus infections in humans cause a wide spectrum of illnesses, ranging from inapparent infection to severe and sometimes fatal hemorrhagic disease [3,4]. The majority of infections present as a mild to severe febrile illness, while only a small percentage of patients progress to hemorrhagic disease. The case fatality rate for dengue hemorrhagic fever is about 5% compared to about 20% for yellow fever.

2.1 A case of effective adaptation

Yellow fever has never been documented in Asia, and dengue fever has never been effectively controlled in that region. This discussion focuses primarily on the Americas, where both yellow fever and dengue fever were effectively controlled by controlling

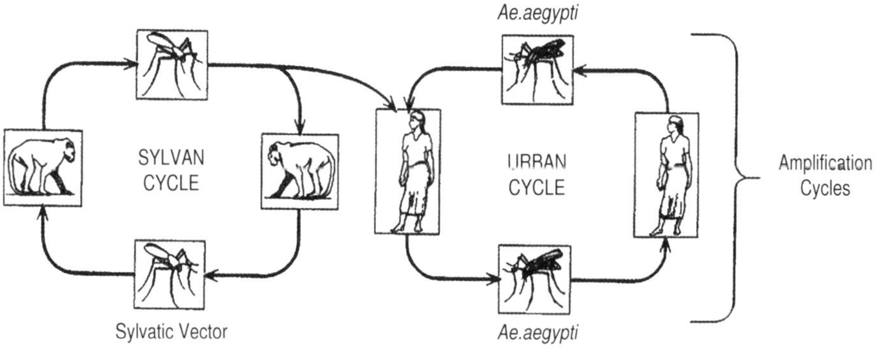

Figure 3.1 Yellow fever and dengue virus transmission cycle

the principal urban mosquito vector, *Ae. aegypti*. Yellow fever was controlled in West Africa by vaccination.

Yellow fever and dengue fever are old diseases: their initial geographic expansion was closely tied to the global spread of *Ae. aegypti* mosquitoes from Africa to other parts of the tropics as shipping and commerce expanded in the 17th and 18th centuries [7]. Yellow fever viruses almost certainly had an African origin, and dengue viruses may have evolved in Asia. Regardless of their origin, the viruses were brought out of the rainforest to villages by humans or lower primates, where a domestic cycle was propagated by *Ae. aegypti* (yellow fever/Africa) or *Ae. albopictus* (dengue/Asia). From the villages, the viruses were introduced into port cities where *Ae. aegypti* became established, and from there they were transported to port cities all over the tropics by sailing vessels.

Yellow fever was introduced into the Western Hemisphere in the early 1600s; the first recorded epidemic occurred in 1648 [8]. From then until the middle of the 20th century, yellow fever was the scourge of the Americas, with major epidemics occurring regularly in the tropics and as far north as New York City. The tropical *Ae. aegypti* mosquito vector would expand and contract its geographic distribution annually with seasonal commerce and shipping.

The term yellow fever was used first during the 1750 epidemic in Barbados. It was not until 1900, however, that it was documented that the virus was transmitted to humans by the bite of an *Ae. aegypti* mosquito [9] (Table 3.1). Once the mosquito connection was established, control measures aimed primarily at the larval stage of the mosquito were implemented and yellow fever was effectively controlled, first in Havana, Cuba, in 1901 and then in Panama in 1904. *Ae. aegypti* had adapted closely to humans and preferred to live in houses with humans, blood-feed on humans, and lay their eggs in artificial containers in the domestic environment [3,7]. While this made them highly efficient epidemic vectors, it also made them easy to control because they were localized in the domestic environment, breeding primarily in artificial containers that were used for water storage or that collected water during rainfall.

The highly effective *Ae. aegypti* control programs that eliminated this mosquito in the Americas [10] were based on controlling larval mosquitoes in domestic water storage

Table 3.1 Yellow fever, a case study of effective adaptation (prevention and control)

1648–1900	Yellow fever was the scourge of the Western Hemisphere.
1900	Walter Reed discovers yellow fever virus is transmitted by *Ae. aegypti* mosquitoes, Havana, Cuba.
1901	William Gorgas develops and implements an effective *Ae. aegypti* control program and controls yellow fever in Havana, Cuba.
1904	Gorgas applies similar strategy in Panama, controlling yellow fever – allows Panama Canal to be completed.
1904–1919	Elimination of yellow fever from key urban centers (Panama, 1905; Rio de Janeiro, 1906; Veracruz, 1907; Guayaquil, 1919).
1937	A live-attenuated yellow fever virus vaccine developed.
1938	Fred Soper leads a team of Brazilians to eliminate *Ae. aegypti* from Brazil.
1941	Widespread vaccination for yellow fever implemented in West Africa.
1942	Last major urban epidemic of yellow fever in the Americas.
1946	Pan American Health Organization implements hemispheric *Ae. aegypti* eradication program.

and collection containers with emphasis on both container elimination and treatment with insecticides, especially DDT. The last major urban epidemic of yellow fever in the New World occurred in 1942 in Brazil [4], and of dengue fever in 1945 in the Caribbean islands; a small outbreak also occurred in New Orleans [7]. In 1946, the Pan American Health Organization initiated a hemispheric *Ae. aegypti* eradication program that was highly successful using these methods [10], eliminating the mosquito from all but a few countries in northern South America, the Caribbean islands, and the southern United States, and effectively controlling epidemic yellow fever and dengue fever in the Americas [11] (Fig. 3.2). An effective, safe, and economical live virus vaccine was developed for yellow fever in 1937, and a French version of the vaccine was used to control the disease in West Africa beginning in 1941 [5] (Table 3.1). However, the vaccine was never used as a primary prevention tool in the Americas. Fortunately, urban epidemics of yellow fever have not recurred in the Americas, but in 1963, after an absence of 18 years, epidemics of dengue fever recurred in Jamaica and Puerto Rico, two countries that had not been successful in eliminating *Ae. aegypti* [12]. In the late 1970s, DEN-1 was introduced to these same islands, beginning a dramatic resurgence of epidemic disease that continues today [7,12].

2.2 The resurgence of disease

The elimination of epidemic dengue fever and yellow fever as public health problems in the American tropics during the 1950s and 1960s led to complacency. Malaria had also been controlled in much of the American region during this time (see below). The result was a redirection of resources from *Ae. aegypti* eradication and control programs to other competing priority public health programs [2,11,12] and the discontinuation of the eradication programs for malaria. These programs were merged in many countries to make the best use of very limited mosquito control resources in the early 1970s.

Another important change in the approach to *Ae. aegypti* control occurred in the early 1970s: ultra low volume (ULV) application of insecticides became widespread as the recommended method for controlling *Ae. aegypti*. Instead of controlling larval mosquitoes in water containers, this method targeted adult mosquitoes using space sprays of very small droplets of insecticides that remained suspended in the air for a

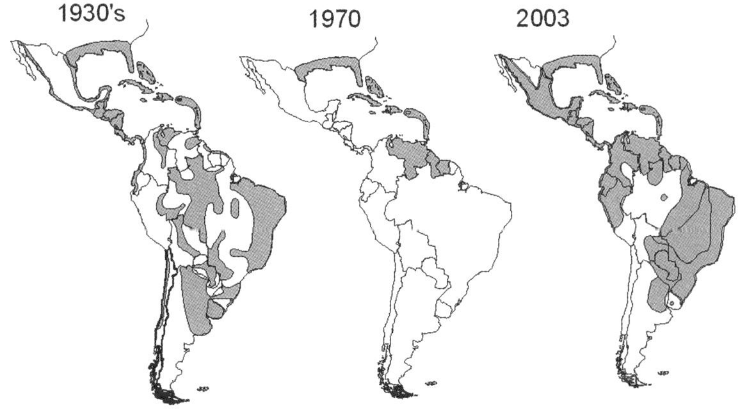

Figure 3.2 Aedes aegypti distribution in the Americas 1930, 1970, and 2003

period of time; the adult mosquito had to be impacted by an insecticide droplet to be killed. This change in control methods led to a major change in health policy: routine mosquito control to prevent epidemics was abandoned in favor of a policy of emergency response with adulticides after epidemics had been reported [11]. This approach to *Ae. aegypti* control was flawed for three reasons. First, space spraying was ineffective in controlling adult *Ae. aegypti* because the spray did not penetrate inside houses where the mosquitoes were most frequently found [11,13]. Second, the ULV sprays did not control immature mosquitoes in water containers. Thus, within hours of the spray, newly emerged adult mosquitoes were feeding on infected humans, continuing the epidemic transmission, even if existing adults had been killed, which they were usually not. Third, surveillance for dengue was poor because it relied on physicians to report dengue-like illness. During interepidemic periods, the index of suspicion for dengue fever was low and cases were usually not recognized until the epidemic was near peak transmission [11]. Thus, the emergency adult mosquito control, which was ineffective to begin with, was always too little and too late to have any impact on epidemic dengue fever transmission [14].

Beginning in the 1970s, after the eradication program had been abandoned, *Ae. aegypti* began to reinfest the tropical countries of Central and South America [11]. Because the mosquito surveillance and control programs for malaria, yellow fever, and dengue fever had been merged, they were more focused on anopheline mosquitoes and malaria, and the new infestations of *Ae. aegypti* were frequently not recognized for months, or even until a dengue epidemic occurred [7]. The invasion and spread of *Ae. aegypti* in the American tropics continued unabated during the 1970s, 1980s, and 1990s [2,11,12]. At the beginning of the 21st century, *Ae. aegypti* has a greater geographic distribution in the Americas, and occurs in higher population densities, than at any other time in history (Fig. 3.2) [14].

Other trends that were occurring at the same time included a dramatic, uncontrolled urbanization of the region and increased frequency of movement of people both within and between regions of the world via jet airplane. This increased the frequency of dengue virus introductions of all four serotypes into large urban human populations that were infested with *Ae. aegypti* [3,7,12]. Beginning in the late 1970s, and intensifying in the last two decades of the 20th century, there were repeatedly larger epidemics of dengue fever caused by all four virus serotypes, the development of hyperendemicity (the co-circulation of multiple virus serotypes in a community), and the emergence of

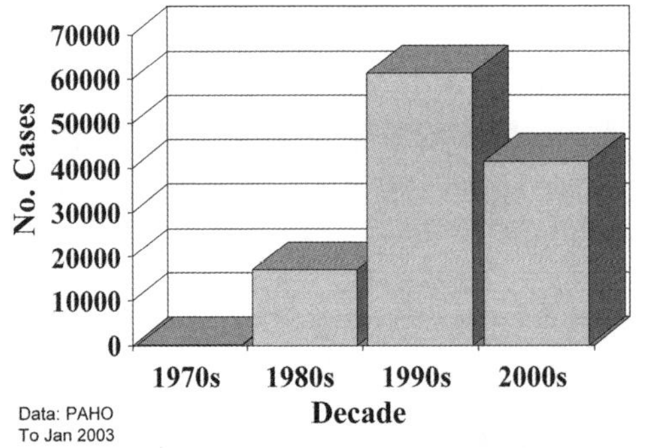

Figure 3.3 The emergence of dengue hemorrhagic fever in the Americas, by decade

the severe and fatal form of disease, dengue hemorrhagic fever, for the first time in history [11,12,14]. In a short period of 20 years, dengue hemorrhagic fever evolved into one of the most important public health problems of the region (Fig. 3.3). In 2003, 28 years after the first sporadic cases were documented, 28 American countries have reported confirmed dengue hemorrhagic fever cases, and major epidemics have occurred repeatedly in many of those countries.

2.3 Prospects for the future

Fortunately, urban yellow fever has not re-emerged in epidemic form in the American tropics, but the region is at the highest risk for urban transmission in over 50 years [15]. Urban human population densities are unprecedented, with more than 150 million people living in urban areas infested with *Ae. aegypti.* Yellow fever is still enzootic in the Amazon Basin, and numerous cities with over 1 million population in the enzootic area are infested with *Ae. aegypti* (Fig. 3.4) [2]. In the last decade, a number of incidents of urban or peri-urban yellow fever transmission have been documented in areas of tropical America [15]. The concern of public health officials is that if epidemic urban yellow fever transmission begins in any of these American cities, it will rapidly spread to other cities in the region, and from there potentially to the Asian-Pacific region, where yellow fever has never occurred, much as dengue has spread around the world via air travel in the past 20 years. This scenario would cause an unprecedented global public health emergency because over 2 billion people live in areas of risk in the Asia-Pacific region.

Yellow fever has never been documented in Asia [16]. It is not known whether it was ever introduced to the region in past centuries when the disease was causing massive epidemics in the Americas. In that era, the virus, the mosquito, and humans moved much more slowly than they do in today's era of jet airplanes. The viruses (yellow fever and dengue) essentially maintained a transmission cycle among the mosquitoes, which were breeding in stored water on the ships, and humans aboard those ships [7]. Clearly the slave trade, with large numbers of humans on any individual ship, increased the chances of virus survival during the voyage across the Atlantic. The logistics of travel

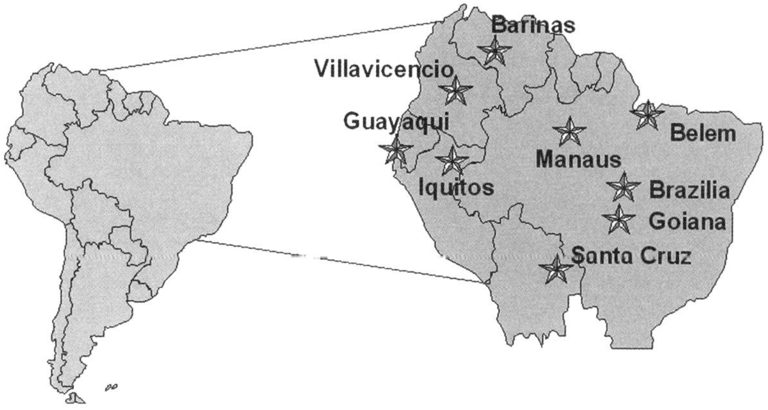

Figure 3.4 Cities in the Amazon yellow fever enzootic region that have human populations of ≥1 million

to Asia from the Americas or from West Africa were not as conducive to movement of the viruses, especially before 1912, when the Panama Canal opened.

Other possible explanations for the lack of yellow fever transmission in Asia include the high prevalence of heterotypic flavivirus antibody (DENV-1, DENV-2, DENV-3, DENV-4, Japanese encephalitis, and others) in those populations, which although not protective for yellow fever infection, may have a modulating effect on the infection, down-regulating it to a milder illness with lower viremia levels and thus decreasing the probability of secondary transmission [17]. Also, there has been some suggestion that Asian strains of *Ae. aegypti* are less competent vectors of yellow fever virus than their American counterparts [16,18]. Finally, evolutionary exclusion may simply prevent yellow fever virus from becoming established in areas where multiple, closely related flaviviruses occur. Most likely, a combination of all of these factors has contributed to preventing yellow fever from being transmitted in Asia in the past.

The logistic and demographic factors that influence virus spread at the beginning of the 21st century, however, are quite different from the past. First, tens of millions of people travelling by jet airplane provides the ideal mechanism to transport pathogens to exotic locations in people incubating the diseases. It is estimated that in 1998, nearly 16 million tourists traveled to yellow fever endemic countries, 75% of them to countries in the Americas (Fig. 3.5).

Ecotourism has increased in recent years, and since 1996, six tourists have died in the United States and Europe as a result of infection with yellow fever virus acquired during travel to endemic countries without vaccination. The point is that if urban epidemic transmission of yellow fever begins in the Americas, there could be thousands of infected people traveling to Asian-Pacific countries where *Ae. aegypti* exposure is high, thus dramatically increasing the probability that yellow fever transmission will occur in Asia. Moreover, if yellow fever viruses were successful in becoming established in the Asia-Pacific region, it would most likely be misdiagnosed as dengue hemorrhagic fever, leptospirosis, rickettsiosis, hantavirus disease, or malaria, thus allowing it to spread to numerous countries before an effective control program could be mounted.

There is a vaccine for yellow fever; it is a live attenuated virus vaccine that is probably the most effective, safe, and economical vaccine available [4]. Yet it is not used in yellow fever endemic countries for primary prevention; instead is used as an emergency response

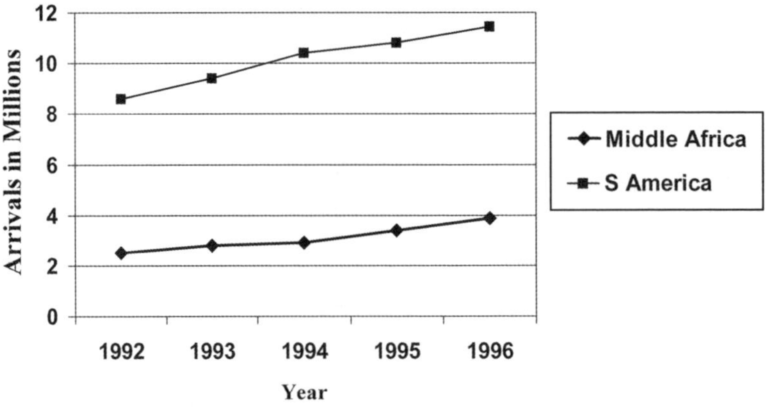

Figure 3.5 US tourist arrivals to yellow fever endemic countries (data taken from Ref. 19)

tool to control epidemics after they have been reported. This is a flawed public health policy because yellow fever surveillance, like dengue surveillance, is poor and insensitive; by the time an immunization program is mounted, the epidemic has usually peaked and is beginning to wane. A classic example was the yellow fever outbreak in Abidjan, Côte d'Ivoire in 2001; the last confirmed case of yellow fever occurred a month before the emergency vaccination program was implemented [20]. Also, yellow fever vaccine stocks are limited; there would not be enough vaccine for American cities, let alone Asia and Pacific cities, if urban transmission were to begin. Moreover, few countries in either region have the mosquito control infrastructure in place to mount an effective prevention program based on *Ae. aegypti* control. Thus, whether or not there is documented yellow fever transmission in the Asian-Pacific region, there will be an "epidemic of panic" that will make the 1994 Indian plague and the 2003 SARS incidents pale by comparison [21].

3. MALARIA

Whether measured by the number of cases, deaths, or any index of economic impact, malaria represents the most important vector-borne disease in the world today [5,22]. This history-altering disease has been important for much of recorded time, and like all vector-borne diseases, including dengue and yellow fever, it is strongly influenced by climatic factors. Malaria is mainly found in tropical and subtropical regions today, but in the past it was common in temperate regions during the summer months; it often has a highly seasonal pattern [23,24]. The distributions and abundances of mosquito vector species are climate-sensitive. Although demographic, social, behavioral, economic, biological, and ecological factors are the most important, there has been much speculation about climate as an important determinant in the recent resurgence of malaria [25–27].

Malaria has been recognized for thousands of years, beginning with Hippocrates (b. 460 BC), who determined that "fevers" occurring as quartan, tertian, and quotidian were associated with this illness [28]. Earlier, Herodotus (b. 485 BC) considered swamps in Egypt to be linked to malaria, although this association had not identified a biophysical mechanism. Others recognized the importance of this *maladie*, but none were able to develop effective preventions or cures. The contemporary word *malaria* was derived in the 1700s from the Italian for "bad air" because of the long association with swamps, marshes, and other mosquito-breeding water. Later, Napoleon and Volta developed a "bad smell" map during invasions of Egypt, to reduce the risk of malaria among French troops. It was not until 1880 that Laveran first observed microscopically the development of the parasite in blood. Shortly thereafter (1897), Ross demonstrated that specific *Anopheles* mosquitoes could transmit malaria. The association of *Plasmodium* species as the cause of malaria in humans is thus a long one that has affected human economic and social development in every reach of the world.

3.1 Human disease

The parasites that cause human malaria include four protozoan species belonging to the genus *Plasmodium*: *P. falciparum*, *P. vivax*, *P. ovale*, and *P. malariae* [28]. Severity of disease is species-specific, and most fatalities are due to *P. falciparum* infection, which can involve renal failure or cerebral disease; disease severity varies with age, exposure, and immune status. Infections with other *Plasmodium* species are typically less severe, rarely fatal, but potentially important because they can be acquired in areas of transmission and remain latent in humans who move to uninfected areas where mosquito vectors are present. This recrudescence of malaria was a major factor in seasonal epidemics of malaria in temperate regions of the world in past centuries.

Malaria treatment, just like the disease itself, also has a long history, dating back more than 2,000 years when the Chinese herbalists used infusions of Qinghao, or sweet wormwood (*Artemisia annua*), as a cure [28]. In the 1600s, bark from the Cinchona tree was used, and by 1820, quinine was identified as the active ingredient. Many synthetic compounds have since been developed, but resistance to most has become increasingly widespread. Thus, the role of parasite evolution complicates analyses of how climate change and other demographic and societal factors are associated with the current reemergence of malaria.

3.2 Mosquito vectors

All of the roughly 60 mosquito species that transmit human malaria parasites are in the genus *Anopheles* [29]. Various environmental conditions play a critical role in the distribution of each species. The four stages of these mosquitoes (egg, larva, pupa, adult) have ecological requirements that determine their survival and abundance. Environmental variables also affect adult feeding frequency, host choice, longevity, and reproduction, factors that are critical to pathogen transmission. As a result, malaria prevention had historically focused on avoiding or eliminating habitats that permitted *Anopheles* egg laying and larval development. As early as 1892, larval control was reported after applying kerosene to water where mosquito larvae were present (this disrupts breathing through the spiracle). Beginning in the 1920s, arsenic and Paris green (a copper acetoarsenite) were used as a chemical larvicide against *Anopheles* mosquitoes. But it was not until the discovery of DDT in 1942 and its successful use in Italy during the mid-1940s that effective chemical control of mosquitoes was realized.

3.3 Malaria eradication successes and failures

By the end of World War II, not only was DDT proving to be a highly effective insecticide, but also chloroquine was available as a very effective treatment for malaria. In 1947, the U.S. Public Health Service began a US$7 million, 5-year campaign aimed at eliminating malaria from the United States, and by 1952 the program ended as a complete success. This encouraged the World Health Organization (WHO) in 1957 to embark on an 8-year program aimed at the global eradication of malaria, with DDT and chloroquine as the primary weapons, and active case surveillance to identify areas of focus [30]. During the initial period (1958–1963), US$430 million (US$1.914 billion in today's currency) was spent on the malaria eradication effort. The program eliminated or dramatically reduced malaria in North America, most of Europe, the former Soviet Union, and parts of Latin America and Asia. Interestingly, only a few countries in Africa that were considered to have had sufficient infrastructure and resources in place were included in the program. Ironically, in 1976, G.W. Jeffrey [31] commented that "the science of malaria control, developed slowly and painfully from the beginning of the century to a relatively high state of sophistication, was almost overnight converted to a rather simplistic technology of malaria eradication, which basically required that one knew how to deliver 2 grams of something to 2 square meters of a sometimes elusive interior wall, and to manage a hopefully ever-diminishing Kardex file of cases". The magic bullets had been successful, but for how long?

The initial euphoria over local elimination quickly turned to anxious concern as *Anopheles* resistance to DDT became increasingly widespread. This should not have come as a surprise, because DDT resistance had already been documented in some areas by the early 1950s, and evolutionary theory combined with previous experience could have anticipated expanding resistance. In addition, as early as the 1960s, chloroquine was found to be increasingly less effective at curing malaria in some areas. It was soon accepted that *P. falciparum* strains in South America and Southeast Asia had become resistant to treatment with chloroquine, and that resistance spread to other parts of the world. During the Viet Nam conflict, chloroquine resistance became widespread in Asia. Loss of confidence in the eradication campaign strategy became widespread, and WHO officially ended the program in 1972, with most funding phased out a few years later. Unfortunately, the eradication program was not replaced with a new control strategy that could maintain the progress made in the previous decade.

At the same time, and in part due to the success of the eradication campaign, international funding for malaria control and basic research declined dramatically in the 1970s. Furthermore, health and social sector spending during the 1970s and 1980s was also declining in many countries, exacerbating the resurgence of malaria in many areas where this disease previously had declined or disappeared. In some countries, global economic changes, civil strife, and war led to deteriorating health infrastructure, migration, or forced movement of people, and changes in land use or settlement patterns, adding more to the growing malaria problem. In many countries, malaria had returned to produce an even greater burden of disease than before the eradication campaign.

Beginning in the 1980s, the international community began to recognize the human toll that malaria was once again taking, and began to fund research aimed at prevention and treatment (new prophylactic drugs, human vaccines, mosquito-based transmission inhibitors). This effort culminated in the 1997 launching of the WHO Roll Back Malaria Campaign, a program that is conceived, designed, and structured very differently than the malaria eradication campaign of four decades earlier.

3.4 Malaria today

Despite enormous efforts during the past century, roughly 40% of the world's population is still at risk of malaria, with case estimates ranging from 120 to 300 million people per year. Regardless of which numbers are correct, malaria is unquestionably one of the world's most important infectious diseases. Misdiagnosis and over- and under-reporting make it difficult to estimate the true impact of malaria, particularly in underdeveloped countries of the tropics. Estimates of malaria deaths range from 800,000 to 3 million per year. Most (roughly 90%) occur in Africa, but such statistics vary because of inaccurate surveillance and the complicated definition of "cause" where many other infectious agents that cause a malaria-like illness, co-infections, or other factors (HIV/AIDS, nutrition). Even if malaria incidence is overestimated by these confounding etiologies, there is little doubt that is has reemerged in much of Africa and other parts of the world [32–34]. Speculation and some data suggest that climate change may be involved in this upsurge, but this is undocumented and controversial. Yet, many other factors are also incriminated as region-specific causes of this increase, including decreased availability of medical services and anti-malarial drugs, greater resistance to drug treatment (chloroquine and other anti-malarials), reduced anti-vector campaigns, lack of insecticides, increased insecticide resistance, movement of infected people into areas of low transmission, population growth and changes in landuse/land cover through deforestation and agricultural development, and other regional factors. Certainly, any or all of these changes could contribute to increased malaria risk, but the role of each will vary among regions, ecological conditions, and intervention efforts. The determinants of increased malaria in many parts of the world have not been adequately studied, making it difficult for development and public health policy makers to develop and implement effective prevention strategies. Nevertheless, basic knowledge of vector ecology, exposure risk, and landuse changes could be applied to reverse the expansion of malaria risk, and decrease transmission in holoendemic areas [35].

Interestingly, in some areas of the world, malaria has declined or remained roughly stable during recent times. Incidence in parts of Asia, for example, has generally been decreasing during the past decade (in China, annual malaria cases numbered perhaps 4 million in the 1970s and diminished to less than 100,000 in the 1990s). At the same time, only slight increases were observed in the Americas during this period. Thus, although malaria continues to be an increasingly serious health concern throughout much of the tropics, there is considerable regional variation and changing incidence. It appears that local conditions (vegetation, human behavior, economic conditions, and effective control programs) kept malaria from returning to some subtropical regions, but it has become an even greater burden in other parts of the world such as Africa [34] than it was before the eradication campaign of a few decades ago.

4. DETERMINANTS OF ADAPTIVE CAPACITY AND REQUIREMENTS FOR PUBLIC HEALTH PREVENTION

The history of successes and failures of interventions against these vector-borne diseases provides important insights concerning our ability to adapt to future conditions, and reduce or prevent future problems (Table 3.2). First, the range of available technological options is considerable, comprising many "low-tech" interventions that have been effective in reducing abundance of mosquito vectors or limiting their contact with people. These include removing breeding sites or reducing vector-human contact,

Table 3.2 Ranking of determinants of adaptive capacity and requirements for public health prevention[a]

Determinants of adaptive capacity	*Requirements for public health prevention*
Range of available technological options (4)	Awareness that problem exists (5)
Availability and distribution of resources (3)	Sense that the problem matters (5)
Structure of critical institutions (4)	Understanding of the causes (3)
Human capital (5)	Capability to intervene (2)
Social capital (4)	Political will to influence (5)
Access to risk spreading (1)	
Ability to manage information (3)	
Public's perceived attribution (4)	

Note
a Scale = 1 (low) to 5 (high).

for example. For *Ae. aegypti*, containers suitable for egg laying can be eliminated, and for *Anopheles* vectors of human *Plasmodium* species, it is possible to remove or treat breeding sites or limit their creation. Other basic changes in housing (screens, air conditioners, indoor residual spraying) that would reduce human exposure to vectors could be employed. Antimicrobial treatment of malaria, and supportive therapy and vaccines for dengue and yellow fever, will reduce disease severity and prevent disease transmission. Resources for this kind of adaptive capacity, however, are less available to the poor and will require financial assistance from developed countries. A greater commitment from foundations and international funding agencies is needed to address the serious problems that these vector-borne diseases represent.

Critical institutions include not only those that treat cases (hospitals and clinics) but also those that provide continual disease surveillance and control and laboratories to support these efforts. Again, these are not sophisticated or difficult tasks, but they do require coordination and funding. These are vital services that yield considerable benefits for the resources that are spent.

Human and social capital is critical to the success of interventions against these vector-borne diseases. Trained professionals are in critically short supply. People will be unable to implement effective coping strategies unless there is understanding and political will. This includes knowledge of foci of elevated transmission, local conditions that lead to greater vector abundance or human contact, and appropriate interventions for specific conditions or groups. Local governmental agencies are critical to implementing prevention and control, and should work closely with nongovernmental organizations (NGOs) and citizens' groups to ensure adequate surveillance and compliance. Individual citizens play an extremely important role in enacting measures that a government establishes.

For implementation of public health prevention measures to be effective, it is vital that the community recognize the existence of various vector-borne diseases and consider them to be important. In many contexts where dengue or malaria are present, people have other concerns that often are more basic (employment, food, shelter, etc.). Thus, it can be difficult to encourage people to see these diseases as significant factors that affect their well-being. Yet, if it is made clear that many simple interventions can reduce risk at minimal cost of time or resources, and that participating in efforts to reduce risk will affect the lives of many others as well, public health prevention is likely to be more effective. Even if people do not understand the detailed biological processes, simple interventions are more easily accepted if they recognize the potential benefits that will result, and if it is done in partnership with responsible government agencies.

5. IMPLICATIONS FOR THE FUTURE

What lessons can be learned from the recent histories of malaria, dengue, and yellow fever epidemiology that might guide future public health practice? In the context of climate change and increasing climate variability, what interventions will have both short-term and long-term benefits regardless? What strategic goals should guide policy and shape practice?

The first lesson learned from these historical experiences is that most vector-borne diseases can be effectively controlled or prevented if adequate trained personnel and resources are made available to endemic countries. This has important implications for effective adaptation to potential adverse events as a result of climate change. Second, the strategy of *eradication* will never succeed. The goals of *control*, and possibly of *local elimination*, are the only logical strategies to consider. Eradicating these vector-borne diseases means killing every individual vector or all of the associated microbes throughout the world. Global eradication of competent *Anopheles* and *Aedes* mosquitoes is technologically impossible. Similarly, global eradication of *Plasmodium* spp., dengue, and yellow fever viruses would not be possible because all occur in enzootic primate cycles. No vaccine exists for dengue or malaria, and it is unlikely that effective immunization will be possible in the foreseeable future. Even if effective vaccines were available, the prospect of immunizing billions of people in endemic areas is unrealistic. Despite having an effective yellow fever vaccine, it has not been used effectively for primary prevention in the past 30 years [35].

Third, control efforts will ultimately fail unless they are designed to be *sustainable*. This is a challenge and a problem because short-term successes often lead to complacency and a redirection of resources. The effort required to maintain effective control will always be less than that needed to re-establish it following a lapse. Methods for controlling *Ae. aegypti* and various *Anopheles* mosquitoes are well understood, and many are technically straightforward. Community education and participation, local mosquito surveillance, and development and introduction of new, effective, antivector technologies are all critical. Successful interventions will be vector-specific and different for dengue/yellow fever than for malaria. Nevertheless, we have the capacity to dramatically reduce *Anopheles*- and *Aedes*-associated diseases provided adequate trained personnel and resources are made available. Until such efforts are undertaken, morbidity and mortality are likely to continue to rise. Programs to reduce any mosquito-borne disease will benefit from removing or controlling breeding sites. For malaria, renewed and expanded use of indoor residual spraying of DDT is warranted, because it remains environmentally safe when used indoors, and is effective against most *Anopheles* and *Aedes* adults. Expanded use of insecticide-treated bednets would also reduce malaria incidence, particularly among children, but care should be taken to forestall insecticide resistance. For dengue and yellow fever, community-based, integrated approaches to eliminate or control *Ae. aegypti* breeding sites could bring back the successes of just a few decades ago. And yellow fever vaccine should be incorporated into the WHO Expanded Immunization Program in endemic/enzootic countries of Africa and the Americas.

Whether climate change will permit dengue, yellow fever, and malaria to gradually expand into regions where they occurred in the past, but are currently absent, is being debated, but it should be remembered that many of these regions already have the mosquito vectors that have been present for hundreds of years. Current adaptive measures have effectively controlled these diseases for many years, an important lesson to remember as policy makers consider adaptive responses to climate change. Even if climatic conditions favor expansion of competent mosquito vectors, sustained and improved adaptive measures could limit, prevent, or even reverse disease transmission in new geographic areas.

Fourth, control or adaptation strategies require appropriate application of available tools. That many effective interventions exist is not enough if people lack knowledge of how and where they should be deployed. Not only can financial constraints limit access to interventions, but also lack of adequately trained personnel, ignorance, or lack of coordination can reduce effectiveness. Community-based interventions tend to be most sustainable and can be very cost-effective, because they allow for coordinated program implementation by members of the community. Appropriate use of antimalarials will reduce the rate at which resistance develops, while reducing transmission to others. The effectiveness of controlling breeding sites will depend on the cooperation of many people in a community. Implied in this is the principle that education of people, partnerships with governmental agencies, and coordination of action are critical to the success of disease reduction strategies. Technical innovation may be necessary, but it is certainly not sufficient when used in isolation.

Fifth, it is critical to enhance surveillance for both the diseases and the mosquito vectors. In a world of modern transportation where rapid movement of people and goods increasingly allows for rapid transport of exotic microbes and vectors to new geographic regions, public health surveillance becomes critical to prevention and control. Emergency response to outbreaks will reduce morbidity and mortality only if it is implemented early in the epidemic and is effectively targeted. A strong and proactive public health infrastructure that constantly gathers and distributes information, and responds to surveillance data in an appropriate and timely manner, is essential to rapid response. This type of enhanced surveillance and response is even more important where climate-related changes may occur.

Sixth, we must guard against complacency that leads to reduced resources and public health preparedness. Ironically, successful vector-borne disease control programs tend to create their own demise. As fewer cases of disease occur, resources are re-directed because there is less public concern over maintaining effective programs. The risk of disease is ever-changing because of the dynamical nature of societies today. Demographic and ecologic shifts, evolution of parasites and vectors, and rapidly changing interactions of humans and their environments all suggest that we should expect the epidemiology of diseases to be in constant flux. The principal lesson learned from the past, and the challenge for the future, is to not let down our guard in the face of successful disease reduction; if we do, we can anticipate a resurgence that often is more severe or widespread.

Finally, experience that produces new knowledge or effective intervention cannot be sustained in the absence of solid political will, adequate adaptive resources, and committed institutional buy-in. Long-term, win/win, or no-regrets strategies cannot be linked to the typically short terms of politicians. A solid public health system must be apolitical and supported by popular understanding, highly trained professionals, and a strong governmental mandate. Such a public health infrastructure will meet more than just the needs of the reemerging diseases that we have discussed here. As has been demonstrated with the introduction of West Nile virus to the United States, we can anticipate exotic threats to health in the future, and a strong infrastructure will prepare societies to better respond to the unknown, even that which may be climate based. To paraphrase the Cheshire Cat in Lewis Carroll's *Alice in Wonderland*: "If you don't know where you're going, then any road will take you there". We need to carefully consider where we want to go, and which of the many roads is most likely to get us there.

REFERENCES

1. Lederberg, J., Shope, R.E., and Oaks, S.C. Jr. (eds): Emerging Infections: Microbial Threats to Health in the United States. Institute of Medicine, National Academy of Sciences, Washington D.C., 1991.

2. Gubler, D.J.: Resurgent Vector-Borne Diseases as a Global Health Problem. *Emerg. Infect. Dis.* 4 (1998), pp.442–450.
3. Gubler, D.J.: Dengue and Dengue Hemorrhagic Fever. *Clin. Microbiol. Rev.* 11 (1998), pp.480–496.
4. Monath, T.P.: The Arboviruses: Epidemiology and Ecology. In: *Yellow Fever*. Vol. V. CRC Press, Boca Raton, FL, USA, 1988, pp.139–231.
5. WHO: *Communicable Diseases 2002: Global Defence Against the Infectious Disease Threat.* World Health Organization, Geneva, Switzerland, 2003.
6. Westaway, E.G., Brinton, M.A., Gaidamovich, S., Horzinek, M.C., Igarashi, A., Kääriäinen, L., Lvov, D.K., Porterfield, J.S., Russell, P.K., and Trent, D.W.: Flaviviridae. *Interviology* 24 (1985), pp.183–192.
7. Gubler, D.J.: Dengue and Dengue Hemorrhagic Fever: Its History and Resurgence as a Global Public Health Problem. In: D.J. Gubler and G. Kuno (eds): *Dengue and Dengue Hemorrhagic Fever*. CAB International Press, London, 1997, pp.1–22.
8. Carter, H.R.: Yellow Fever. In: L.A. Carter and W.H. Frost (eds): *An Epidemiological and Historical Study of its Place of Origin.* Waverly Press, Baltimore, MD, 1931.
9. Reed, W., Carroll, J., and Agramonte, A.: The Etiology of Yellow Fever. *JAMA* 86 (1901), pp.431–440.
10. Schliessman, D.J. and Calheiros, L.B.: A Review of the Status of Yellow Fever and *Aedes aegypti* Eradication Programs in the Americas. *Mosq. News* 34 (1974), pp.1–9.
11. Gubler, D.J.: *Aedes aegypti* and *Aedes aegypti*-Borne Disease Control in the 1990s: Top Down or Bottom Up. *Am. J. Trop. Med. Hyg.* 40 (1989), pp.571–578.
12. Gubler, D.J. and Trent, D.W.: Emergence of Epidemic Dengue/Dengue Hemorrhagic Fever as a Public Health Problem in the Americas. *Infect. Agents Dis.* 2 (1994), pp.383–393.
13. Newton, E.A.C. and Reiter, P.: A Model of the Transmission of Dengue Fever with Evaluation of the Impact of Ultra-Low Volume (ULV) Insecticide Applications on Dengue Epidemics. *Am. J. Trop. Med. Hyg.* 47 (1992), pp.709–720.
14. Gubler, D.J.: Epidemic Dengue/Dengue Hemorrhagic Fever as a Public Health, Social and Economic Problem in the 21st Century. *Trends Microbiol.* 10 (2002), pp.100–103.
15. Gubler, D.J.: Yellow Fever. In: R.D. Feigin and J.D. Chery (eds): *Textbook of Pediatric Infectious Diseases.* 4th ed. WB Saunders, Philadelphia, PA, 1998, pp.1981–1984.
16. Monath, T.P.: The Absence of Yellow Fever in Asia Hypotheses: A Cause for Concern? *Virus Info. Exch. Newsl.* 6 (1989), pp.106–107.
17. Theiler, M. and Anderson, C.R.: The Relative Resistance of Dengue-Immune Monkeys to Yellow Fever Virus. *Am. J. Trop. Med. Hyg.* 24 (1975), pp.115–117.
18. Gubler, D.J., Novak, R., and Mitchell, C.J.: Arthropod Vector Competence – Epidemiological, Genetic, and Biological Considerations. In: *Proceedings of International Conference on Genetics of Insect Disease Vectors, Bellagio, Italy.* Stipes Publishing, Champaign, IL, USA, 1982, pp.343–378.
19. WTO: *Yearbook of Tourism Statistics*. World Trade Organization, Geneva, 1998.
20. Marfin, A.A. and Burkett, D.: Investigation of Urban Yellow Fever (YF), Abidjan, Cote d'Ivoire. Unpublished CDC trip report, 2001.
21. Fritz, C.L., Dennis, D.T., Tipple, M.A., Campbell, G.L., McCance, C.R., and Gubler, D.J.: Surveillance for Pneumonic Plague in the United States during an International Emergency: A Model for Control of Imported Emerging Diseases. *Emerg. Infect. Dis.* 2 (1996), pp.30–36.
22. Oaks, S.C. Jr. (ed): *Malaria: Obstacles and Opportunities.* National Academy Press, Washington, D.C., 1991.
23. Molineaux, L.: The Epidemiology of Human Malaria as an Explanation of its Distribution Including Some Implications for its Control. In: W.H. Wernsdorfer and I. McGregor (eds): *Malaria: Principles and Practice of Malariology.* Churchill Livingstone, Edinburgh, 1988.

24. Haworth, J.: The Global Distribution of Malaria and the Present Control Effort. In: W.H. Wernsdorfer and I. McGregor (eds): *Malaria: Principles and Practice of Malariology.* Churchill Livingstone, Edinburgh, 1988.
25. Kovats, R.S., Campbell-Lendrum, D.H., McMichael, A.J., Woodward, A., and Cox, J.S.: Early Effects of Climate Change: Do they Include Changes in Vector-Borne Disease? *Philos. Trans. Roy. Soc. B: Biol. Sci.* 356 (2001), pp.1057–1068.
26. Gubler, D.J., Reiter, P., Ebi, K.L., Yap, W., Nasci, R., and Patz, J.A.: Climate Variability and Change in the United States: Potential Impacts on Vector- and Rodent-Borne Diseases. *Environ. Health Persp.* 109 (2001), pp.223–233.
27. Reiter, P.: Climate Change and Mosquito-Borne Disease. *Environ. Health Persp.* 109 (2001), pp.141–161.
28. Bruce-Chwatt, L.J.: History of Malaria from Prehistory to Eradication. In: W.H. Wernsdorfer and I. McGregor (eds): *Malaria: Principles and Practice of Malariology.* Churchill Livingstone, Edinburgh, 1988, pp.1–69.
29. Russell P.F., West, L.S., and Manwell, R.D.: *Practical Malariology.* 2nd ed. Oxford University Press, New York, 1963.
30. International Development Advisory Board: *Malaria Eradication: Report and Recommendations of the International Development Advisory Board.* International Development Advisory Board, Washington, D.C., 1956.
31. Jeffery, G.M.: Malaria Control in the Twentieth Century. *Am. J. Trop. Med. Hyg.* 25 (1976), pp.361–371.
32. Campbell, C.C.: Malaria: An Emerging and Re-emerging Global Plague. *FEMS Immunol. Med. Mic.* 18 (1997), pp.325–331.
33. Wongsrichanalai, C., Pickard, A.L., Wernsdorfer, W.H., and Meshnick, S.R.: Epidemiology of Drug-Resistant Malaria. *Lancet Infect. Dis.* 2 (2002), pp.209–218.
34. Kitron, U. and Spielman, A.: Suppression of Transmission of Malaria through Source Reduction: Antianopheline Measures Applied in Israel, the United States, and Italy. *Rev. Infect. Dis.* 11 (1989), pp.391–406.
35. Gubler, D.J.: The President's Address. Prevention and Control of Tropical Diseases in the 21st Century: Back to the Field. *Am. J. Trop. Med. Hyg.* 65 (2001), pp.v–xi.

4 Ecology, climate, and campylobacteriosis in New Zealand

Philip Weinstein and Alistair Woodward

ABSTRACT: Campylobacteriosis is a gastrointestinal disease that can be spread through water, and its transmission cycle can by influenced by climate in a variety of ways. We argue that in New Zealand, climate change is likely to increase the disease burden from campylobacteriosis, and adaptation will be necessary to minimize such public health impacts. Infection rates with *Campylobacter* are high in New Zealand by international standards (up to 300 notified cases per 100,000 population), and a significant proportion of cases is believed to relate to exposure to fecally contaminated drinking or recreational waters. New Zealand has numerous, diverse, and often untreated freshwater resources, as well as high stocking densities that result in a fecal load on the environment 40 times that attributable to the human population. The rate at which *Campylobacter* enters freshwater ecosystems and its survival once established are dependent on the "health" of those ecosystems, which is subject to factors such as rainfall, temperature, and ultraviolet exposure. Climate change could affect campylobacteriosis rates through its effects both on the health of freshwater ecosystems and on human and animal behaviors. These include patterns of freshwater use for recreation and exposure to untreated water in other forms. Unlike many countries, New Zealand has peaks in fresh water pathogen levels and peaks in human disease rates that sometimes coincide in summer. Although the current state of knowledge does not allow us to model the potential increase in amplitude and coincidence of these peaks with climate change, there are strong reasons to expect an increasing disease burden, and hence a need for adaptation. Possible responses range from the revegetation of catchments (thus increasing natural water filtration), through improvements in surveillance systems, to altering risk perception and behavior in those communities most likely to be affected (often the socioeconomically disadvantaged). The range of options illustrates the problem of "targeted" versus "universal" adaptation, and the need for interventions that produce benefits in both the short and long term. An accelerated program of reforestation in water catchments is one such "no regrets" intervention, likely to improve both ecosystem health and human health regardless of which climate changes emerge in the future.

1. INTRODUCTION

In recent decades it has become apparent that many "traditional" environmental health problems cannot be solved by "traditional" approaches alone. Rather, we need broader approaches to analyze interactions between humans and biotic and abiotic factors, often drawing on the science of ecology. Nowhere is this more obviously so than in vector- and water-borne diseases, which are influenced by many factors in the physical, social, and biological environments. One important contributor to this complexity is climate. Because ecosystem health and human health are inextricably linked [1], and because climate is a major driver of ecosystem function, it should be no surprise that the incidence of many diseases varies with short- and long-term changes in climate.

In this chapter, we use campylobacteriosis, a gastrointestinal disease that can be transmitted by water, to illustrate the relationship between the ecology of a disease and climate, and between climate change and adaptation. We start by describing the ecology of the disease in New Zealand and its relationship to climatic variables. We describe how climate change may affect this disease ecology (in the absence of adaptation) and review what steps can be taken to prevent illness due to *Campylobacter* infection. Finally, we explore the implications for adaptation to climate and climate change.

2. BACKGROUND: THE ECOLOGY OF CAMPYLOBACTERIOSIS IN NEW ZEALAND

Campylobacteriosis was first recognized as an "emerging" human gastrointestinal disease in the late 1970s [2], and is now the most commonly notified disease in the western world. It accounts for about 10% of all diarrhea worldwide [3], and New Zealand has notification rates amongst the highest recorded anywhere, peaking at over 350 cases per 100,000 population in 1998 (Table 4.1). The increase in monthly numbers of cases over the last two decades is remarkable (Fig. 4.1). Other gastrointestinal diseases have not demonstrated such marked changes, suggesting that the rise in campylobacteriosis is greater than what could be expected as a result of improved ascertainment alone. The question arises, therefore, as to what unique aspects of disease ecology have led to this pattern in New Zealand.

Campylobacter jejuni is a bacterium that invades the intestinal lining. The consequences range from asymptomatic infection, through diarrhea with abdominal pain, to rare but severe complications that include arthritis and nerve inflammation. The characteristic acute diarrhea arises 2–5 days after exposure (ingestion) and is usually associated with abdominal pain, malaise, fever, and nausea. An acute episode is usually over within 2–5 days, but sometimes lasts for up to 10 days and longer when there are complications [3].

Table 4.1 Rates of campylobacteriosis in selected countries

Country	*Rate (per 100,000)*	*Original source (year)*
Australia[a]	106.3	http://www.health.gov.au/pubhlth/cdi/nndss/nndss2.htm (2002)
Canada	37.7	Centre for Infectious Disease Prevention and Control, Health Canada (1999)
England and Wales	108.4	http://www.phls.co.uk/topics_az/campylo/data.htm (2001)
New Zealand	279.8	Annual Surveillance Summary. ESR Population and Environmental Health Group (2001)

Source: Data from Ref. 4.

Note
a Excludes New South Wales (campylobacteriosis is not notifiable in New South Wales).

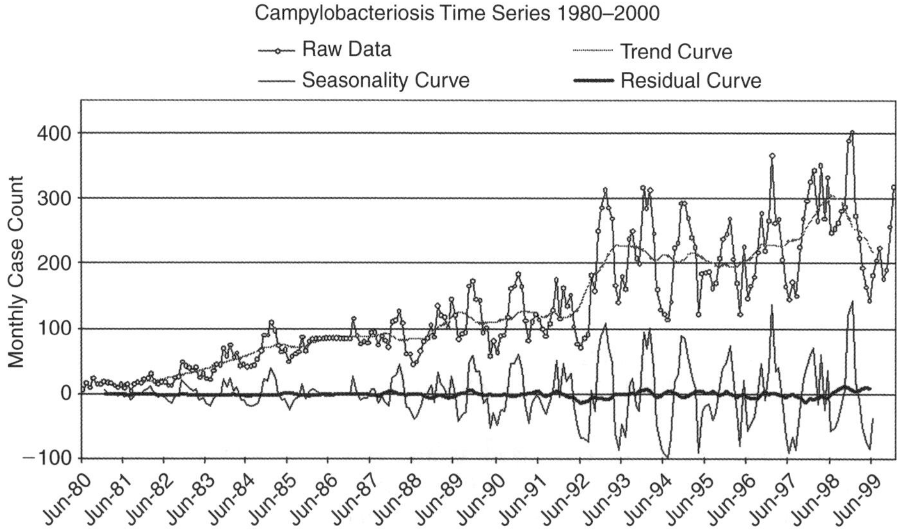

Figure 4.1 Monthly number of notified cases of campylobacteriosis in New Zealand from 1980 to 1990. Source: Data from C. Skelly, New Zealand Ministry of Health

Campylobacteriosis is a food- and water-borne zoonosis with a reservoir often consisting of domesticated animals, including poultry, sheep, and cattle. Only in these reservoirs, and in humans, will *Campylobacter* multiply, and it is therefore known as a thermophilic bacterium. Its transmission in food and water does not depend on growth rates in these media as is the case with some other gastrointestinal pathogens, but is rather a matter of "survival trajectories" between excretion by the reservoir and ingestion by the case [5]. The survival of this organism in the environment is subject to the influence of a variety of abiotic factors. To understand the variety of independent variables acting at different levels of organization, and the interactions between host agent and environment factors, it is necessary to broaden the conventional epidemiological perspective into one more closely aligned with the science of ecology [6,7].

2.1 Environmental prevalence of *Campylobacter*

New Zealand is a fertile, mountainous country, separated by thousands of miles of ocean from neighboring land masses. The country has over 770 lakes, 70 major rivers, and thousands of streams that provide about 60% of the water consumed by the human population of 3.8 million [8]. Pastoral farming has a major impact on both water flow and quality. Over the last 150 years, deep-rooted vegetation has been removed from hillsides and riverbanks, increasing the volume and speed of runoff during heavy rains that in some places reach 11,000 mm per year. Much of the farming land is heavily stocked with sheep and cattle, and over 50 million head of stock excrete about 40 times the mass of feces produced by the human population [8]. Some of this excrement is washed into waterways with heavy rains, where it may come into contact with humans both directly (drinking, recreation) and indirectly (by "seeding" secondary cases; see following section).

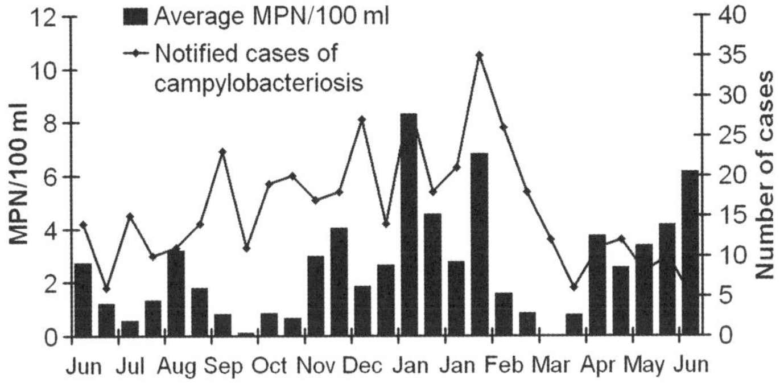

Figure 4.2 Average maximum probable number of *Campylobacter* by month, for one site (Outram Glen) on the Lower Taieri River combined, January 2000–June 2001. Source: Data from ref. 4

The natural self-purification of water percolating through soil and vegetation is reduced as a result of changes in land cover. This exposes both stock and humans downstream to a variety of zoonotic pathogens, including *Campylobacter*, *Cryptosporidium*, and *Giardia* [9]. Recent surveys have found over 50% of New Zealand's surface waters appear to be contaminated with *Giardia* (T. Brown, Massey University, personal communication, 2000). Even in undeveloped catchments, *Campylobacter* occurs in over 50% of samples [10]. *Campylobacter* levels in the Taieri River near the southern city of Dunedin ranged from 0 to over 11/100 mL maximum probable number (MPN) in a recent study [11] (see Fig. 4.2), frequently exceeding both drinking water standards and recreational water standards.

Given the prevalence of *Campylobacter* in the environment, drinking water is clearly one possible source of exposure for the human population. The organism is sensitive to chlorine, but the quality of water treatment across New Zealand varies widely. About 20% of the population is served by supplies that have not been graded as "satisfactory". This includes supplies for small rural populations, which are not covered by the public health water grading system. Exposure may also occur through an outdoor lifestyle that includes hiking (and drinking from untreated surface waters) and recreational water use (in untreated surface waters). Thus, the New Zealand environment contains a widespread hazard in the form of contaminated surface water, to which a significant proportion of the population is regularly exposed through drinking water and outdoor activities. It is perhaps not surprising then that New Zealand has very high notification rates, by international standards, of most zoonotic and potentially water-borne gastrointestinal diseases [9,12].

What of other sources of infection? A New Zealand case control study of campylobacteriosis found an association between notified illness and consumption of certain foods, notably raw or undercooked chicken [13]. There was also an increased risk with consumption of raw dairy products, and contact with puppies and cattle, particularly calves, but no association with contact with young children. Interestingly, there was an increased risk for people with rainwater as a source of water at home, presumably from contamination of roof water catchments and storage tanks with feces of animals such as cats, birds, and possums.

However, case control and other within-population studies cannot provide an insight into differences between populations. There is no evidence or *a priori* reason to indicate that New Zealanders have poorer hygiene or significantly different cooking practices than other countries with a similar sociocultural makeup, such as Australia, the UK, or Canada, where campylobacteriosis rates are as much as an order of magnitude lower (see Table 4.1). Variations in health care services and disease notifications are also unlikely to produce such marked differences in disease rates, nor can agricultural practices (e.g., use of veterinary antibiotics) or demography (population age structure) account for these variations. As we have indicated already, agricultural development directly influences biodiversity, ecosystem health, and hence the risk of disease transmission in freshwater ecosystems [14]. Therefore the high rates of campylobacteriosis may be influenced by the unusual ecology of the causative organism in New Zealand's uniquely modified ecosystem.

2.2 Drinking water quality and gastroenteritis

Although most New Zealand drinking water supplies are treated and monitored, infection in a small number of supplies could theoretically generate a disproportionately large number of cases. In addition to direct transmission, pathogens in water may affect humans through occupational exposures to infected stock, contamination of the food chain, or contamination of a variety of fomites by feces from both infected humans and infected stock. Generation of cases in this way could obscure the importance of potential "seeding" sources of infection: for example, recreational use of an untreated rural fresh water body may lead to fecal-oral transmission in an urban environment, accounting for the majority of cases. It is difficult to track the development of outbreaks, or even to make urban-rural comparisons of incidence rates, because of historical limitations in the accuracy of georeferencing cases (the uncertainty in rates between rural areas exceeds the difference in rates between rural and urban areas) [15]. But regardless of data issues, the potential for seeding of secondary cases means that simple spatial comparisons are not necessarily helpful in assessing the extent of waterborne transmission.

The frequency of infection with another zoonotic water-borne pathogen, *Cryptosporidium*, has been shown to relate directly to drinking water quality in New Zealand [12]: mean rates of notified cryptosporidiosis in drinking water zones that complied with the New Zealand drinking water standards were lower than rates in zones that did not comply with the standards. A similar study could not be carried out for campylobacteriosis because of the data problems alluded to in the preceding paragraph [15]. However, given the similarities in the epidemiology of these organisms, it would be surprising if campylobacteriosis rates were not also associated with water quality. Although the introduction of modern standards of water treatment over the last century has reduced the overall burden of enteric infections in New Zealand in parallel with the rest of the developed countries, we suggest that water-borne *Campylobacter* may still account for a large number of cases. This may occur by direct contact with the bacterium in waterways and by human-to-human transmission of infections originally picked up from water.

2.3 Socioeconomic deprivation and gastroenteritis

Why have the New Zealand rates increased so rapidly over the last two decades (Fig. 4.3)? Improved ascertainment may have played a part in the early years, but cannot be the full explanation. It is unlikely that the prevalence of the organism in the

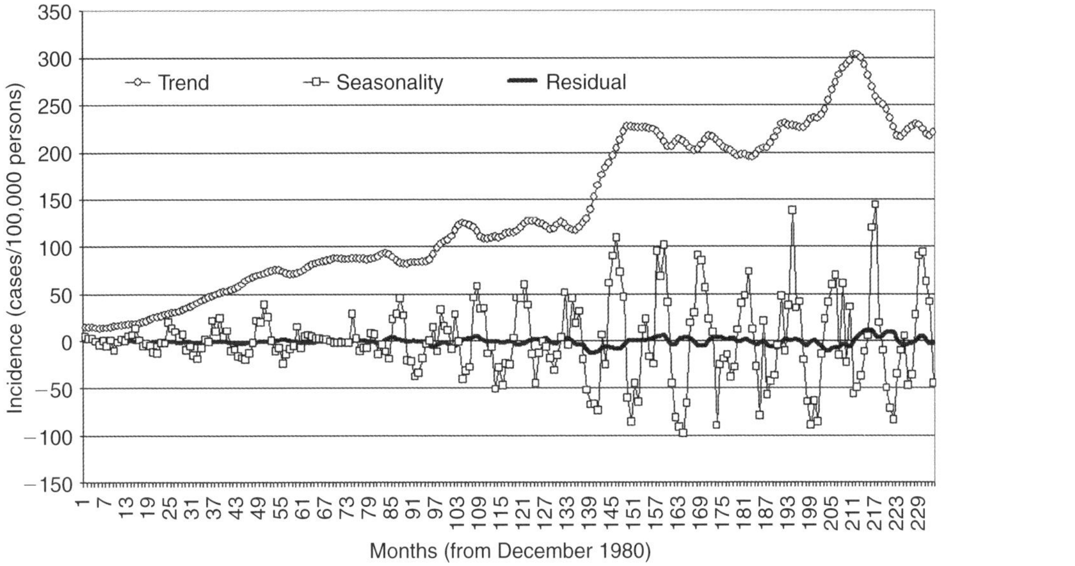

Figure 4.3 Time series decomposition of New Zealand's Campylobacteriosis notifications into trend and seasonality, with a residual component. Curves sum to expected crude rate calculated from notifications. Monthly data were unavailable for 1986, and annual total was instead evenly divided amongst the months

environment has changed significantly with time, other than in some specific areas where the dairy industry has intensified. Might socioeconomic factors explain the rapid increase in numbers of cases? The incidence of diarrheal diseases is sensitive to factors such as housing conditions that make it difficult to maintain a high standard of hygiene, crowding, and access to primary health care. Moreover, in New Zealand social disadvantage is associated with poor quality drinking water.

The New Zealand Deprivation Index [16] is a measure of socioeconomic deprivation by mesh block (smallest area unit used in the New Zealand Census, average population about 100). The index is derived from nine variables reported in the national census, including household income, crowding, and educational achievement. By comparing the deprivation index of a mesh block to the quality of water supplied to the same area, Hales et al. [17] established that the most deprived communities in New Zealand are most likely to receive the poorest water quality. The odds of a community using drinking water that is not of a satisfactory standard is about three times higher in communities with the highest decile of deprivation scores compared to those in the lowest decile (that is, the least disadvantaged).

In New Zealand, as in many other countries, the last two decades have seen a "stretching" of society, with a small number of people getting very much richer and a large number of people staying much the same in terms of income, or getting poorer [18]. We know of no studies that have tested the hypothesis that growing social inequalities have influenced trends in water-borne diseases. However, social trends such as greater household crowding and reduced disposable income, which have been concentrated in the most disadvantaged sections of the population, may have made these groups more susceptible to infection with agents such as *Campylobacter*.

3. THE RELATIONSHIP BETWEEN CAMPYLOBACTERIOSIS ECOLOGY, RAINFALL, AND TEMPERATURE

The combination of intensive sheep and cattle farming and the large variety of fresh water sources used for drinking and recreation in New Zealand is distinctive. In Australia, for instance, where water is a far scarcer commodity, the profile of campylobacteriosis is more likely to relate to the ecology of food-borne disease than to the ecology of water-borne disease.

Some of the features of the epidemiology of campylobacteriosis are consistent with significant water-borne transmission; others are not. For example, one might expect that both environmental pathogen levels and human illness would peak in winter, when runoff from pastoral landscapes is at its maximum. However, notifications of illness show a consistent summer peak in New Zealand (Fig. 4.1), as is commonly reported from other countries [5]. On the other hand, many New Zealanders lead an active, outdoor lifestyle, with exposure to pathogens in surface waters through both drinking and recreational water use. These behaviors peak during the warmer months of summer, which is also the time when environmental levels of *Campylobacter* are highest.

In the United States and Europe, winter tends to be the peak time for environmental prevalence of viable *Campylobacter* [5]. This seasonal pattern is explained in terms of increased survival of the organism at lower temperatures, a decreased *Campylobacter* mortality with lower exposure to ultraviolet radiation, and increased runoff with winter rainfall. But this is not the case in New Zealand: *Campylobacter* is found in greater numbers in New Zealand freshwater ecosystems in summer (Fig. 4.2). This pattern is still unexplained. Perhaps it relates to thirsty stock directly accessing fresh water bodies more frequently in summer, although the peak occurs 1–2 months before the warmest period of the year (February–March). But whatever the reason, the fact that environmental contamination and human disease rates peak concurrently in New Zealand is consistent with a strong role for water-borne transmission of *Campylobacter*.

4. CLIMATE CHANGE AND CAMPYLOBACTERIOSIS

Human-induced climate change is expected to lead to higher temperatures across New Zealand. Warming will be slightly greater in the north of the country than in the south, with increases of 1.2 to 1.6°C in mean summer temperatures by 2080 [19]. Such changes are less marked than those forecast for large land masses, especially those in the Northern Hemisphere.

Increased temperatures are likely to affect not only the prevalence of *Campylobacter* in the environment but also behaviors that lead to human exposures to the organism. Although the half-life of *Campylobacter* in freshwater is shorter at higher temperatures, we have seen that in New Zealand pathogen levels nevertheless tend to be higher in fresh waters in summer (Fig. 4.2). To understand this, we need to consider possible effects of climate at other levels of the ecosystem. For example, higher temperatures may influence the behavior of stock, such as more frequent or prolonged visits to fresh water sources. At the level of human populations, exposure to contaminated surface waters, through both drinking and recreational use, is also likely to increase with higher average temperatures and more frequent high temperature extremes. These are suppositions, and need testing with empirical data. But in the absence of specific research in this area, we believe it is likely that some of the fundamental ecological factors driving the incidence of disease in New Zealand would be enhanced by warming, with effects on both hazard and exposure in the cycle of campylobacteriosis.

Precipitation changes are more difficult than temperature to forecast. However, current scenarios suggest there will be more frequent high rainfall events in association with changes in the El Nino/Southern Oscillation. Conditions are likely to be generally wetter (i.e., higher total rainfall) on the west coasts of both main islands, although there is a high degree of uncertainty [19]. Noting these uncertainties, it still seems likely that runoff from many areas of pastoral land use will be both more rapid and greater in magnitude. The relevance of total rainfall to human health lies in the sheer volume of feces and fecally contaminated water that enters surface waters.

Rapidity of runoff is important because of the limited survival of fecal pathogens, whose half-lives are less likely to be exceeded before human exposure occurs. The risk of viable pathogens passing through water treatment plants is a function of both total runoff and rainfall intensity. Furthermore, a greater frequency of extreme rainfall events is likely to exacerbate the problem of fecal pathogens from point sources such as dairy farms and human sewage treatment plants. Other countries, with more extensive data, have been able to quantify the impact. For example, using data from 1948 to 1994, U.S. researchers showed a close correlation between outbreaks of water-borne disease and extreme rainfall events within the preceding 2 months [20]. There is insufficient information in New Zealand to model the effects of increased rainfall on campylobacteriosis rates, but it is likely that disease transmission will be enhanced.

It was noted earlier that cattle and sheep in New Zealand excrete about 40 times more waste than the 3.8 million human population, giving a human fecal load "equivalence" of a population of 160 million, most of which is untreated. Bearing in mind this microbiological pressure on the local environment, climate change may indirectly magnify the risk of enteric infections among humans. For example, in New Zealand higher temperatures and increased rainfall will be conducive to the growth of richer and more extensive pasture, which may well lead to further increases in stock ranges and density.

The New Zealand economy is likely to remain agriculturally based for the foreseeable future, and the effects of midlatitude, midcontinental drying on global food production may increase New Zealand's competitive advantage in this sector. As a consequence there may be powerful economic incentives to expand and intensify agricultural activity and thus, indirectly, increase the environmental prevalence and human exposure to *Campylobacter*.

5. ADAPTATION TO AN "EMERGING" DISEASE BURDEN

An assessment of the ecology of campylobacteriosis in New Zealand and possible effects of climate change on this disease (this chapter) is the first step in the process of developing effective public health interventions to reduce the damage that might be caused by rapid changes in climate. We have suggested a number of contributing factors to the "emergence" and maintenance of this disease. It is not possible, with the current state of knowledge, to determine the relative contribution of each factor (or combination of factors), or to predict the increase in disease burden in New Zealand with higher temperatures and changed rainfall patterns. Nevertheless, what we do know currently about the ecology of *Campylobacter* in this country suggests a number of potential adaptive changes.

Current preventive measures for controlling transmission and infection with *Campylobacter* include food and farm hygiene, thorough cooking (or irradiation) of food, use of pasteurized milk and chlorinated water supplies, and control of the disease in domestic and domesticated animals [3]. Whatever their impact on the transmission of *Campylobacter*, these basic public health strategies should be maintained for many other reasons. However, they have failed to arrest the "emergence" of campylobacteriosis in most developed countries, particularly in New Zealand, and therefore cannot, on their own, be counted on to prevent expansion of the disease in a world in which the climate may generally favor further spread of the pathogen.

Historically in New Zealand, catchment areas providing water destined for human consumption have been closed to agricultural use or recreation. The advent of powerful, high volume water treatment plants has made the protection of catchments a less pressing concern. From the perspective of campylobacteriosis and other enteric infections such as cryptosporidosis and giardiasis, however, there are good reasons to review catchment management. For instance, many catchment areas have been cleared and left as grassland or planted in exotics (*Pinus radiata* in particular), but native vegetation in New Zealand has the highest natural filtration capacity of any land use, slowing runoff and increasing percolation [21].

Water from catchments with native vegetation is less likely to contain viable pathogens than water from catchments with no native cover (G. McBride, National Institute of Water and Atmospheric Research, personal communication, 2002). Replanting and protecting native vegetation in these areas reduces erosion and enhances conservation, and may bring public health benefits also. Importantly, it is not only the direct transmission of *Campylobacter* in drinking or recreational water exposure that will be affected. If stock infections are also decreased as a result of regrowth of native plants in water catchments, then the number of human infections acquired occupationally (in farms and abattoirs) and by the food-borne route (via animal products) will also be reduced.

6. IMPLICATIONS FOR ADAPTATION TO CLIMATE CHANGE

This example demonstrates how the range of possible adaptation options widens when moving from a clinical perspective on disease control to an ecological view. For campylobacteriosis this shift in perspective involves (metaphorically and literally) looking "upstream" for significant, alterable causes of disease. There are many other specific instances in which this general principle applies. But the example of campylobacteriosis in New Zealand also shows the importance of local factors. The particular, and often unique, features of the New Zealand landscape need to be incorporated in the ecological analysis. As we have already mentioned, New Zealand's closest neighbor, Australia, has a very different geological and biological history and quite different settings for organisms such as *Campylobacter*. As shown in Table 4.2, which summarizes the basic requirements for public health action, we have little understanding of the primary causes of campylobacteriosis, and there is only modest public engagement with the problem. It is highly likely that the predominant sources of infection, and the vectors by which infection is transmitted, will vary from one country to another. Furthermore, the patterns of susceptibility depend on many factors, including socioeconomic circumstances. As a

Table 4.2 Campylobacteriosis in New Zealand – what factors affect the capacity to adapt and what are the most important requirements for prevention: Ranking of determinants of adaptive capacity and requirements for public health prevention[a]

Determinants of adaptive capacity	*Prerequisites for public health prevention*
Range of available technological options (2 – relatively few options)	Awareness that problem exists (3)
Availability and distribution of resources (4 – ample resources)	Sense that the problem matters (3)
Structure of critical institutions (4 – regional government critical)	Understanding of the causes (2)
Human capital (4)	Capability to intervene (2)
Social capital (3)	Political will to influence (3)
Access to risk spreading (2)	
Ability to manage information (4)	
Public's perceived attribution (2)	

Note
a Scale = 1 (low) to 5 (high).

consequence, the specific steps that need to be taken to control enteric infections and respond to the pressures of changing climate will also to a large extent be country-specific. But whatever the setting, we suggest that an ecological analysis of the kind we have outlined for New Zealand is a useful means to identify options for adaptation.

The *Campylobacter* story illustrates the compounding effects of social disadvantage. The communities at greatest risk of diarrheal diseases because of factors such as low income, poor housing, and low standards of education are also most likely to be served by ungraded or unsatisfactory water supplies. These communities are also over-represented in parts of New Zealand where the most marked changes in climate are forecast to occur (the far north, where warming will be greatest, and the west coast of the South Island, where rainfall increases are likely to be heaviest). Adaptation must take account in some way of this general phenomenon – that those most likely to be harmed by climate variability, now and in the future, are those with the fewest resources available to respond to change.

One challenge that arises from the social distribution of risk is the tension of targeting versus universalism. Targeting means directing resources to the individuals and communities at greatest risk. Universal approaches tend to favor the larger population centers, where the greatest gains are likely to be made, in absolute terms. But the most serious problems with the quality of drinking water tend to occur in small communities in rural areas, often remote from service centers. This means that interventions are expensive, and are frequently more than the local rate-payers can afford. In narrow "efficiency" terms, it is difficult to justify subsidizing these services. But on equity grounds, and in terms of making the greatest difference for individuals, the improvement of small water supplies deserves to receive a high priority.

It is frequently pointed out that the best options for adaptation are those that produce benefits in both the short and long term. These are "no regrets" interventions, which can be justified whatever climate changes emerge in the future. Better care of water catchments, including revegetation with an emphasis on planting native species, would have multiple benefits, independent of climate trajectories. As mentioned already, some of these would be apparent relatively quickly (reduced runoff, improved water quality), while others (such as the promotion of biodiversity and carbon sequestration) would occur rather later.

Recommendations for adaptation to climate change often stress the importance of public health infrastructure without specifying the elements that matter most. In this instance, desirable changes in the health system would include improvements in disease surveillance and reporting, improvements in environmental hazard surveillance and reporting, and closer links between the professionals involved in both (including human and veterinary epidemiologists, environmental scientists, and ecologists).

7. CONCLUSION

New Zealand is a small country with a high dependence on primary agricultural production. Its "clean green" image adds value both to this production and to a significant ecotourism industry. Whereas agricultural production demands ecosystem change, a clean green image demands conservation and restoration of native ecosystems. The present balance between these two competing demands is tilted toward agricultural production, and we suggest this has been a contributing factor in the emergence of campylobacteriosis. Rising temperatures and changes in rainfall patterns are likely to favor the transmission of enteric pathogens such as *Campylobacter* in this environment, both directly and indirectly.

One response to this challenge would be an accelerated vegetation restoration program in water catchment areas. Such a program not only would retard runoff and improve the self-purification of water from appropriately reforested catchments, but also would (albeit in a small way) help slow the rise in greenhouse gases.

For a variety of vector-borne, water-borne, and other "environmental" diseases, appropriate, scientifically based public health interventions can be devised only with an understanding of the ecology of the disease. By extension, effective adaptation to the potential effects of climate change is more likely to occur if a sound understanding of the relationship between ecosystem health and human health exists. An accelerated program of reforestation in water catchments would serve as a means of reducing simultaneously greenhouse gases and the burden of gastrointestinal disease. Adaptation to climate change in this case provides the opportunity to improve both ecosystem health and human health concurrently.

ACKNOWLEDGMENTS

Work contributing to this paper was funded, in part, by the Health Research Council of New Zealand, the University of Otago, and the Wellington Medical Research Foundation. We thank Rebekah Eyles, Simon Hales, Des O'Dea, Margot Parkes, Chris Skelly, and Colin Townsend for their contributions to the develop-ment of some of the ideas expressed herein, and Agnes Weinstein for typing the manuscript.

REFERENCES

1. McMichael, A.J.: *Human Frontiers, Environments, and Disease*. Cambridge University Press, Cambridge, UK, 2001.

2. Skirrow, M.: Campylobacter Enteritis: A 'New' Disease. *BMJ* 2 (1977), pp.9–11.
3. Chin, J.: *Control of Communicable Diseases Manual*. American Public Health Association, Washington, DC, 2000.
4. Eyles, R.: Linking Aquatic Ecosystem Health and Human Health: The Ecology of Campylobacter in Freshwaters. PhD Thesis, University of Otago.
5. Skelly, C. and Weinstein, P.: Pathogen Survival Trajectories: An Eco-environmental Approach to the Modelling of Human Campylobacteriosis Ecology. *Environ. Health Persp*. 111 (2003), pp.19–28.
6. Weinstein, P.: An Ecological Approach to Public Health Intervention: Ross River Virus in Australia. *Environ. Health Persp*. 105 (1997), pp.364–366.
7. Aron, J.L. and Patz, J.A.: *Ecosystem Change and Public Health*. John Hopkins University Press, Baltimore, MD, USA, 2001.
8. Ministry for the Environment: The State of New Zealand's Environment. New Zealand Ministry for the Environment, Wellington, 1997.
9. Weinstein, P., Russell, N., and Woodward, A.: Drinking Water, Ecology, and Gastroenteritis in New Zealand. Interdisciplinary Perspectives on Drinking Water. *IAHS Publ*. 260 (2000), pp.41–47.
10. McBride, G., Till, D., Ryan, T., Ball, A., Lewis, G., Palmer, S., and Weinstein, P.: Freshwater Microbiology Research Programme Report: Pathogen Occurrence and Human Health Risk Assessment Analysis. Ministry for the Environment, Wellington, NZ, 2002.
11. Eyles, R., Niyogi, D., Townsend, C., Benwell, G., and Weinstein, P.: Spatial and Temporal Patterns of *Campylobacter* Contamination Underlying Public Health Risk in the Taieri River, New Zealand. *J. Environ. Qual*. 32 (2003), pp.1820–1828.
12. Duncanson, M., Russell, N., Weinstein, P., Baker, M., Skelly, C., Hearnden, M., and Woodward, A.: Rates of Notified Cryptosporidiosis and Quality of Drinking Water Supplies in Aotearoa New Zealand. *Water Res*. 34 (2000), pp.3804–3812.
13. Eberhart-Phillips, J., Walker, N., Garrett, N., Bell, D., Sinclair, D., Rainger, W., and Bates, M.: Campylobacteriosis in New Zealand: Results of a Case-Control Study. *J. Epidemiol. Commun. Health* 51 (1997), pp.686–691.
14. Townsend, C., Eyles, R., Niyogi, D.K., Riley, R., and Weinstein, P.: Agricultural Development Can Influence Biodiversity, Ecosystem Health and the Risk of Disease Transmission in Streams. Presented at Healthy Ecosystems, Healthy People, Washington, DC, 6–11 June 2002.
15. Skelly, C., Black, W., Hearnden, M., Eyles, R., and Weinstein, P.: Disease Surveillance in Rural Communities is Compromised by Address Geocoding Uncertainty: A Case Study of Campylobacteriosis. *Aust. J. Rural Health* 10 (2002), pp.87–93.
16. Salmond, C., Crampton, P., and Sutton, F.: NZDep96 Index of Deprivation: Report 8. Health Services Research Centre, Wellington, 1988.
17. Hales, S., Black, W., Skelly, C., Salmond, C., and Weinstein, P.: Social Deprivation and the Public Health Risks of Community Water Supplies in New Zealand. *J. Epidemiol. Commun. Health* 57 (2003), pp.581–583.
18. Mowbray, M.: *Distribution and Disparity: New Zealand Household Incomes*. New Zealand Ministry of Social Policy, Wellington, 2001.
19. McCarthy, J.J., Canziani, O.F., Leary, N.A., Dokken, D.J., and White, K.S.: *Climate Change 2001: Impacts, Adaptation, and Vulnerability. Contribution of Working Group II to the Third Assessment Report of the Intergovernmental Panel on Climate Change*. Cambridge University Press, Cambridge, UK, 2001.
20. Curriero, F.C., Patz, J.A., Rose, J.B., and Lele, S.: The Association Between Extreme Precipitation and Waterborne Disease Outbreaks in the United States, 1948–1994. *Am. J. Public Health* 91 (2001), pp.1194–1199.
21. Mosley, M.P.: *Waters of New Zealand*. New Zealand Hydrological Society, Wellington, 1992.

5 A case study of unintended consequences: arsenic in drinking water in Bangladesh

Kristie L. Ebi, David Mills, and Joel B. Smith

ABSTRACT: Beginning in the 1970s, tubewells were widely installed in Bangladesh in an effort to provide a "safe" source of drinking water to a population that was experiencing high morbidity and mortality, especially among children, from water-related diarrheal diseases (e.g., cholera, dysentery, and other intestinal diseases). However, in many regions of Bangladesh, the groundwater accessed by these wells has naturally occurring high concentrations of arsenic, a known carcinogen. The result has been called the largest mass poisoning of a population in history. Possibly 30 million out of the 125 million inhabitants of Bangladesh drink arsenic-contaminated water. Health consequences of exposure range from skin lesions to a variety of cancers. Because of the latency of arsenic-related cancers, morbidity and mortality from historical and current exposures may continue for approximately 20 years after exposures are discontinued. A number of international initiatives are under way to help resolve this problem; however, solutions will most likely take a decade or more. This case study illustrates a number of important lessons, including adequately analyzing interventions before they are implemented, recognizing problems that may arise with relying on a single solution, intervening initially on a small scale to test results, considering other problems when designing an intervention, and having an effective monitoring system to identify problems.

1. INTRODUCTION

Public health interventions are designed to improve the health of a community. However, in some cases, interventions are not as successful as originally planned and unforeseen consequences of implementing the intervention can result in a new, significant public health problem that replaces or compounds the original problem. The installation of tubewells and subsequent increase in use of groundwater in Bangladesh provide an extreme example [1,2].

The story of this public health intervention is not directly related to climate change. However, this case study is relevant for those currently involved in climate change adaptation efforts in that it illuminates the risk of undertaking intervention programs without adequately assessing both benefits and risks. It cautions those implementing adaptation measures to consider the scale of the intervention and the consequences of any associated risk when it is magnified across a large population. It illustrates the problem of relying on a single technical solution to a problem instead of taking an integrated, multidisciplinary approach; an effective technology does not always solve the problem. And it shows the importance of implementing adaptation measures in an incremental or staged fashion instead of instituting a wide-scale deployment before all information is available. Flexible and responsive approaches are needed in which new information and experience are properly monitored and evaluated, and then used to appropriately modify interventions.

2. ORIGINAL HEALTH ISSUE – BURDEN OF DIARRHEAL DISEASE

Surface water in Bangladesh is readily available because of a combination of the country's geography (it is chiefly an alluvial delta interlaced by tidal waterways) and annual flooding during the monsoon rains. Surface water sources include irrigation and transport canals, rivers, lakes, and ditches. Historically, water sources included this surface water (Fig. 5.1 shows the two main rivers [3]), even though it is heavily polluted with fecal and other materials, and rainwater collected in "tanks". Typical rural villages are clusters of homes built above flood level on earthen mounds, and the surface depressions excavated for mound construction are called tanks. A 1975 survey of water use noted urination and defecation into the water, and manual bathing of the perianal region in surface water sources also used for other purposes, including washing rice seedlings, food, utensils, household items, and other parts of the body (mouth, teeth); drinking (humans and animals); taking water; taking mud; gathering windfall fruit from water; collecting water hyacinth; playing; fishing; and cooling buckets of milk [4].

Diarrheal diseases are caused by a variety of bacterial, viral, and parasitic agents, including *Vibrio chlorea*, rotavirus, enterotoxic *Escherichia coli*, *Camphylobacter*,

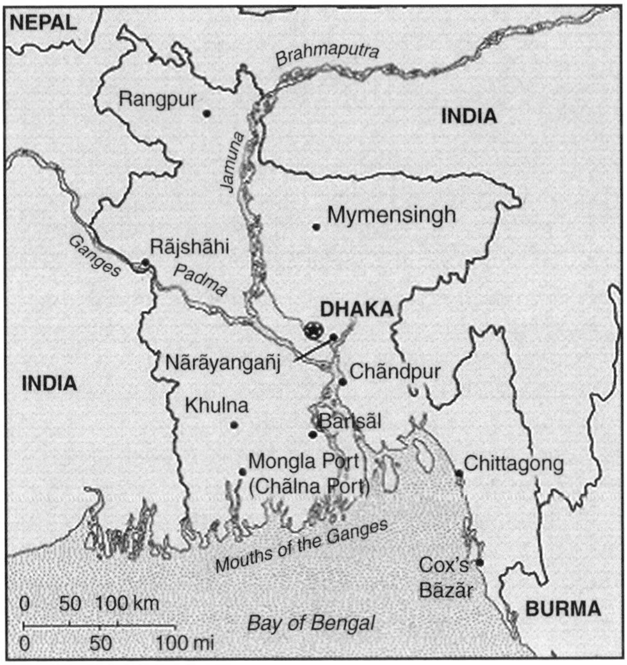

Figure 5.1 Major rivers of Bangladesh. Source: Ref. 3

Table 5.1 Adjusted number and rate of death (per 1,000) among children under age 5 according to cause in Matlab, Bangladesh (1975–1977)

Cause of death	*Number*	*Percentage of deaths*	*Rate*
Diarrhea	2,099	26.7	80.0
Watery	1,258	16.0	46.6
Dysentery	841	10.7	33.4
Tetanus	1,233	15.7	39.8
Measles	536	6.8	21.1
Fever	520	6.6	18.9
Respiratory	488	6.2	16.8
Drowning	237	3.0	9.4
Skin	99	1.1	3.5
Others	2,646	33.7	90.2
All	7,858	100	279.7

Source: Data from Ref. 5.

Shigella, *Cryptosporidium*, *Aeromonas*, and *Giardia*. Diarrheal diseases occur most frequently in conditions of poor environmental sanitation and hygiene, inadequate water supplies, poverty, and limited education [5]. Water-borne transmission has been documented for most enteropathogens, including not only the consumption of contaminated water but also the use of contaminated water for bathing, washing, swimming, cleaning, and washing eating utensils. Person-to-person and food-borne transmission also occurs for most enteropathogens.

Diarrhea is a symptom complex characterized by stools of decreased consistency and increased number. Defining an episode of diarrhea is a challenge in the face of multiple causative agents and different personal stool patterns. The standard working definition of diarrhea is three or more loose stools in 24 hours; an episode is considered to have ended after two diarrhea-free days [5]. A typical episode lasts for eight days. Dysentery is a diarrheal disease defined by the presence of blood in liquid stools.

Diarrheal diseases result in reduced food intake, and malabsorption and wastage of essential nutrients. The consequences include morbidity, retarded growth and development, and mortality. Diarrheal diseases are a leading cause of childhood mortality in developing countries. Table 5.1 provides the results of a longitudinal surveillance program conducted from 1975 to 1977 in a rural Bangladesh population of 263,000 [6]. The data are for children under the age of 5, and are adjusted for the mortality-reducing effects of the health services provided by the International Centre for Diarrhoeal Disease Research, Bangladesh (ICDDR,B). The leading cause of death was diarrhea (watery and dysentery). The top eight causes of death combined resulted in an infant mortality rate of 142.6 per 1,000 live births, and an overall childhood (under age 5) morality rate of 279.7 per 1,000 live births. The study was conducted in the Matlab *thana*, in the Comilla district, where the ICDDR,B has operated a surveillance system of regular cross-sectional censuses and longitudinal registration of vital events among 228 villages since 1963. As a consequence, there are better data for this area than for most of the country. However, the mortality rates in this area may not be typical of other areas in Bangladesh because Matlab was chosen for research purposes based on the hyperendemicity of watery diarrhea.

Another way to express these data is as follows: the 15.9% of the Matlab population under the age of 5 experienced 53.1% of all deaths [6]. Over the 3-year study

Table 5.2 Diarrheal morbidity by age group, Bangladesh

<table>
<tr><th rowspan="2">Age</th><th colspan="2">Incidence (per 100)</th><th rowspan="2">Prevalence (per 100)</th><th colspan="3">Rural</th></tr>
<tr><th>Urban (1964–66)</th><th>Rural (1975)</th><th>Population (%)</th><th>Episodes (per 100)</th><th>Morbidity (1,000 person days)</th></tr>
<tr><td>>1</td><td>1.62</td><td rowspan="2">1.86</td><td rowspan="2">4.1</td><td>4.3</td><td rowspan="2">32.9</td><td rowspan="2">244</td></tr>
<tr><td>1–4</td><td>0.90</td><td>13.4</td></tr>
<tr><td>5–9</td><td>0.38</td><td>1.07</td><td>2.4</td><td>13.7</td><td>14.6</td><td>117</td></tr>
<tr><td>10–14</td><td>0.10</td><td>0.52</td><td>1.1</td><td>15.7</td><td>8.2</td><td>65</td></tr>
<tr><td>15+</td><td>0.09</td><td>0.56</td><td>1.2</td><td>53.0</td><td>29.7</td><td>238</td></tr>
<tr><td>All</td><td>0.31</td><td>0.90</td><td>2.0</td><td>100.1</td><td>85.4</td><td>684</td></tr>
</table>

Source: Data from Ref. 6.

period, the average life expectancy at birth was 49.9 years. Had a child survived to age 1, life expectancy increased to 57.2 years. Survivors to age 5 had an average life expectancy of 63.1 years.

Table 5.2 shows the burden of diarrheal morbidity by age group in Bangladesh. Although the urban and rural data are not strictly comparable because of different diagnostic criteria, time periods, and sampling strategies, the data suggest higher rates in rural than urban areas [7]. Assuming an average duration of eight days per episode, the overall Bangladesh incidence rate implies a prevalence of 2% for the entire population, with a prevalence of 4.1% among children under the age of 5. When converted to the number of diarrheal episodes in a rural population of 100,000, Sunoto calculated an annual attack rate of 85.4% (85,400 cases annually), with 39% of those attacks in children under the age of 5 [7]. Among about 53,000 adults, the 238,000 person-days of diarrhea experienced translated into an annual loss (per person) of 4.5 workdays. The potential economic impact of this morbidity may be higher than the data suggest because the peaks of morbidity coincided with the fall harvest of the major rice crop in November and December, and with the planting of the winter crop in February [8].

Diarrheal diseases do not occur in isolation from other health problems. For example, Sunoto and Markum found that other diseases, particularly malnutrition and upper respiratory tract infections, accompanied about 80% of the diarrheal cases [9].

The severity of the international burden of diarrheal diseases led the World Health Organization (WHO) to declare 1981–1990 to be the International Drinking Water Supply and Sanitation Decade [10]. WHO estimated that during this time period US$133.9 billion was invested in water supply and sanitation, with 55% spent on water and 45% on sanitation. Urban areas received 74% of the total. This investment resulted in about 1.60 billion people being served with safe water and about 0.75 billion with adequate sanitation. However, because of population growth, by 1990 there remained 1.015 billion people without safe water and 1.764 billion without adequate sanitation.

3. SOLUTION – INSTALLATION OF TUBEWELLS

Based on experience, the public health and other sectors approached solutions to the diarrheal disease problem on the premise that providing safe water to rural areas would drastically reduce the high burden of diarrheal diseases. It was assumed that accessing groundwater via tubewells would provide water free from the pathogens causing diarrheal disease.

In Bangladesh, shallow tubewells are 5 cm tubes that are inserted into the ground at depths of usually less than 200 m [1]. The tubes are then capped with a pump. Deep tubewells tap water sources deeper than 200 m. Some tubewells existed during the British colonial era [11]. When the partition of India took place in 1947, an estimated 50,000 tubewells were in what was to become Bangladesh. By 1972, the Government of Bangladesh had installed approximately 135,000 tubewells. At that time, there was about one tubewell for every 400 people.

Following a devastating cyclone that hit Bangladesh in November 1970, the United Nations Children's Fund (UNICEF) provided assistance to repair some 10,000 damaged tubewells. UNICEF continued to assist the Bangladesh Department of Public Health Engineering in initiatives to provide a safe source of drinking water, primarily over concern for diarrheal diseases. These sources included shallow and deep tubewells, pond sand filtration systems, and rainwater harvesting. UNICEF support for shallow tubewell installation was discontinued in the late 1980s when the private sector became successful at installation. Today, approximately 97% of the rural population relies on shallow tubewells for drinking water [12]. On average, 75% of tubewells are privately owned [11]. It is estimated that there are 6–11 million tubewells in Bangladesh today [13].

At the time the wells were installed, standard water testing did not include tests for arsenic; unfortunately, it is still not a standard or routine test for rural water supplies. However, arsenic in drinking water was recognized as a problem before the installation of tubewells. The U.S. Environmental Protection Agency set the current standard of 50 ppb in 1975, based on a Public Health Service standard originally established in 1942 [14].

4. IMPACT OF TUBEWELLS ON DIARRHEAL DISEASE MORBIDITY AND MORTALITY

The installation and use of tubewells to control diarrheal diseases in Bangladesh appear to have elements of both success and failure.

4.1 The good news

A 1995 survey measured water quality in one rural area of Bangladesh [15]. This area had three potential sources of water: hand operated tubewells, rain-fed water storage ponds (tanks) and surface water. From whatever source it was collected, water was generally stored in the household in traditional earthenware pots. All households in the district had access to tubewell water for drinking purposes. Tank or river water was used for personal bathing, for washing of clothes, and occasionally for cooking. Water was collected from each water source and each storage pot used by the study families, and the number of fecal coliforms per 100 mL was measured. Tubewell water had low fecal coliform counts (<10 coliforms per 100 mL). This and other studies demonstrated that tubewells provide clean water uncontaminated with pathogens.

However, tubewell water was initially not well accepted in some areas [16]. As cited in [17], a 1975 WHO/UNICEF study showed that only 52% of households with access to tubewells used tubewell water, at least for drinking purposes. As a result, health education campaigns were launched that delivered information about personal hygiene, sanitation, and use of tubewell water. The installation of tubewells, educational campaigns about use of tubewell water, and other measures were so effective that UNICEF stated in its 1997 country report that Bangladesh had surpassed its goal of providing 80% of the population with access to "safe" drinking water by 2000 in the form of tubewells, ring wells, and taps [11]. According to Hoque et al. [18], during the 1980s, when the population of Bangladesh increased from about 90 to 100 million, access to safe water increased in rural areas from 37% to 96%. That percentage is now decreasing because of arsenic contamination of tubewell water.

4.2 The bad news

A number of studies were conducted in Bangladesh during the 1970s to determine the efficacy of tubewell installation in reducing diarrheal diseases. The results were not expected – most studies found that providing clean water from tubewells did not significantly decrease diarrheal diseases [4,19,20]. These results even led some authors to question whether cholera was primarily a waterborne disease [e.g., 21]. Based on similar results from studies conducted in India and Africa, research efforts were launched to understand why the provision of clean water was not sufficient to decrease diarrheal diseases. These studies identified other environmental, personal hygiene, and host factors associated with diarrheal disease.

For example, a survey of household sanitation in rural communities in Bangladesh found that although 89% of households used tubewell water for drinking, only 11.5% had a safe water supply and usage (defined as satisfactory if there was a tubewell within 400 m of the household and the area within 15 m of the well was free from any source of pollution, if all the members of the household used this water for drinking, and this or other well water was used for *all* domestic purposes, including cooking, washing utensils, and bathing) [17]. Table 5.3 summarizes the data collected on the other water sources in the 1995 survey [15]. When measured at the point of collection, tubewells provided water of good quality; however, 62% of household storage pots contained water with moderate (11–50 colony-forming units/100 mL) to high (>50 colony-forming units/100 mL) contamination [15].

Other factors were found to influence water usage patterns. In the Matlab area, tubewell water was not attractive for a variety of reasons, including being less accessible than surface water, requiring considerable effort to pump, and being generally

Table 5.3 Occurrence of fecal coliforms in different waters

Source	*Coliform count (colony-forming units/100 mL)*			
	N	*≤10*	*11–50*	*>50*
Tubewells	10	10	0	0
Storage pots	29	11	4	14
Surface waters	7	0	2	5

Source: Data from Ref. 16.

poor in quality [20]. Tubewell water was reported to turn turbid and to form brown scum and precipitates during overnight storage. In addition, it causes discoloration of teeth and food, and tastes of iron. Cultural beliefs also are important factors in water usage patterns. For example, as cited in [20], "Children below the age of about one year are said never to be given plain cold water to drink as its temperature is considered too chilling for them ... [and] ... some people say that [tubewell water] temperature is too cold and causes them to catch cold and lose their voices".

Studies that reported reduction of diarrheal disease with tubewell water use also reported other interventions. One of the studies reported a reduction in diarrheal disease when latrines were installed in 92% of the households and hygiene education in relation to water use and sanitation practices was provided [22]. In that study, episodes of diarrhea per child per year decreased 25% in the intervention area; environmental risk factors included distance from a handpump tubewell, latrine use by children, and exclusive use of handpumped water for all major domestic activities in the wet season.

However, Shaikh et al. [23] concluded in 1990 that although the pattern of diarrheal mortality showed a significant decline from 1976 to 1986, the decline did not appear to be commensurate with intervention efforts and considerable progress still needed to be made. One of the activities of the ICDDR,B is a Demographic Surveillance System in Matlab. Table 5.4 summarizes diarrheal disease mortality per 1,000 population for Matlab from 1966 to 1987 [23]. The only major difference across the 5-year periods was the large increase in mortality for 1971–1975; this time period included the 1971 War of Liberation that led to the creation of Bangladesh and the 1974–1975 famine. Another peak observed in 1983–1984 was due to an epidemic of shigellosis. Over the 22-year study period, 20% of all deaths were related to diarrheal diseases.

One factor not understood at the time of tubewell installation was the relative importance of water quality and quantity in the transmission of diarrheal diseases. Since the 19th century, most public health engineers believed, based on historical evidence, that poor water quality was the most significant factor in transmission of diarrheal disease [24]. This was based on the impression that diarrheal diseases were generally water-borne. Since the 1970s, the ways of thinking about water and health shifted with the realization that water-related infectious diseases have four means of transmission: infections spread through water supplies (water-borne); infections spread because of lack of water (whether clean or contaminated) for personal hygiene (water-washed); infections spread through an aquatic invertebrate host (water-based); and infections spread by insects that depend on water [25]. These categories are not mutually exclusive; many diarrheal diseases have more than one means of transmission.

Table 5.4 Diarrheal disease mortality per 1,000 population by consecutive 5-year periods and by gender during 1966–1987 in the Demographic Surveillance System area in Matlab, Bangladesh

	1966–1970	*1971–1975*	*1976–1980*	*1981–1985*	*1986–1987*
Males	2.1	4.3	2.4	2.7	1.5
Females	2.4	4.7	2.6	3.5	2.0
All	2.2	4.5	2.5	3.1	1.8
% of all deaths	14.5	24.9	17.6	22.5	16.5

Source: Data from Ref. 23.

Where overall levels of fecal contamination and endemic diarrhea are low, and resources necessary for good water quality and sanitation are generally available, water quality is an important determinant of diarrheal disease [24]. However, much of the developing world suffers from both relatively high levels of fecal contamination in water and high incidences of diarrheal disease. In these situations, water quantity may be more important than water quality in the transmission of several important diarrheal diseases [26], because water availability is important for hygienic behaviors that would prevent much of the transmission by person to person or through food [27].

For example, a comparison of transmission of diarrhea in two crowded areas with different sanitary facilities in Dhaka found that although contamination of household drinking water (defined as >10,000 colony forming units per gram) was more prevalent in one area (58% versus 35% of households), there was no significant correlation between water contamination and childhood diarrheal incidence [28]. Both areas had a significant correlation between the incidence of diarrhea and the degree of contamination of the hands. The authors concluded that "these findings imply that for slum communities, where heavy contamination abounds, using resources to provide increased access to a plentiful supply of water is likely to be more beneficial than providing a source which is absolutely pure". Another survey found that the hands of 70% of slum mothers were contaminated with fecal coliform bacteria [29].

The risk of diarrhea increases with inadequate handwashing, particularly after defecation or cleaning a child. Handwashing with soap is effective in eliminating fecal contamination. However, in rural Bangladesh, soap is rarely used for handwashing; it is costly and it is perceived more as a beautifying agent [29]. The health risk related to hand contamination is not clearly understood; the need for handwashing in rural areas is explained in religious terms. In areas where soap may not be available or not used, handwashing with mud, ash, or other agents is more effective than solely rinsing hands with water [29].

In addition, the presence of multiple routes of disease transmission in environments with heavy fecal contamination limits the effectiveness of controlling only one or two routes. Esrey et al. [26] reviewed 144 intervention studies to assess the impact of water and sanitation interventions on diarrheal disease among young children; the results are summarized in Table 5.5. Although important, improving water quality alone reduced diarrheal disease morbidity by only 17%. A more recent study analyzed data collected in the late 1980s from eight countries in sub-Saharan Africa,

Table 5.5 Median reductions in diarrheal disease morbidity from improvements in one or more components of water and sanitation

	All studies		*Rigorous studies*	
	N	*Median reduction (%)*	*N*	*Median reduction (%)*
Water and sanitation	7	20	2	30
Sanitation	11	22	5	36
Water quality and quantity	22	16	2	17
Water quality	7	17	4	15
Water quantity	7	27	5	20
Hygiene	6	33	6	33

Source: Data from Ref. 26.

Asia/North Africa, and the Americas to test whether improved health effects regarding diarrhea and nutritional status resulted from incremental improvements in water or sanitation conditions [30]. Nationally representative random samples of ever-married women 15–49 years of age were interviewed in all countries, and the weight and height of children 3–36 months of age were recorded. Multiple linear regressions controlled for household, maternal, and child-level variables. Improvements in sanitation resulted in fewer cases of diarrhea and in taller and heavier children within each level of water supply (on the premises, improved public water supplies, and unimproved). Health benefits from improved water were less pronounced. Benefits from improved water occurred only when sanitation was improved and only when water was available on the premises.

Interventions now accepted to have high effectiveness and strong feasibility in prevention of childhood diarrhea in developing countries include breast-feeding, improved weaning practices, improved water supplies and sanitation, promotion of personal and domestic hygiene, vitamin A supplementation, and prevention of low birth weight [31]. Effective implementation of these preventive strategies requires the involvement of not only the health sector but also agriculture, water supply, and sanitation sectors.

5. UNINTENDED CONSEQUENCES OF TUBEWELLS IN BANGLADESH: EXPOSURE TO ARSENIC

The discovery of arsenic contamination of groundwater in neighboring West Bengal led to the testing of tubewell water in Bangladesh, with contamination confirmed in 1993 [1,13]. The source of the arsenic appears to be geological because none of the explanations for anthropogenic contamination can account for the regional extent of groundwater contamination. The sediments in West Bengal and Bangladesh are rich in iron pyrites that contain arsenic. The process by which arsenic is released into the groundwater is currently under investigation.

Arsenic is a metalloid abundant in the earth's crust, with an average concentration of 2 mg/kg [32]. It occurs in trace quantities in rock, soil, water, and air. Arsenic and its compounds occur in crystalline, powder, amorphous, or vitreous forms. Arsenic is the main constituent of more than 200 mineral species.

Arsenic is widely distributed in surface freshwater, with concentrations in rivers and lakes generally below 10 μg/L. Arsenic levels in groundwater average about 1–2 μg/L, except in areas with volcanic rock and sulfide mineral deposits where arsenic concentrations can range up to 3 mg/L [32]. Elevated concentrations (>1 mg/L) in groundwater from geochemical origins have been found in Taiwan, West Bengal (India), and Bangladesh. Elevated concentrations of arsenic in drinking water also have been found in Chile, northern Mexico, and several areas of Argentina. Arsenic contaminated groundwater has been found in parts of China, the United States (California, Utah, Nevada, Washington, and Alaska), and Finland [32]. Arsenic in water cannot be tasted or smelled.

5.1 Contamination of groundwater

In 1998, the British Geological Survey in collaboration with the Bangladesh Department of Public Health Engineering conducted a survey of 3,534 well waters from 61 of the 64 districts of Bangladesh and from 433 of the 496 *upazilas* [13]. Approximately one sample per 37 km^2 was taken, representing perhaps 0.03–0.05% of all tubewells. Stratified random sampling was attempted, but not fully realized for practical reasons (i.e., flooded areas, lack of roads for vehicular access, and the local lack of familiarity with randomized sampling schemes). The majority of the tubewells sampled were government-installed wells that are believed to be representative of all wells. Except for four wells, all samples also were analyzed for a wide variety of other elements. In addition, special study areas were established in three *upazilas* to undertake greater sampling density. A survey was conducted to examine the variation of arsenic at the village level.

Figure 5.2 [13] is a smoothed map showing the regional trends in groundwater arsenic concentrations in shallow tubewells (defined at less than 150 m deep) based on the national survey. The map shows clear regional differences in arsenic concentrations, with the southern and southeastern areas having the highest contamination.

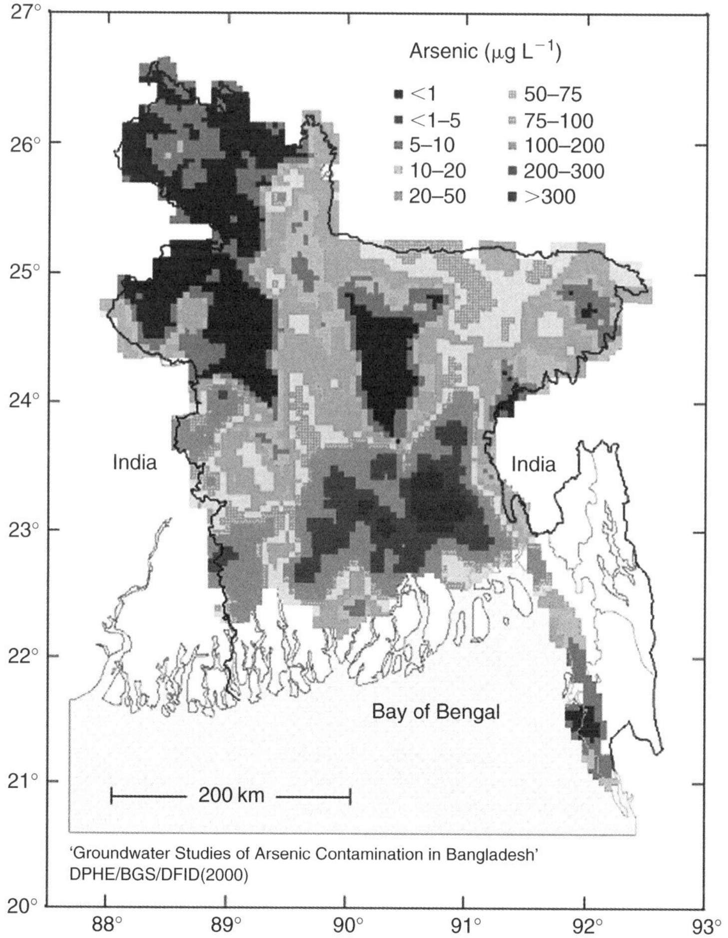

Figure 5.2 Regional trends in Bangladesh arsenic concentrations in tubewell groundwater. Source: Ref. 13. Reproduced by permission of the British Geological Survey. Copyright NERC. All rights reserved. IPR/46-11C

The survey showed that 25% of all tubewells sampled contained more than 50 μg/L of arsenic, which is the Bangladesh drinking water standard. In addition, 9% of the tubewells exceeded 200 μg/L, 1.8% exceeded 500 μg/L, and 0.1% exceeded 1,000 μg/L. On the other hand, 24% of samples fell below the detection limit for arsenic (0.25–0.5 μg/L). Of the shallow tubewells, 27% contained more than 50 μg/L and 46% contained more than 10 μg/L, the WHO guideline value for arsenic. Only 3 out of 327 deep wells (defined as equal to or greater than 150 m in depth) exceeded 50 μg/L and 16 exceeded 10 μg/L. Based on an estimate of 6–11 million tubewells in Bangladesh, these results suggest that 1.5–2.5 million wells are contaminated with arsenic concentrations above the Bangladesh standard.

Another important finding was that there was considerable well-to-well variability over the scale of a few kilometers. The variability was so high that it was difficult to predict whether a particular well was contaminated based on measurements taken in neighboring wells. In the 243 samples collected, arsenic concentrations varied over four orders of magnitude. Even within an area as small as a village, no pattern of contamination emerged. This pointed to the necessity of establishing a program of testing nearly all shallow tubewells if they are to be used for drinking water. Even areas of generally low concentrations had occasional hot spots with a cluster of tubewells with unusually high concentrations.

In addition, there appears to be no correlation between percentage of tubewells contaminated and the incidence of arsenicosis [11]. Table 5.6 shows the result of one survey; the incidence of arsenicosis is not highest in areas where the contamination of tubewells is most widespread. Although preliminary, these findings suggest that factors other than contaminated water are contributing to the occurrence of arsenicosis. Understanding the key factors associated with arsenicosis will be important for identifying and implementing effective interventions.

The survey used two different methods to estimate the number of people exposed to drinking water with arsenic concentrations in excess of the Bangladesh standard; these methods gave estimates of 28 million and 35 million people in a 1999 population of 125.5 million for the whole of Bangladesh. Using the WHO guidelines, the numbers were 46 and 57 million. The authors suggested that the larger estimates were more reliable.

The survey also clearly identified that arsenic is not the only problem: 35% of the national survey samples exceeded the WHO guideline for manganese, and some significantly so. The spatial patterns of the arsenic and manganese problem areas differed significantly; only 33% of shallow tubewell water was in compliance with the

Table 5.6 Tubewell contamination and incidence of arsenicosis

Thana	*Contaminated tubewells (%)*	*Number of patients*	*Surveyed population ('000)*	*Patients per 10,000 people*
Bera	55	86	197	4.3
Jhikargacha	59	96	267	3.6
Kachua	97	1	83	0.1
Sonargaon	62	213	301	7.1
Total	61	396	848	4.7

Source: Data from Ref. 11.

WHO guidelines for both. Elevated concentrations were found for other inorganic constituents of potential health concern, including boron and uranium. Iron and ammonium also were present in high concentrations, particularly in the southern part of Bangladesh.

5.2 Arsenic-related diseases

As noted earlier, arsenic is an abundant, naturally occurring metalloid; as such, everyone is exposed to some degree. Environmental exposure to arsenic is primarily through the ingestion of food and water, and food is the principal contributor to the average daily intake of 20–300 μg [32]. The human body has evolved mechanisms for removal of minerals, including arsenic. Adverse health effects from arsenic exposure begin once an individual's threshold body burden is exceeded; there is variability in the individual body burden at which symptoms begin to occur.

Soluble inorganic arsenic is acutely toxic, and ingestion of large doses leads to gastrointestinal symptoms, disturbances of cardiovascular and nervous system functions, and eventually death. Survivors may experience bone marrow depression, haemolysis, hepatomegaly, melanosis, polyneuropathy, and encephalopathy.

Early symptoms from long-term exposure to arsenic in drinking water can range from development of dark spots on the skin to a hardening of the skin into nodules, often on the palms of the hands and the soles of the feet. One study reported the prevalence of skin lesions in Bangladesh villages drinking from arsenic contaminated tubewells was 26–30% [33]. A survey of 18,000 people found 20.6% had arsenical dermatological symptoms [34]. Children appeared to have higher body burdens than adults despite fewer dermatological symptoms. Symptoms developed after 6 months to 2 years or more of exposure, and were related to the amount of arsenic ingested, the concentration of arsenic in the water, and nutritional status. With higher arsenic concentration in water and greater daily intake, clinical symptoms develop earlier. Chronic arsenicosis can affect multiple organ systems, including the lungs, gastrointestinal system, liver, spleen, genitourinary system, hemopoietic system, eyes, nervous system, and cardiovascular system. In the survey mentioned above, of 413 cases of chronic arsenicosis, 154 (37.3%) had evidence of neuropathy [34]. There was evidence of damage to both the peripheral and the central nervous system.

Arsenic exposure also is causally related to increased risks of cancers in the skin, lungs, bladder, and kidney [32]. Increased risks of lung and bladder cancer and of arsenic-associated skin lesions have been reported to be associated with ingestion of drinking water with arsenic concentrations of ≤50 μg/L.

Morales et al. [35] reanalyzed data from a study in Taiwan and concluded that although the shape of the exposure-response curve is uncertain at low levels of arsenic exposure, over a lifetime, one out of every 100–300 people who consume drinking water containing 50 μg/L arsenic may suffer an arsenic-related cancer (lung, bladder, or liver cancer) death. Smith et al. predicted similar levels of risk [36]. Given the British Geological Survey estimates of 28–35 million people consuming water above this concentration, and given cancer latency periods of about 20 years, about 93,000 to 350,000 excess cancer deaths could be expected unless effective and safe interventions are implemented quickly, including the provision of safe drinking water.

Conclusions on the causality of the relationships between arsenic exposure and other health effects are less clear-cut. The evidence is strongest for hypertension and cardiovascular disease, suggestive for diabetes and reproductive effects, and weak for cerebrovascular disease, long-term neurological effects, and cancer at sites other than the lung, bladder, kidney, and skin [32].

There are few effective treatments for arsenicosis. A number of arsenic chelators have been tested, with 2,3-dimercapto-1-propanesulfonate (DMPS) showing the most promise [37]. In addition, recent evidence suggests that discontinuing exposure to contaminated drinking water can reduce body burdens of arsenic [38,39]. Ahmad et al. [38] concluded that many patients with first and second stages of arsenicosis would recover if further exposures were prevented. Supplying safe water appears to be a major step to reversing as well as preventing further arsenic poisoning.

In addition to health effects, there are social consequences of arsenicosis in Bangladesh. Because of illiteracy and lack of information, many confuse the skin lesions of arsenicosis with leprosy [12]. Within a community, individuals with skin lesions have been barred from social activities and often face rejection, even by family members. Women have been unable to get married, wives have been abandoned by their husbands, men have lost jobs, and children have not gone to school in an effort to hide the problem.

5.3 Arsenic control efforts

The problem of arsenic in drinking water in Bangladesh is huge and complex; an estimated 60,000 villages are at risk in 59 out of Bangladesh's 64 districts [12]. These numbers are expected to increase as more testing is done. Major constraints to addressing the problem include lack of information on the extent, causes, effective treatment, and long-term sustainable remedies. Uncertainties range from why some people develop skin lesions while other family members do not to low-cost technological options for removal of arsenic from drinking water. Because of these uncertainties, the primary approach must be prevention.

The initial response to this crisis was slow because little was known about the extent of the problem, about how to test and treat contaminated water, and about how to identify and treat patients. A variety of international agencies began initiating programs in 1996/1997 to respond to the arsenic crisis, including the World Bank, the United Nations Development Programme, the WHO, UNICEF, the Food and Agriculture Organization, the UK Department for International Development, the Swiss Development Corporation, the Canadian International Development Agency, the Netherlands Ministry of Development Corporation, and others. Activities generally fall into five areas: testing existing tubewells to identify those that are safe sources for drinking water; providing alternative safe water sources; enhancing the health sector capacity for the surveillance, prevention, diagnosis, and treatment of individuals with arsenic related conditions; promoting research to enable the development of effective policies and measures to address the various dimensions of the problem; and initiating communication campaigns to educate the public about arsenic contamination, its health effects, how to avoid exposure, and treatment options.

The largest single project is the US$44 million Bangladesh Arsenic Mitigation and Water Supply Project, which aims to identify affected tubewells, to provide alternative sources of safe drinking water, and to build capacity at the national and local levels to address the challenge in the longer term [12]. Because of the lack of a pattern in tubewell contamination, every tubewell needs to be tested. The limited evidence available on the need for continuous monitoring is inconsistent; some tubewells that tests showed to be safe were contaminated on subsequent monitoring [13,34]. Laboratory testing is time consuming, expensive, and resource intensive (about US$10 for a single arsenic analysis) [11]. A reliable field test kit that can accurately detect arsenic at 50 ppb, the current Bangladesh standard, is not yet available commercially. The field test kit being used in UNICEF-funded projects is relatively reliable in indicating whether a water sample has arsenic below 20 ppb or above 100 ppb. Currently, all tubewells that contain arsenic above 20 ppb are painted red to indicate that they are unsafe.

One of the difficulties that has been encountered is that by 1999, a Community-Based Action Research Project found the perception that tubewells are a safe source of drinking water to be deep-seated [11]. The project found that even when tubewells were shown to contain unacceptable levels of arsenic, people continued to drink from them. This was partly because villagers are reluctant to fetch water from a source that is not close to their dwellings, and because the urgency to switch to a safe source was not readily felt. Communication programs are working to change this perception.

Another difficulty is that it is not simple for a family to switch from their unsafe well to someone else's safe well [40]. Most wells are privately owned. Women are traditionally not expected to leave their *bari* (cluster of related households) unaccompanied. The need for privacy can be an obstacle because many of the wells are located close to the family latrine.

Alternative options for provision of safe drinking water include installing deep tubewells (deeper than 200 m), dug wells, and pond sand filters; harvesting rainwater; and removing arsenic from contaminated water [11]. Deep tubewells tap into an aquifer that is less contaminated than the one from which shallow tubewells draw. Pond sand filters treat surface water to make it safe for drinking and cooking. Rainwater harvesting systems have been used in the coastal districts for years, and are now being introduced in arsenic-affected areas. Effective, safe, and inexpensive low technology methods for removal of arsenic from contaminated water in rural settings are under development.

Developing more comprehensive and sustainable water supply options is a challenge. These options will require a means of addressing the lack of sanitation and a commitment for operation and maintenance of implemented options. One successful arsenic mitigation project in rural Bangladesh achieved this through integrated research and development of appropriate water supply options and through community participation [2]. Political leaders and women played key roles in the success of the project. The main recommendations included integration of tubewell screening with supply of safe water; research on technological and social aspects; community participation by women and local governments; education and training of all stakeholders; immediate and appropriate use of the available knowledge; links between intermediate and long-term investment; effective coordination; and immediate attention by health, nutrition, agriculture, education, and other programs.

Until recently, the focus has been on the challenge of providing safe drinking water. In addition, besides prevention of new cases, there is a need for medical management of persons already suffering from arsenicosis. More definitive information is needed on the magnitude, distribution, and epidemiology of the health impacts in Bangladesh; the clinical progression of arsenicosis, both with and without continuing exposure (including whether changes in pigmentation and hyperkerotosis are good markers of more serious health effects); determinants and risk factors that explain variations in outcomes for individuals with similar levels of exposure; and the food safety issues related to crops irrigated with water contaminated with arsenic [41].

At this time, most of the population is unaware of the hazard and ways to deal with it, their water sources have not been tested, most health care providers have not been trained to recognize and treat affected patients, and crucial epidemiologic questions about incidence, prevalence, and effective medical treatments have not been answered. Obviously, educating the public about the potential hazards of arsenic exposure is ineffective under these conditions.

Addressing these challenges will be difficult given other issues facing Bangladesh. Bangladesh has some of the poorest people in the world; most earn less than US$1 per day [42]. Bangladesh has among the highest levels of malnutrition and the fourth largest concentration of tuberculosis cases in the world [41], in addition to the high rates of diarrheal and acute respiratory diseases. Contributing to the health challenges are marked gender disparities in health care, low levels of education, poor sanitation, and inadequate health care services.

6. CLIMATE CHANGE PROJECTIONS

About one-fifth of Bangladesh consists of low-lying coastal zones that are currently vulnerable to climate variability. The significant changes in climate projected for Asia are expected to increase vulnerability in Bangladesh [43]. Major risks include the following:

- Large deltas and coastal low-lying areas could be inundated by sea level rise.
- Climate change and its variability could exacerbate current vulnerability to extreme climate events.
- Increased precipitation intensity, particularly during the summer monsoon, could increase flood-prone areas.
- Tropical cyclones could become more intense. Tropical cyclones combined with sea level rise could result in enhanced risk of loss of life and properties in low-lying coastal areas.

The impacts of climate change could exacerbate the existing safe drinking water problems in Bangladesh. Flooding not only could result in loss of life but also could affect water supplies in terms of amount and distribution of rainwater. It could also cause problems in isolating "clean" surface water supplies from contaminated ones. Climate change projections should be taken into consideration when planning, designing, and implementing options for providing access to clean water and sanitation. Because infrastructure decisions have a long lifetime, prudent decision-making today could increase future adaptive capacity. Particular attention should be paid to the low-lying coastal areas, which are vulnerable to storm surges and sea level rise.

7. LESSONS FOR ADAPTATION TO CLIMATE CHANGE

As discussed in Chapter 2, the determinants of adaptive capacity and the requirements for public health prevention are measures of the ability of a community, nation, or region to cope with and respond effectively to an external stress. Table 5.7 provides a rough ranking of the dimensions of each measure for the current response to the problem of

Table 5.7 Ranking of determinants of adaptive capacity and requirements for public health prevention[a]

Determinants of adaptive capacity	*Requirements for public health prevention*
Range of available technological options (2)	Awareness that problem exists (5)
Availability and distribution of resources (2 – aid)	Sense that the problem matters (5)
Structure of critical institutions (2)	Understanding of the causes (1)
Human capital (2)	Capability to intervene (2)
Social capital (3)	Political will to influence (4)
Access to risk spreading (1)	
Ability to manage information (2)	
Public's perceived attribution (1)	

Note
a Scale = 1 (low) to 5 (high).

arsenic in drinking water in Bangladesh; the ranking is on a scale of 1 (low) to 5 (high). For both measures, the current response is limited on a number of dimensions, from an understanding of the cause of the elevated concentrations of arsenic to the human and social capital required to address the problem. Understanding where limitations occur can help focus activities to facilitate the development of effective response strategies. The consequences of the installation of tubewells to reduce the burden of diarrheal diseases offer a variety of lessons; more lessons are likely to be learned as the resolution of this problem unfolds over time. This case study clearly illustrates the risk of undertaking massive intervention programs without determining both benefits and risks. A systematic and thorough assessment is needed of the dynamics of a particular issue (including the risks and benefits of doing nothing), of the extent to which there are key uncertainties, and of the magnitude of any potential adverse impacts. Without an assessment of the potential consequences of an intervention, beneficial steps may be inadvertently presented as cures. Acute problems such as arsenic in drinking water in Bangladesh create pressure to find quick solutions. Programs should evaluate short-term responses while finding long-term solutions, within the context of the underlying causes. Claims for simple cures to complex problems from those proposing interventions often increase the skepticism of both those funding and those receiving the intervention.

In this case, arsenic in drinking water was not a recognized problem by those funding tubewell installation, so testing and monitoring were not recommended in the 1970s. However, the problem in Bangladesh could have been recognized sooner, particularly after health problems began to arise in Taiwan and India. Once problems were identified elsewhere, prompt testing of wells would have prevented many current and future adverse health effects. That, of course, is easy to say now that the size of the arsenic problem in Bangladesh is apparent.

Imbedded in this lesson is the issue of scale, the difference of absolute versus relative risks. A solution associated with a small risk implemented on a wide scale (such as the installation of tubewells) is likely to have a far more significant adverse health impact than a solution with a larger risk implemented on a small scale.

This case study also reinforces the problem of relying on a single technical solution to a problem instead of taking an integrated, multidisciplinary approach. An effective technology does not always solve the problem. A more holistic and experience-based approach is needed to ensure that the mistakes of the past are not repeated. During the tubewell installation, it was discovered that diarrheal diseases result from numerous contamination pathways. Supplying clean drinking water from tubewells was only part of the solution; equally important were changes in hygienic practices and sanitation.

In addition, the arsenic problem shows the danger of designing interventions in isolation from problems being addressed by other sectors. In this case, tubewell installation was undertaken without developing effective watershed management plans. Watershed management and villager participation are needed to assure the appropriate utilization of the large surface water resources available in Bangladesh [34]. Effective coordination and collaboration across disciplines and across donors are needed to address large-scale problems.

This problem also illustrates the importance and the challenges of community-based interventions. Technological solutions such as the initial installation of tubewells may not achieve the success envisioned because social, cultural, and behavioral factors also are critical determinants of success. That is, technology often tells us what can be done, but does not tell us how to do it [8]. In addition, as with other public health interventions, the process of intervention is intended not just to improve physical well-being but also to be a positive force for social change. This is the case in the response to the current crisis in Bangladesh. There are concerns that testing of tubewells without concurrent education will not result in appropriate behavioral changes in a community, particularly if none of the community members has signs of arsenic-related disease. Habits are hard to break; one visit will not be convincing when the villagers look at the clear, clean water and compare it with the murkier surface water. As a consequence, follow-up monitoring and education programs are being established as part of the tubewell testing program.

Another lesson relates to the process of intervention. Because tubewells were viewed as technological fix, installation was implemented on a broad scale as rapidly as possible, not in an incremental or staged fashion that would incorporate regular evaluation of success. Flexible and responsive approaches are needed in which new information and experience are properly evaluated and then used to appropriately modify interventions. Arsenic contamination of drinking water is a classic second-generation problem, with the contamination discovered many years after the initial tubewells were installed. The government and the donor agencies took time to accept and then respond to the problem; meanwhile, millions of people were being exposed to hazardous concentrations of arsenic. Taking a staged approach to implementation could have had a much different result.

REFERENCES

1. Smith, A.H., Lingas, E.O., and Rahman, M.: Contamination of Drinking Water by Arsenic in Bangladesh: A Public Health Emergency. *Bull. WHO* 78 (2000), pp.1093–1103.
2. Hoque, B.A., Mahmood, A.A., Quadiruzzaman, M., Khan, F., Ahmed, S.A. et al.: Recommendations for Water Supply in Arsenic Mitigation: A Case Study from Bangladesh. *Public Health* 114 (2000), pp.488–494.
3. U.S. Central Intelligence Agency. World Factbook. www.cia.gov/cia/publications/factbook.

4. Levine, R.J., Khan, M.R., D'Souza, S., and Nalin, D.R.: Failure of Sanitary Wells to Protect Against Cholera and Other Diarrhoeas in Bangladesh. *Lancet* 2 (1976), pp.86–89.
5. Black, R.E.: Diarrheal Diseases. In: K.E. Nelson, C. Masters-Williams, and N.M.H. Graham (eds): *Infectious Disease Epidemiology: Theory and Practice*. Jones and Bartlett Publishers, Boston, MA, USA, 2001, pp.497–517.
6. Chen, L.C., Rahman, M., and Sarder, A.M.: Epidemiology and Causes of Death Among Children in a Rural Area of Bangladesh. *Int. J. Epidemiol.* 9 (1980), pp.25–33.
7. Sunoto: Diarrhoeal Problems in Southeast Asia. *Southeast Asian J. Trop. Med. Public Health* 13 (1982), pp.306–318.
8. Chen, L.C.: Control of Diarrheal Disease Morbidity and Mortality: Some Strategic Issues. *Am. J. Clin. Nutr.* 31 (1978), pp.2284–2291.
9. Sunoto and Markum, H.A.H.: Child Health Problems in Indonesia. *Indian Paediatr.* 14 (1977), p.763.
10. World Health Organization: Water and Sanitation. Fact Sheet 112, 1998. http://www.who.int.inf-fs/en/fact112.html.
11. UNICEF Media Brief: Arsenic Mitigation in Bangladesh. Updated Version of January 2000. 2000. Dhaka@unicef.org.
12. World Bank: The Bangladesh Arsenic Mitigation Water Supply Project: Addressing a Massive Public Health Crisis. 1998. http://www.worldbank.org.
13. British Geological Survey: Groundwater Studies for Arsenic Contamination in Bangladesh. Final Report. 1999. Mott MacDonald Ltd., London, UK.
14. U.S. EPA: Fact Sheet: Drinking Water Standard for Arsenic. September 20, 2002. U.S. Environmental Protection Agency, Washington, DC. http://www.epa.gov/safewater/ars/ars_rule_factsheet.html.
15. Shears, P., Hussein, M.A., Chowdhury, A.H., and Mamun, K.Z.: Water Sources and Environmental Transmission of Multiply Resistant Enteric Bacteria in Rural Bangladesh. *Ann. Trop. Med. Parasit.* 89 (1995), pp.297–303.
16. Taha, A.Z., Sebai, Z.A., Shahidullah, M., Hanif, M., and Ahmed, H.O.: Assessment of Water Use and Sanitation Behavior in a Rural Area of Bangladesh. *Arch. Environ. Health* 55 (2000), pp.51–57.
17. Yusuf, M. and Zakir Hussain, A.M.: Sanitation in Rural Communities in Bangladesh. *Bull. WHO* 68 (1990), pp.619–624.
18. Hoque, B.A., Juncker, T., Sack, R.B., Ali, M., and Aziz, K.M.A.: Sustainability of a Water, Sanitation and Hygiene Education Project in Rural Bangladesh: A 5-Year Follow-Up. *Bull. WHO* 74 (1996), pp.431–437.
19. Sommer, A. and Woodward, W.E.: The Influence of Protected Water Supplies on the Spread of Classical/Inaba and El Tor/Ogawa Cholera in Rural East Bengal. *Lancet* 11 Nov (1972), pp.985–987.
20. Briscoe, J.: The Role of Water Supply in Improving Health in Poor Countries (with Special Reference to Bangla Desh). *Am. J. Clin. Nutr.* 31 (1978), pp.2100–2113.
21. Feachem, R.G.: Is Cholera Primarily Water-Borne? *Lancet* 2 (1976), p.957.
22. Aziz, K.M.A., Hoque, B.A., Hasan, K.Z., Patwary, M.Y., Huttly, S.R., Rahaman, M.M., and Feachem, R.G.: Reduction in Diarrhoeal Diseases in Children in Rural Bangladesh by Environmental and Behavioural Modifications. *T. Roy. Soc. Trop. Med. Hyg.* 84 (1990), pp.133 138.
23. Shaikh, K., Wojtyniak, B., Mostafa, G., and Khan, M.U.: Pattern of Diarrhoeal Deaths During 1966–1987 in a Demographic Surveillance Area in Rural Bangladesh. *J. Diarrhoeal Dis. Res.* 8 (1990), pp.147–154.
24. Kolsky, P.J.: Diarrhoeal Disease: Current Concepts and Future Challenges. *T. Roy. Soc. Trop. Med. Hyg.* 87 (1993), pp.43–46.
25. Bradley, D.: Health Aspects of Water Supplies in Tropical Countries. In: R. Feachem, M. McGarry, and D. Mara: *Water, Wastes and Health in Hot Climates*. John Wiley & Sons, Ltd., London, UK, 1977, pp.6–7.

26. Esrey, S.A., Potash, J.B., Roberts, L., and Shiff, C.: Effects of Improved Water Supply and Sanitation on Ascariasis, Diarrhoea, Dracunculiasis, Hookworm Infection, Schistosomiasis, and Trachoma. *Bull. WHO* 69 (1991), pp.609–621.
27. Feachem, R.G.: Interventions for the Control of Diarrhoeal Diseases Among Young Children: Promotion of Personal and Domestic Hygiene. *Bull. WHO* 62 (1984), pp.467–476.
28. Henry, F.J. and Rahim, Z.: Transmission of Diarrhoea in Two Crowded Areas with Different Sanitary Facilities in Dhaka, Bangladesh. *J. Trop. Med. Hyg.* 93 (1990), pp.121–126.
29. Hoque, B.A. and Briend, A.: A Comparison of the Local Hand-Washing Agents in Bangladesh. *J. Trop. Med. Hyg.* 94 (1991), pp.61–64.
30. Esrey, S.A.: Water, Waste, and Well-Being: A Multicountry Study. *Am. J. Epidemiol.* 143 (1996), pp.608–623.
31. Huttly, S.R.A., Morris, S.S., and Pisani, V.: Prevention of Diarrhoea in Young Children in Developing Countries. *Bull. WHO* 75 (1997), pp.163–174.
32. Gomez-Caminero, A., Howe, P., Hughes, M., Kenyon, E., Lewis, D.R., Moore, M., Ng, J., Aitio, A., and Becking, G.: *Environmental Health Criteria 224: Arsenic and Arsenic Compounds*, 2nd edition. International Programme on Chemical Safety, World Health Organization, Geneva, 2001.
33. Tondel, M., Rahman, M., Magnuson, A., Chowdhury, I-A., Faruquee, M-H., and Ahmad, S-A.: The Relationship of Arsenic Levels in Drinking Water and the Prevalence Rate of Skin Lesions in Bangladesh. *Environ. Health Persp.* 107 (1999), pp.727–729.
34. Rahman, M.M., Chowdhury, U.K., Mukherjee, S.C., Mondal, B.M., Paul, K., Lodh, D. et al.: Chronic Arsenic Toxicity in Bangladesh and West Bengal, India – A Review and Commentary. *Clin. Toxicol.* 39 (2001), pp.683–700.
35. Morales, K.H., Ryan, L., Kuo, T-L., Wu, M-M., and Chen, C-J.: Risk of Internal Cancers from Arsenic in Drinking Water. *Environ. Health Persp.* 108 (2000), pp.655–661.
36. Smith, A.H., Hopenhayn-Rich, C., Bates, M.N., Goeden, H.M., Hertz-Picciotto, I., Duggan, H.M., Wood, R., Kosnett, M.J., and Smith, M.T.: Cancer Risks from Arsenic in Drinking Water. *Environ. Health Persp.* 97 (1992), pp.259–267.
37. Guha-Mazumder, D.N., De, B.K., Santra, A., Ghosh, N., Das, S., Lahiri, S., and Das, T.: Randomized Placebo-Controlled Trial of 2,3-Dimercapto-1-Propanesulfonate (DMPS) in Therapy of Chronic Arsenicosis Due to Drinking Arsenic-Contaminated Water. *Clin. Toxicol.* 39 (2001), pp.665–674.
38. Ahmad, S.A., Sayed, M.H.S.C., Hadi, S.A., Faruquee, M.H., Khan, M.H., Jalil, M.A., Ahmed, R., and Khan, A.W.: Arsenicosis in a Village in Bangladesh. *Int. J. Environ. Health Res.* 9 (1999), pp.187–195.
39. Mandal, B.K., Chowdhury, T.R., Samanta, G., Mukherjee, D.P., Chanda, C.R., Saha, K.C., and Chakraborti, D.: Impact of Safe Drinking Water and Cooking on Five Arsenic-Affected Families for 2 Years in West Bengal, India. *Sci. Tot. Environ.* 218 (1998), pp.185–201.
40. Van Geen, A., Ashan, H., Horneman, A.H., Dhar, R.K., Zheng, Y., Hussain, I. et al.: Promotion of Well-Switching to Mitigate the Current Arsenic Crisis in Bangladesh. *Bull. WHO* 80 (2002), pp.732–737.
41. World Bank: *Bangladesh Arsenic Public Health Project.* 2002. http://www.worldbank.org/pics/pid/bd76693.txt. Accessed July 2002.
42. Huq, S.: Climate Change and Bangladesh. *Science* 294 (2001), p.1617.
43. Lal, M., Harasawa, H., and Murdiyarso, D.: Asia. In: J.J. McCarthy, O.F. Canziani, N.A. Leary, D.J. Dokken, and K.S. White (eds): *Climate Change 2001: Impacts, Adaptation, and Vulnerability. Contribution of Working Group II to the Third Assessment Report of the Intergovernmental Panel on Climate Change.* Cambridge University Press, Cambridge, UK, 2001, pp.533–590.

6 Epidemic early warning systems: Ross River virus disease in Australia

Rosalie E. Woodruff

ABSTRACT: An early warning system brings together research and policy into a response plan, which aims to minimize the risks to population health of a threat (such as an outbreak of mosquito-borne disease, a natural disaster, or a famine). This chapter presents the core requirements for early warning systems. It considers the feasibility of establishing early warning systems for mosquito-borne diseases, the dynamics of which are strongly influenced by climate. A case study of Ross River virus disease in Australia is used to explore the practical issues involved. If effectively designed and implemented, early warning systems provide the opportunity to prevent cases and reduce public health expenditure.

1. INTRODUCTION

The importance of adequate, timely response to the success of an early warning system is often discussed only as an addendum to prediction modeling [1]. Yet, the purpose of an early warning system is to guide decision-making regarding disease preventive actions. The common division of responsibility between those who generate disease prediction techniques and those who organize a response based on warnings (the policy-makers) means that the two primary components – prediction and response – are usually developed in isolation. This split is a potential limitation of many early warning systems.

This chapter considers the practicalities of implementing an early warning system for Ross River virus (RRv) disease, the vector-borne disease with the highest caseload in Australia (see Box 6.1 for a description of the disease).

In previous work [9], two models were developed to predict epidemics of RRv disease in the Murray region. The purpose of the study was to see if climate factors could be used to build a model to predict epidemics of RRv disease with sufficient accuracy and timeliness for use as part of an early warning system. To this end, geographic information system (GIS) techniques were used to choose two climatically similar study regions located around the Murray River, the largest river system in southeastern Australia. One region was semi-arid, the other temperate. Each region comprised a number of local government areas, the unit of analysis. Ross River notification data over an 8-year period (1991 to 1999) were used to define epidemic years. Climatic variables and epidemic years were analyzed in a logistic regression model to determine those factors that would best predict epidemics. The methods and variables used, along with the results and validation, have been published elsewhere [9].

Box 6.1 What is Ross River virus disease?

Ross River virus (RRv) is a mosquito-borne virus that causes epidemic polyarthritis in people. RRv disease is widely distributed in Australia, and outbreaks have been reported in many Pacific island countries. The disease syndrome is characterized by the sudden onset of headache, fever, rash, and muscle and joint pain. The arthritic symptoms can be severe and debilitating, and for most people last between three and six months [2]. Although not fatal or permanently disabling, RRv disease can cause considerable pain, distress, and loss of productivity. There is no treatment for the disease, and, in the absence of a vaccine, prevention remains the sole public health strategy. In Australia, there were 51,761 notifications from 1991 to 2002 (an average of 4,500 a year) [3].

Macropods (kangaroos and wallabies) are considered the primary vertebrate hosts, although a wide range of vertebrates (typically marsupials) can be infected, including humans in certain circumstances [4]. The natural cycle of the virus is between the vertebrate host and the mosquito vector. A large number of species of mosquitoes have been implicated as vectors of RRv. When immunity is low in the host population and climatic conditions are suitable, it is theorized that massive virus amplification occurs, resulting in a "spill-over" of infection into exposed human populations.

The epidemiology of the disease varies across Australia [e.g., 5,6,7], reflecting the many vector and host species involved in the transmission cycle and the impact of diverse climatic and environmental conditions. In tropical and subtropical regions, temperature and precipitation enable adult vectors to remain active all year round [5], a continuous transmission cycle results, and the disease is endemic. In colder, temperate regions, mosquitoes are active only during the warmer months [8], and the disease is usually epidemic. The many vector species and vertebrate hosts enhance survival and persistence of the virus. Vertical transmission and the "overwintering" of female mosquitoes are the two primary mechanisms by which the virus is believed to persist in a region.

A prerequisite condition for an epidemic, relating to the host and virus dynamics, was lower than average spring rainfall in the pre-epidemic year. A two-stage predictive model was developed. The early warning model (Table 6.1) includes weather conditions for July-November (late Southern Hemisphere winter to end of spring). RRv cases typically commence around November, and control measures at this point would be effective in reducing mosquito breeding. The late warning model (Table 6.2) includes variables for December-February (the Southern Hemisphere summer). In most years, the bulk of cases do not appear until March-April. Since the interval between onset and symptoms is low (7–9 days), it would still be useful to check the probability of an epidemic at this time, and to issue public alerts if the results were conclusive. Table 6.3 gives the sensitivity and specificity of the early warning and late warning models across all 8 years for both regions.

The timing, intensity, duration, and distribution of outbreaks of vector-borne diseases are expected to alter because of climate change. This case study of RRv disease in the Murray River region provides a number of lessons for both the public health and adaptation communities. It demonstrates the value of early warning as a planning tool for climate-related health risks. It argues in favor of focusing, in an integrated fashion, on both the prediction and response components of early warning systems.

Table 6.1 Early warning (variable months Jul-Nov) logistic regression models for the two regions, predicting the occurrence of an epidemic year

Region	*Variable*	*Odds ratio*	*95% CI*
1[a]	Rain days (preceding spring)	0.25	0.07–0.91
	Rain days (July)	2.92	1.21–7.03
	Rainfall (August-September ave.)	1.87	1.02–3.42
	Relative humidity (November)	0.27	0.09–0.81
	Max temperature (November)	0.01	0.00–0.37
2[b]	Rain days (preceding spring)	0.36	0.22–0.59
	Rain days (July)	1.40	1.10–1.78
	Sea surface temperature (August)	27.9	5.05–154
	Rain days (September)	1.49	1.05–2.12
	Max temperature (November)	0.37	0.24–0.57

Source: Data from Ref. 9.

Notes

a Region 1 diagnostics: R^2 (percentage of variance explained by model) = 84%.
b Region 2 diagnostics: R^2 (percentage of variance explained by model) = 78%.

Table 6.2 Late warning (variable months Jul-Feb) logistic regression models for the two regions, predicting the occurrence of an epidemic year

Region	*Variable*	*Odds ratio*	*95% CI*
1[a]	Rain days (preceding spring)	0.15	0.0–0.9
	Rain days (July)	1.88	1.1–3.3
	Rainfall (August-September ave.)	1.34	1.1–1.6
	Max temperature (November)	0.37	0.2–0.8
	Relative humidity (February)	0.25	0.1–0.8
	Min temperature (February)	4.19	1.2–14.0
2[b]	Rain days (preceding spring)	0.60	0.41–0.89
	Rain days (September)	1.52	1.02–2.27
	Max temperature (November)	0.45	0.20–0.99
	Max temperature (December)	0.38	0.19–0.78
	Vapour pressure (January-February)	5.73	1.46–22.4
	Max temperature (February)	2.14	1.38–3.31

Source: Data from Ref. 9.

Notes

a Region 1 diagnostics: R^2 (percentage of variance explained by model) = 79%.
b Region 2 diagnostics: R^2 (percentage of variance explained by model) = 80%.

Table 6.3 Sensitivity and specificity of the early warning and late warning models, all years

	Early warning model		*Late warning model*	
	Sensitivity	*Specificity*	*Sensitivity*	*Specificity*
Region 1	62%	95%	96%	93%
Region 2	73%	81%	66%	98%

Source: Data from Ref. 9.

And it proposes a re-orienting of research and policy toward the individual and population factors that reduce infection rather than continuing our reliance on the more conventional methods of vector control.

2. HISTORICAL DEVELOPMENT OF EARLY WARNING SYSTEMS

Our ability to use our knowledge of Earth's natural systems to predict future risks to health has taken a quantum leap in the past 30 years. In this period, weather forecasting has gradually increased from same-day forecasting to the 3 day advance prediction, and our understanding of the mechanics and teleconnections of the El Niño/Southern Oscillation (ENSO) now provides the capacity for seasonal and annual rainfall forecasting – assumed as recently as the 1970s to be the stuff of fancy, not science [10]. The Southern Oscillation Index (SOI) is now a standard tool used by farmers worldwide for drought monitoring and crop yields. The need for famine early warning systems was first recognized in the early 1970s, following droughts in Africa [11], and initial systems were established at the global level in 1974 by the Food and Agriculture Organization (FAO) and the US Agency for International Development (USAID) [12]. But it was not until after widespread famines in Africa during the mid-1980s that the importance of climate and other information in determining the timeliness of response was recognized, which prompted a boom in the famine early warning system industry [13]. The majority of these systems operate at the national level, although there are also global, regional, and local systems designed to monitor famine and food security.

In Italy and other European countries (also, United States and China), heat-watch warning systems are being established to give advance warning of heat waves. In the area of infectious diseases, recent work in the United States has focused on the use of dead bird surveillance as an early warning of West Nile virus activity [14]. An early warning system for Rift Valley fever based on rainfall has been proposed for Kenya [15], and the FAO is investigating the value of developing one in the east African region [16].

3. REQUIREMENTS FOR AN EFFECTIVE EARLY WARNING SYSTEM

There is wide variation in the meaning of the term "early warning system". Most early warning systems focus on developing or describing the prediction component of the system [1,17,18]. For example, Myers et al. [1] identified the principal components of infectious disease early warning systems to be the routine surveillance of disease and the modeling of current and future risk.

Although they provide the basis for an early warning system, disease prediction models do not on their own constitute one. Even when predictive models are comprehensive and able to account for physical or biological feedback, climate change, and human adaptation [19], they are still of limited value until they are actually used to direct disease control efforts. A "system" (in the context of vector-borne diseases) suggests an interrelation of predictive models, data, procedures, interventions, and communication between groups that combine to accomplish a specific task – i.e., the reduction in cases of disease in a region. An early warning "system", to be effective, has to elicit an appropriate societal response [20–24].

The purpose of an early warning system is to reduce vulnerability and increase preparedness [19]. To do this effectively, early warning systems should integrate the two principal components of disease prediction and response, with the situation-specific constraints for response designed into the development of the prediction model. Specific requirements include:

Prediction
- knowledge of disease transmission,
- reliable and up-to-date information on the exposure and the health outcome, and
- a model that is accurate, specific, and timely.

Response
- a response plan, detailing thresholds for action,
- available and effective interventions,
- economic assessment of the benefit and affordability of the system,
- communication strategy, and
- involvement of all relevant stakeholders in the process.

3.1 Disease prediction

3.1.1 *Knowledge of the disease transmission process*

Knowledge of the disease cycle, including the influence of climate and environmental factors in modifying this cycle, and availability of quality exposure and health data are essential preconditions for an early warning system, because no system can be developed in their absence [16]. It follows that "operational early warning systems are not yet generally feasible, due to our limited understanding of most relationships between climate and disease" [19].

Not all regions currently affected by RRv disease would be suitable for the development of an early warning system. RRv has a complex transmission cycle that occupies several different ecological zones [7]. Developing a predictive model requires data on the timing of past outbreaks, regional information on the vector and host species involved in transmission, and knowledge of the environmental and climatic factors that influence their breeding and survival. In many parts of Australia this information is entirely lacking or not sufficiently detailed. Mosquito trapping and virus isolation, for example, occur only in few parts of the country, and to date no surveillance has been conducted of RRv host populations. To maximize the use of resources, trap positions are often selected with the aim of obtaining information on multiple disease vectors (usually with a priority for collecting information on the potentially fatal Murray Valley encephalitis), and so traps may not necessarily be located in areas of highest RRv disease risk. In the RRv case study region, there has been a long history of outbreaks, and the transmission cycle, vector species etc., are well documented.

3.1.2 *Reliable and up-to-date information on the exposure and health outcome*

Good disease data (historical and current) are needed to model the current disease risk and to predict future risk. "Preparedness" implies that potentially emerging events are known [25]. For this to occur, public health and disease-related infrastructure (e.g., pathology laboratories) need the capacity for good clinical or serological surveillance, and clear diagnostic criteria to confirm cases [16]. Many industrialized countries have national disease surveillance systems, and the reporting of infectious diseases is mandatory [1]. Even so, depending on the scale of the early warning system, these data may not be sufficiently accurate for prediction in all regions. For RRv, one of the limitations of the notification data (whereby cases are reported by place of postcode rather than place of presumed infection) dictated that modeling could not be reliably conducted in metropolitan areas.

Meteorological data (e.g., rainfall, temperature, humidity, ENSO indices) are available at various resolutions and quality, depending on the region and country. Access to data is limited by, among other things, costs associated with its provision, limited availability in terms of poor station coverage, and missing values [26]. In many countries remote sensing data provide the opportunity for weather data not available from observation sites, and these are often free.

For many vector-borne diseases, the ecology of the pathogen varies widely between regions, and a suitable ecological zonal classification would be required to develop predictive models [e.g., see 16]. For the RRv study described above, GIS software was used to combine the major Australian climate and vegetation zones into regions within which conditions were assumed to be approximately similar [9].

3.1.3 *A model that is accurate, specific, and timely*

Any formal attempt to predict the future occurrence of disease requires the use of modeling [27]. Myers et al. [1] identify two basic methods for modeling the disease and environment relationship: statistical and biological, which is also called "process-based" [27]. The RRv disease prediction model described above was a statistical exploration of the relationship between climatic predictors and past epidemic events. The following section describes the generic elements of a prediction model for an early warning system.

Model accuracy

The accuracy of a model includes the elements of precision (consistent measurement, or reduced standard error) and validity (e.g., the number of epidemics/nonepidemics that are correctly identified) [28]. Predictions of disease epidemics can be made at several levels of precision [16]. As a rule, the value of greater precision in study results must be matched against the greater cost required to achieve it. For a disease with a high fatality rate and infectiousness such as Ebola, detailed local analyses may be justified, even required, to determine the mechanism of infection and generate forecasts of outbreaks. RRv disease is not such a high-priority disease: although there are on average some 5,000 cases a year, it is a nonfatal disease, patients typically recover without long-term sequelae, and treatment costs are low. In this case, resource expenditure for surveillance and prediction at the local level would have been disproportionate to the perceived public health burden. As such, a relatively simple, low-cost early warning system was appropriate.

Highly accurate models may be neither possible nor warranted. For example, early warning systems can also use qualitative assessment criteria for decision-making. Where the availability of data and the understanding of the dynamics of transmission in a region are minimal, a simple scoring system could be employed to assess the potential for virus activity (based on general knowledge acquired from other regions) [16]. This approach would be relevant for regions that are geographically close or climatically similar.

Early warning systems for vector-borne diseases are better suited to regions that record epidemic patterns of disease with a strong interannual pattern of events rather than endemic ones in which a steady background rate of cases are reported, year in and year out, with slightly higher peaks in some years than others.[1] Statistical modeling by its nature is more suited to regions that record high variability in the exposure variables (climate or environmental factors) and outcome variables (occurrence of epidemic). These regions are easier to model than those with low variability, in which it would be hard to distinguish the epidemic "signal" from the background "noise".

Timeliness and specificity

Developing an early warning system involves an inevitable compromise between the timeliness of the system in terms of the lag between prediction and event and the technical accuracy and reliability of the indicators. Long-range forecasts give the least specific warnings, but have the benefit of providing planners with long lead times [1]. This is especially the case with the use of ENSO as a prediction variable: there are powerful advantages of a prediction 6–12 months in advance, although the nature of ENSO teleconnections means there is no guarantee that rainfall (or lack of it) will necessarily arrive.

The WHO [29,30] has proposed a useful framework for conceptualizing the prediction stage of early warning systems. They distinguish three elements of "epidemic forecasting", "early warning", and "early detection". Epidemic forecasting predicts conditions in the future, usually using medium-range weather forecasts. It provides the longest lead time, but will also be the least specific and reliable. Early warning involves monitoring local factors (such as weather or mosquito activity) to provide greater reliability and local specificity, although this allows less time for response. The early detection stage includes monitoring of case data to provide highly specific and reliable information on where an outbreak is occurring. Obviously, this has very reduced lead time (although, depending on the shape of the epidemic curve, action at this point may still effectively reduce a substantial number of cases).

The timeliness of warnings relates to more than choosing an indicator that provides sufficient lag between the prediction and cases. It also relates to the "ability of the system to collect and analyse data quickly" [31] and to the time taken to achieve a response. A warning two months before an outbreak is useful only if communication with relevant stakeholders and appropriate control measures can take place within that period.

Informal systems

Although the benefits of formally describing and forecasting epidemics are many, the role of existing informal early warning systems should not be overlooked [20]. In relation to RRv disease, the knowledge base of some field entomologists and public health

1 In this chapter the terms "endemic" and "epidemic" refer to the human epidemiological patterns of disease occurrence rather than the ecological pattern of the persistence of the RR virus. Thus, the case study region is described as epidemic for RRv disease. Even though it is suspected that RR virus can be maintained year-round in the region, the pattern of human cases is sporadic: typically large numbers occur every 3 to 4 years, with few cases recorded in between.

officers – who rely on their own observations of local events as a basis for decision-making, and who have established information sharing networks with colleagues from other regions – can be considerable and highly sophisticated. This expertise should be incorporated, where possible, into the development of the early warning system. The disadvantage of an informal system of this type is that it concentrates expertise among very few people.

Appropriate scale of risk analysis

Buchanan-Smith categorizes early warning systems into local, subnational, national, regional, and global levels [20]. In principle, the appropriate scale of an effective early warning system will be determined by the needs of the public health authority responsible for disease prevention, guided by consideration of the cost and benefit of developing an early warning system at different levels, the complexity of the disease transmission dynamics, and the resolution at which disease and exposure data are available. The first of these points is discussed later in this chapter. In relation to the second, prediction modeling at the national- or state-level scale is not appropriate for RRv and many other arboviral diseases, given the different ecologies of the virus across Australia. Analyses are more successfully conducted within bioclimatic zones [9] in which temperature, humidity, rainfall, flora, fauna, and land use are more uniform [32].

Regarding the available resolution of data, although some macro-level data can be disaggregated, others cannot necessarily be scaled down for use at the micro-level [31]. In the RRv study, the weather data were available as an interpolated surface, and analyses could theoretically have been conducted at any resolution. Modeling at the local scale of, say, 20–50 km^2 could have incorporated substantial detail about the ecology of the vector and host species and the climatic factors driving breeding and abundance. As it turned out, the number of RRv notifications at this scale was too small to determine an epidemic pattern, and the study used climatically homogeneous collections of local government area units (each an average of several hundred square kilometers).

3.2 Disease response

A *feasible* response plan is an essential component of a system, so that the benefit of early warning is realized [19]. Feasibility in this context relates to the ability to generate timely, effective, and affordable prevention activities. Unlike famine early warning systems, where the prediction developers and users are usually separate entities, for vector-borne disease early warning the same agencies (typically public health and local government agencies) are responsible for generating the warning indicators and for providing the response.

3.2.1 *A detailed response plan*

Thresholds for action

Experience from famine early warning systems has demonstrated that decision-making takes time and can interrupt the prompt and effective mobilization of resources in an emergency [13,22]. The process of turning a prediction of an epidemic into an operational response requires agreement in advance on the triggers for activity. How many cases, for example, will constitute an epidemic? A suitable prediction model will "quantify thresholds, sizes and durations of deviations from normal" [16] of the environmental and health indicators. Ideally, these levels should be linked to prepared prevention and control plans [30] that are developed in advance of the epidemic, and preferably "concurrent with the development of the early warning system" [33]. A sensitivity analysis can be used to examine how changes in "soft data" can influence policy decisions [34].

Vulnerability assessment

Vulnerability is variously defined, and a detailed discussion of this term is provided in Chapter 2 of this volume. Briefly, vulnerability has two components: the risk or frequency of being exposed to a hazard, and the ability of an individual, group, or society to protect themselves from a hazard [12,35,36]. Successful prevention requires attention to high-risk groups as well as high-risk areas. In relation to vector-borne diseases, some exposure risk will always remain (given that control techniques will never eradicate mosquitoes), and an attempt must be made to "increase the coping capacity and reduce the consequences of the exposure" [35].

The capacity of a population to respond to the threat of vector-borne disease transmission relates to the quality of public health infrastructure, the availability and effectiveness of response options (i.e., vector control, prevention education campaigns, ongoing evaluation), and the environmental context. At the individual level, vulnerability to RRv is a function of a person's age, immune status, exposure behaviors, and individual practice of personal protection measures, which depends on their perception that the risk is real and significant, the extent of behavioral change that the personal protection requires, and to a lesser extent the individual's financial resources. Thus, groups within populations will be at differential levels of risk. Outdoor workers recorded higher levels of RRv antibody in one study [37], perhaps because of the difficulty of repeatedly applying mosquito repellent in some working conditions and their increased exposure to mosquito bites.

Other early warning system experiences [29,38] have confirmed the importance of locating, with reasonable specificity, the spatial occurrence of a hazardous event so as to enable effective intervention. The RRv prediction model was developed at the regional level, which comprises numbers of local government areas. Within each region, risk probabilities were generated for each area. Opportunities for fine-tuning exist within each local government area by the use of case surveillance and mosquito trapping, where resources permit.

Knowledge of individuals and groups at greatest risk of transmission will help target prevention campaigns. This requires quality baseline data, which are generally collected from ongoing epidemiological surveillance and research, combined with intersectoral collaboration in the region (e.g., with community groups, unions, tourism boards).

3.2.2 *Available and effective interventions*

There is no benefit in prediction information if nothing can be done to prevent cases from occurring. In general, the degree of sophistication of an early warning system should be matched to a country's capacity to take meaningful action to mitigate the hazards being predicted [19]. For a disease like RRv, no treatments are available to mitigate the impact of disease, and prevention is the only intervention. An early warning system could enable more sensitive timing of interventions. For example, education campaigns could be used only in "high risk" years, to avoid fatigue from people exposed to the same message every year. It would also allow for more accurate (and efficient) application of vector control measures (discussed later). An added benefit would be a reduction in the total amount of resources expended.

3.2.3 *Assessment of cost-benefit and affordability*

In public health, and in most areas of policy, prevention is better than a cure for several reasons. First, it avoids human suffering. Second, the approach maximizes the use of limited public health resources. It is helpful to understand the relative costs of action versus inaction. Most public health regions would support the establishment of an early warning system for RRv disease in theory. However, the realities of scarce resources and competing priorities for the health dollar demand that the system costs and benefits be assessed before proceeding. Such an assessment helps in deciding if the system is worth implementing at all, and if it is the most efficient way of achieving the greatest possible social benefit [39].

Mishan [40] defines the "benefits" of a particular disease-control program as the current costs of the cases that are averted by the program compared with the resource costs of the program. Depending on the sophistication of the system required, and the seriousness of the disease, monetary and nonmonetary benefits may both need to be quantified [41]. It is not always appropriate or feasible to fully quantify the benefits of a disease control system. At minimum, the nature of unquantifiable benefits for each strategy should be stated.

In general, the benefit of preventing an additional case of disease (the marginal benefit) should be equal to or greater than the cost of preventing a case of disease (the marginal cost) [42] – although this will depend on the political nature of the disease, and the stage of its establishment in a community or region. The relatively low morbidity of RRv disease, combined with the fact that it has been long established in the Murray region and is one of only many competing health priorities, means the costs of intervention would definitely need to be less than the cost of the status quo.

Detailing the annual costs *averted* by an early warning system program is essential for its ongoing survival, because there is a general problem with maintaining government interest in funding public health prevention activities. In Australia, for example, AIDS education funding for sex worker groups has been constantly threatened since the mid-1990s, when extraordinarily low HIV transmission rates were achieved between sex workers and clients. Sex worker groups attributed this Australian anomaly (HIV rates in neighboring Asian countries are very much higher) to the value of ongoing prevention education and law reform.

In terms of the effectiveness of an early warning system, it is also necessary to count the cost of a type I (false positive prediction) or type II (false negative prediction) error. Even if the system has a high sensitivity (i.e., the number of epidemics correctly predicted is high), the number of "false alarms" may also be high. If this is the case, the cost of attending to these may outweigh the overall benefit that the real alerts provide.

Although RRv disease affects a considerable number of people in Australia each year, a comprehensive assessment of the costs of the disease to the community has yet to be prepared. As is often the case, many data needed for this estimate are unavailable,

Table 6.4 Estimates for annual cost of RRv disease

Type of cost	*Details of cost*	*Cost estimates*
Number of cases	Average number of cases per year (1991–2000)	4745
Testing	Pathology cost for single RRv test	A$38.40[a]
	Percentage of positive tests	5%–25% of tests ordered
	Total number of tests ordered per annum	18,980 – 94,900
	Total cost of testing (positive and negative)	A$728,832–$3,644,160[b]
Medical care	Cost of two consultations with family doctor	A$65.90
	Total cost of medical care	A$217,796[c]
Lost earnings	Average time off work	one week in 50% of cases
	Average Australian weekly earnings before tax	A$908.20[d]
	Total lost earnings	A$2,154,704
Total estimated costs		A$3.1 m to $6 m per annum

Source: Ref. 43, updated where noted.

Notes

a This cost includes the cost of the test (IgG or IgM, both of which are A$23.80) and the cost of a collection fee (A$14.60).
b Does not include cost of testing to exclude differential diagnoses.
c Does not include cost of medications.
d Estimate updated to May 2002 [44].

available only for very small population groups or regions (and therefore not necessarily generalizable), or unvalidated. In addition, even if accurate costs were available, it is likely they would vary between regions, depending on the method of vector control used. Harley and others [43] provide the most detailed estimate to date of annual disease costs, summarized in Table 6.4. They calculate the annual total costs incurred by all cases to lie somewhere between A$3.1 and $6 million (one Australian dollar is approximately equal to US$0.75). The average cost per case (derived from these figures) is between A$650 and A$1,265.

These figures provide only a partial estimate, however, because they are based on data for one state of Australia (Queensland). In addition, the assessment does not account for treatment costs, patient travel to and from a general practitioner, decreased work output of more than one week, and the intangible costs of pain and family support. Recent research [43,45] suggests that previous studies may have overestimated the prevalence and duration of symptoms with RRv disease. For the majority of patients the disease is self-limiting, and progressive resolution over three to six months appears usual.

This assessment also does not account for the costs incurred by the public health system in disease prevention, or by the environment from the negative impacts of control measures. Examples include the cost of vector control, education campaigns, and research, and the possible costs of drug resistance or ecological damage – all of which should be considered in a comprehensive cost-benefit assessment.

3.2.4 *Communication strategy*

Walker [24] identified two principal components of a successful early warning system that relate to the transmission of information. These can be summarized as:

- constructing the warning or response message (in lay terms, detailing what is predicted, and who is most at risk, in a consistent manner); and
- distributing the message (accurately, and through the correct channels).

Even reliable and specific warnings will fail to achieve their objective if the message is inaccurately or improperly transmitted [23].

In the proposed RRv system, prediction is made at the regional level, but monitoring and response are more likely to be managed at the local government level. In the initial years of a system, achieving consistency of response messages may be a simple matter. As time goes on, however, and with additional research into adaptation strategies, divergent views on how best to promote disease prevention (at the public education campaign level as well as at the individual protection level) may emerge. It would be ideal to aim for unanimity among collaborating regions to avoid contradictory warnings (i.e., different self-protection measures).

In a developed country such as Australia, channels of communication are technically excellent. However, good communication consists of more than just technical capacity. Foran and Brosnan [33] recommend the development of a "chain of communication" as part of a response plan. This details the people involved from the point of prediction analysis through to those responsible for transmitting, receiving, and responding to the warning. It should also indicate the maximum time period for communication between collaborators.

3.2.5 *Involvement of all partners in the system*

Thomson and Connor [30] have noted the limited routine use of early warning systems in Africa for malaria control programs due, they argue, to poor intersectoral collaboration between health and the meteorological and agricultural sectors, and to the lack of evidence of the cost-effectiveness of these systems. For a disease like RRv, the primary stakeholders (i.e., those involved in the operation of the system) might include public health regional units (responsible for disease surveillance and public education), local government authorities (responsible for vector control interventions), and staff in the meteorological organizations that provide regular weather data. Other people who might need to be involved in establishing a system could include government staff at state health levels who can direct funds or provide nonfinancial support, and researchers, statisticians, entomologists, and others with expertise in RRv issues.

4. THE CONTEXT FOR RRv EARLY WARNING IN AUSTRALIA

4.1 Current RRv disease management

No vaccine is available to prevent RRv infection. Therefore the principal disease management activities can be categorized as surveillance (of the vector, pathogen, or human populations); source reduction and control of vectors; and prevention education. RRv disease management is a joint responsibility of both health departments and local government authorities in all states and territories, with different jurisdictions adopting different models to apportion costs and responsibility for surveillance, control, and prevention activities. Disease control aims to decrease the incidence or prevalence of a disease, or to eliminate the conditions that contribute to outbreaks. Regarding vector-borne diseases, conventional reliance on adulticiding (the spraying of adult mosquitoes) for the control of mosquitoes has been replaced, in Australia as in many other countries, by an integrated pest management approach to minimizing the risk of disease [46].

4.1.1 *Surveillance*

Anker and Wilson provide a review of public health surveillance issues (Chapter 10 in this volume). Briefly, surveillance data indicate who is affected, where problems are occurring, and where interventions should be directed. The surveillance mechanisms for RRv disease include case data collection, surveillance of vector and virus activity, and meteorological surveillance.

Primary surveillance for RRv disease relies on the collection of health status data such as RRv notifications, case follow-up, and serosurveys to determine population immunity levels. All states and territories in Australia record case notifications from laboratory serological tests and are required to report them to the National Notifiable Disease Surveillance System.

4.1.2 *Vector control*

Vector control strategies consist of reducing mosquito populations through chemical sprays or biological control, or reducing mosquito breeding sites by environmental engineering (e.g., filling in swamps, runnelling, water management, removing vegetation near human populations). Effective vector control relies on knowledge of the ecology of the mosquito species involved (i.e., habitat, biology, and dispersal pattern).

In most states, mosquito control is primarily the responsibility of local governments. Only a few states practice ground or aerial spraying of larvae (Western Australia, Northern Territory, Queensland). The largest mosquito control program in Australia is conducted in Brisbane, Queensland, a city that is surrounded by salt marshes (some 30,000 hectares of salt-marsh mosquito breeding habitat are aerially treated each year). Adulticiding is uncommon in some states, and not practiced at all in most because it is generally considered ineffective. New South Wales and South Australia (where the RRv study was located) place a strong focus on personal responsibility for protection rather than relying on mosquito control [47].

Source reduction involves eliminating larval habitats [46]. Runnelling, a form of habitat modification, allows water to drain from trap pools along natural drainage lines to connect with tidal sources [48] and permits predatory fish to gain access to mosquito larvae. Runnelling has been successfully used in parts of Australia, although it is not suitable for regions that generate large, flat fields of water.

4.1.3 *Prevention education*

Community education to reduce exposure to infection is important, given that control will usually be inadequate. For example, larviciding will never completely knock out mosquito populations in years of heavy rainfall or high tides [46], and it is economically impractical in rural areas where human population density is low.

Two types of prevention are relevant to vector-borne disease management. Primary prevention involves action taken by the individual or community to reduce the risk of being bitten by mosquitoes. Primordial prevention, on the other hand, consists of public health policy measures that seek to inhibit the environmental, economic, social, and behavioral conditions that are known to increase the risk of disease [49]. Such activities include workplace education for workers who are regularly exposed to mosquitoes, or the use of planning instruments to discourage or reduce development around mosquito-breeding sites.

In all states, prevention education is the responsibility of the public health departments, occasionally in concert with local governments. The approach varies slightly, depending on the climate of the region. In the tropical north, "low level" education all year round is combined with media alerts when an increase in cases is detected. In the temperate south, education generally commences only as the weather warms at the start of summer and when mosquito activity is more likely (or, in those areas that have the resources for mosquito surveillance, when activity has been actually observed). States adopt a variety of similar education techniques, including the use of written material (such as pamphlets explaining the recommended protective measures, placement of warning signs in high-risk recreational areas) and health alerts via the media when high numbers of cases have been recorded. Both local residents and tourists are usually targeted, again depending on resources.

Town planning is a consideration for eliminating potential breeding sites. Tai and others [50] found different attack rates between two towns in the Northern Territory, even though both areas had identical rainfall patterns. They attributed the lower incidence of infection in one town to groundwater control systems, which drained water away from the ground surface.

4.2 Why change the current approach?

4.2.1 *Economic and environmental sustainability*

The use of vector control strategies for RRv disease will continue to be appropriate in certain ecologies and at particular times. As a long-term public health strategy to reduce disease, however, mosquito control (particularly spraying to reduce population numbers) has several problems. Weinstein [51] has argued that the use of insecticide sprays to control mosquitoes is appropriate at the interface between endemic and nonendemic areas, but not when nonimmune populations enter an endemic area (e.g., as a result of urban or agricultural expansion or tourism), because (i) endemic areas are likely to be large and spraying is not cost-effective in the long term, (ii) there is growing opposition to use of insecticides from the public, (iii) applying insecticides without complete eradication of the mosquito population increases the chance of building resistance, and (iv) it will delay exposure of children to the RR virus until they are likely to show symptomatic disease. In regard to the last point, the use of insecticides for mosquito control could ultimately lead to an increase in the incidence of disease. Instead, education about anti-mosquito measures would be more appropriate [51].

The steady opposition to the use of insecticide control is for both health and environmental reasons. Although used only at low rates, many insecticides are toxic to insects, birds, fish, bees, and aquatic life [46]. Larviciding is not permitted in fish habitat areas or conservation parks in Queensland [52]. In New South Wales, the Griffith Council did not use insecticides to control mosquito activity in a large outbreak because of local community and industry concern about its effect on consumer confidence of the fruit grown in the area [53], even though the area recorded some of the highest rates of RRv disease in New South Wales (1,070 per 100,000 people in an outbreak in 1984) [37].

4.2.2 *Insensitive targeting of prevention education*

There is accumulated knowledge through international research of other diseases (such as malaria and dengue), and through some Australian research [54–56], that wearing long-sleeved loose clothing, using personal mosquito repellent, insect-proofing buildings and water tanks, and removing domestic breeding sites will reduce infection rates. No studies in Australia have investigated the comparative effectiveness of these different personal protection strategies, however. Research from other countries on vector-borne diseases usually cannot be generalized to the experience of RRv disease in Australia, given differences in vector biology and pathogen transmission, or for social or cultural reasons, etc.

Arbovirus prevention campaigns in Australia currently lack the sociobehavioral information needed to formulate effective messages. Research into the factors that increase or decrease the use of personal protection, and into the appropriate targeting of information, is scarce. More information is needed, for example, about how knowledge of local vector control activities affects people's perception of their risk of infection. Do they become less attentive to self-protection measures? How often should prevention messages be delivered over the course of a season, and by whom? Is it better to provide specific details about the habits of mosquito vectors in each region, or will general (and basic) information be more effective? Should particular groups in communities be targeted, or would a blanket population approach be better? Assertions about the merit of particular activities (such as modifying behavior to avoid peak biting times or areas of high mosquito abundance) have not been accompanied by research focusing on the often mundane difficulties of implementation.

Given the lack of information noted above, regional public health authorities are forced to rely on simplistic prevention messages [47] for what are quite complex behavioral changes. Given minimal resources in most regions in Australia, education campaigns are almost never evaluated.

To improve the quality of data, differences in risk perception between regions and groups should be identified. In rural communities, for example, it has been anecdotally noted that there "is an attitude of apathy or stoicism" to contracting RRv disease [48]. If true, this would obviously affect the level of prevention education uptake: annual repetition of prevention messages may cause people to "switch off" to a risk they perceive to be constant and unavoidable.

4.2.3 *Tourism in Australia*

Tourism is one of Australia's fastest growing and most significant industries: domestic and international tourism accounted for 4.5% of the gross domestic product in 1997–1998. More than 4.8 million people from overseas visited Australia in 2002, 55% of them as tourists [57]. International visitor arrivals are predicted to increase to

8.1 million by 2012, an average annual increase of 4.8% [58]. The growth of tourism in northern tropical and subtropical Australia, predominantly in nonurban areas, suggests the risk of infection among nonimmune visitors will increase [59]. Advance warning of outbreaks at the local government area level provides the opportunity to target information toward tourists entering the region.

4.2.4 *Future pressures*

Future population growth and human settlement patterns are likely to place increasing numbers of people at risk of exposure to RRv disease. The Australian population is expected to grow from 19 million to between 25 and 28 million by 2051 [60]. This represents an increase of 9–38% in the total number of people in the RRv higher risk age groups (i.e., 15 to 65 years of age). Increasing urbanization has been accompanied by a general trend from rural to city living in Australia, as elsewhere. Currently about 86% of the Australian population live in urban areas. This is expected to rise to 96% by 2050 [60]. Peri-urban sprawl has been a persistent feature of Australia population redistribution. Between 1986 and 1991, 28% of Australian population growth occurred in nonmetropolitan coastal regions [61], mostly into productive farm land and bushland. If this future picture holds true, nonimmune humans are progressively likely to be exposed to the "natural cycle" of the virus between the mosquito and reservoir hosts. In regions where the virus is endemic, this presents a significant public health management issue. Because of these changes, the impetus for more precise methods of determining when and where outbreaks of disease will occur is also likely to increase.

4.2.5 *Summary*

The current approach to RRv risk management in Australia, although following the world's best practice in terms of integrated pest management, has a number of intractable problems. Concern for possible ecological and human health impacts, the possibility of vector resistance, and the increasing financial costs of vector control all favor limiting the use of sprays and environmental modification as much as possible. At present, there is no mechanism for public health authorities to discriminate between high-risk, medium-risk, or low-risk seasons, and the community is beset with similar prevention messages each year. Accurate and timely prediction, coupled with timely and effective response, would enable targeting of vector control and education campaigns to those years when there is a high risk of epidemics. Hopefully, this would have the benefit of reducing cases and stretching limited resources further.

5. OVERVIEW OF THE RRv EARLY WARNING SYSTEM

Figure 6.1 is a diagram of the three-stage early warning system developed for the RRv study. Using weather data in a staged prediction model (i.e., where forecasts increase in specificity as the time before the commencement of the epidemic season decreases) increases the response time and allows for improved disease control activities.

The interventions outlined in the diagram are triggered by predefined thresholds. The predicted probability of an epidemic would need to be determined, perhaps differently for each local government area, depending on the result of cost-benefit analyses (including

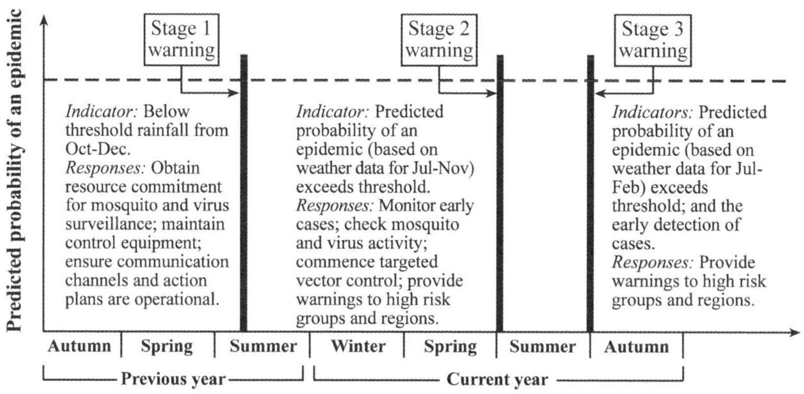

Figure 6.1 The use of early warning indicators in the RRv disease system

sensitivity analyses), a local risk aversion assessment, etc. Some regions might choose a 0.7 probability of an epidemic as a signal for action, whereas others may choose a lower cut-off. Similarly, the mosquito trap density, amount of virus isolated, and number of human cases would need to be quantified. It has been observed that early season population reduction for RRv vectors is more effective than spraying during an outbreak [62]. Quantifying "early" and "late" in this context should be done for the establishment of an early warning system, and would be regionally specific.

5.1 Benefits of the system

The benefits of the proposed RRv system have yet to be tested. Theoretically, advance warning of the likelihood of epidemics enables better communication of risk and more sophisticated and specific prevention information, improved timing of insecticide spraying, fewer cases of disease, and more efficient use of public health resources.

With the end use of the prediction model in mind, the RRv study intentionally used data, software, and statistical techniques that are currently readily available to public health authorities, or that can be obtained with minimum effort and expenditure. The interpolated weather surface, generated daily for the whole of Australia from weather station reports, is available via the World Wide Web. Similarly, the choice of scale (i.e. local government area) reflects the need to keep an RRv prediction model as simple as possible, given the resources available to control outbreaks.

The models were developed at the regional level, but provide an ability to increase the specificity of prediction by incorporating more detailed information about local mosquito-breeding habitat, host density, and epidemiological surveillance at a smaller scale.

Although RRv disease vaccination is not currently available, research is being conducted into the development of a vaccine [63,64], and human trials are being considered. The RRv prediction model could potentially be used to inform decision-making for an immunization schedule. Output from the model could categorize local areas into high (>70% probability), medium (30–70%), or low (<30%) risk each season. This would enable immune-compromised people living, or tourists travelling into an area, to consider the merits of vaccination.

5.2 Limitations of the system

In addition to the constraints on the development of the RRv early warning system that have been raised throughout this chapter, a number of other points are worth making. Theoretically, advance warning of an epidemic gives health authorities enough time to conduct mosquito control and to provide warnings to vulnerable groups. The reality is that success is not guaranteed. Because of a combination of political and institutional factors, as well as logistical and resource ones, there is often a failure to capitalize on an early warning [13]. In the case of RRv, direct costs for aircraft time, chemicals, and wages, plus ongoing infrastructure costs for training (application of chemical and occupational health and safety), are all costs that must be borne by the public health sector. Staff need to be regularly trained to ensure high technical accuracy in the application of chemicals. There is little tolerance with the timing of spraying in relation to the breeding season and local weather conditions. Equipment failure, absent employees [46], and lack of chemical availability and aircraft or ground vehicles when needed are all factors that compromise the effectiveness of larviciding.

A statistical model that attempts to predict future patterns of disease is limited by a reliance on past or current data. Thus it cannot be assumed that the relationships will remain the same in the future [1]. Even if all known confounders were incorporated as variables in the model (which is impossible to achieve), future changes in climate, environmental conditions, and social factors will be likely to influence the disease transmission cycle. While the specific regional impacts remain to be researched, it is reasonable to assume that long-term climate changes may limit the predictive power of current observations at future time points [19].

6. THE IMPACT OF CLIMATE CHANGE

Long-term changes in the abundance, spread, and distribution of all mosquito-borne diseases may occur as the result of changing global climate (depending on the state of public health infrastructure). The future climate changes predicted for Australia indicate that changes will vary locally, and impacts will vary as well, depending on the local mosquito ecology and environmental conditions. Climate projections for 2030 indicate that temperatures may rise by 0.4 to 2°C over most of Australia [65]. By 2070 they may rise by 1 to 6°C. Winter rainfall in the southeast of the country may decrease by up to 10% by 2030 (35% by 2070), and summer rainfall in these regions may change by −10 to +20% by 2030 (−35 to +60% by 2070) [65].

The effect of these changes on the breeding and survival of mosquitoes and vertebrate populations, even in a region as "localized" as the southeast of Australia, is difficult to anticipate. Winter-breeding mosquitoes (thought to be responsible for viral build-up in the natural cycle) may be adversely affected by decreased winter rainfall, possibly resulting in delayed or precluded RRv activity [66]. Conversely, the predicted increase in summer rainfall may increase the availability of mosquito habitat, which, combined with higher average temperatures, may lead to higher humidity, a lengthened season of abundance, and greater transmission levels. In summary, the impact of climate changes on the timing, intensity, and distribution of RRv epidemics in most parts of Australia is still essentially unknown.

7. LESSONS FOR ADAPTATION TO CLIMATE CHANGE

It is useful to consider this case study in relation to the determinants of adaptive capacity and prerequisites for prevention identified in Chapter 2 (Table 6.5 provides a simple ranking for these).

This case study demonstrates the potential value of using an early warning system as a planning tool to increase preparedness of climate-related events. If effectively designed and implemented, early warning provides the opportunity to prevent cases and reduce public health expenditure. Several of the main points raised in this chapter are worth summarizing. First, excellent predictive ability is of little value if the response capacity is not equally as good. Response can be impeded by political, institutional, and administrative factors. In fact, the major work of establishing an early warning system is ensuring that an advance warning can be smoothly responded to. Research and policy need to focus on the less technologically sophisticated (but no less important) details of action thresholds, vulnerability assessments, communication strategies, etc.

Second, the lack of information about the future impact of climate changes on the ecology of RRv – and other complex vector-borne disease transmission cycles – is a problem in terms of devising adaptation policy and measures. At present it is unclear how the incidence of the disease will change. Continued monitoring of the vectors in the current epidemic regions – and particularly at their borders – is needed to provide evidence of changes in species type, distribution, or breeding patterns (information that could be used to adjust prediction models). The resources available for RRv risk management are potentially high, but the current distribution of funds is low because RRv is not perceived to be a significant public health priority at the national level or in the southeast region. This limits the effort directed toward developing adaptive capacity.

Finally, the public health response to RRv disease outbreaks should shift the emphasis from a dominant focus on surveillance and vector control to include more effort in developing an effective prevention capacity. While the former are necessary components of a risk management strategy, they are not sufficient. It is the adequacy of the total public health response that prevents large outbreaks occurring. Research should be taken into the factors that support and prevent people from making positive health behavior changes.

Table 6.5 RRv disease in southeastern Australia[a]

Determinants of adaptive capacity	*Requirements for public health prevention*
Range of available technological options (3)	Awareness that problem exists (5)
Availability and distribution of resources (3)	Sense that the problem matters (3)
Structure of critical institutions (5)	Understanding of the causes (4)
Human capital (3)	Capability to deal with problem (3)
Social capital (4)	Political will to influence (2)
Access to risk spreading (2)	
Ability to manage information (4)	
Public's perceived attribution (3)	

Note

a Scale = 1 (low) to 5 (high).

ACKNOWLEDGMENTS

The author would like to thank Charles Guest and Graeme Garner for their thoughtful comments on this chapter. Research into the RRv disease and climate relationship was conducted with financial support of the Electric Power Research Institute, Palo Alto, California, USA.

REFERENCES

1. Myers, M.F., Rogers, D.J., Cox, J., Flahault, A., and Hay, S.I.: Forecasting Disease Risk for Increased Epidemic Preparedness in Public Health. *Adv. Parasit.* 47 (2000), pp.309–330.
2. Harley, D., Bossingham, D., Purdie, D.M., Pandeya, N., and Sleigh, A.C.: Ross River Virus Disease in Tropical Queensland: Evolution of Rheumatic Manifestations in an Inception Cohort Followed for Six Months. *Med. J. Australia* 177 (2002), pp.352–355.
3. Communicable Diseases Network Australia. National Notifiable Diseases Surveillance System. http://www.health.gov.au/pubhlth/cdi/nndss/year002.htm. Accessed 10 October 2002.
4. Aaskov, J.G., Nair, K., Lawrence, G.W., and Dalglish, D.A.: Evidence for Transplacental Transmission of Ross River Virus in Humans. *Med. J. Australia* 2 (1981), pp.20–21.
5. Kay, B.H. and Aaskov, J.G.: Ross River Virus (Epidemic Polyarthritis). In: T. P. Monath (ed.): *The Arboviruses: Epidemiology and Ecology.* CRC Press, Boca Raton, FL, USA, 1989, pp.93–112.
6. Mackenzie, J.S., Lindsay, M.D., Coelen, R.J., Broom, A.K., Hall, R.A., and Smith, D.W.: Arboviruses Causing Human Disease in the Australasian Zoogeographic Region. *Arch. Virol.* 136 (1994), pp.447–467.
7. Russell, R.C.: Ross River Virus: Disease Trends and Vector Ecology in Australia. *Bull. Soc. Vector Ecol.* 19 (1994), pp.73–81.
8. Dhileepan, K.: Mosquito Seasonality and Arboviral Incidence in Murray Valley, Southeast Australia. *Med. Veterin. Ent.* 10 (1996), pp.375–384.
9. Woodruff, R.E., Guest, C.S., Garner, M.G., Becker, N., Lindesay, J., Carvan, T., and Ebi, K.: Predicting Ross River Virus Epidemics from Regional Weather Data. *Epidemiology* 13 (2002), pp.384–393.
10. Nicholls, N.: Climatic Outlooks: From Revolutionary Science to Orthodoxy. In: *Climate and Culture.* Canberra, 25–27 September 2002. Australian Academy of Science.
11. Glantz, M.H.: Eradicating Famines in Theory and Practice: Thoughts on Early Warning Systems. *Internet J. African Studies* 2 (1997). http://www.brad.ac.uk/research/ijas/ijasno2/glantz.html. Accessed September 2002.
12. Moseley, W.G. and Logan, B.I.: Conceptualising Hunger Dynamics: A Critical Examination of Two Famine Early Warning Methodologies in Zimbabwe. *Appl. Geogr.* 21 (2001), pp.223–248.
13. Buchanan-Smith, M.: Drought and Famine in Africa: Time for Effective Action. *Food Policy* 17 (1992), pp.465–467.
14. Eidson, M., Kramer, L., Stone, W., Hagiwara, Y., Schmit, K., and The New York State West Nile Virus Avian Surveillance Team: Dead Bird Surveillance as an Early Warning System for West Nile Virus. *Emerg. Infect. Dis.* 7 (2001), pp.631–635.
15. Linthicum, K.J., Anyamba, A., Tucker, C.J., Kelley, P.W., Myers, M.F., and Peters, C.L.: Climate and Satellite Indicators to Forecast Rift Valley Fever Epidemics in Kenya. *Science* 285 (1999), pp.397–400.
16. Garner, M.G. and Kalunda, M.: Early Warning Systems for Rift Valley Fever and other Transboundary Diseases in Africa. Report No. TCP/RAF/8821. Food and Agricultural Organization of the United Nations, Rome, 1999.

17. Cox, J.S., Craig, M.H., Seur, D.L., and Sharp, B.: Mapping Malaria Risk in the Highlands of Africa. MARA/HIMAL Technical Report. Durban, South Africa, 1999.
18. Hay, S.I., Rogers, D.J., Shanks, G.D., Myers, M.F., and Snow, R.W.: Malaria Early Warning in Kenya. *Trends Parasitol.* 17 (2001), pp.95–99.
19. Committee on Climate, Ecosystems, Infectious Disease, and Human Health, Board on Atmospheric Sciences and Climate, and National Research Council: *Under the Weather: Climate, Ecosystems, and Infectious Disease.* National Academies Press, Washington, DC, 2001.
20. Buchanan-Smith, M.: What Is a Famine Early Warning System? Can It Prevent Famine? *Internet J. African Studies* 2 (1997). http://www.brad.ac.uk/research/ijas/ijasno2/smith.html. Accessed September 2002.
21. Davies, S., Buchanan-Smith, M., and Lambert, R.: *Early Warning in the Sahel and Horn of Africa: The State of the Art. A Review of the Literature.* Volume One. Research Report No. 20. IDS, Sussex, UK, 1991.
22. Saidy, D.: Early Warning and Response. *Internet J. African Studies* 2 (1997). http://www.brad.ac.uk/research/ijas/ijasno2/saidy.html. Accessed September 2002.
23. Tapscott, C.: Is a Better Forecast the Answer to Food Security? To Better Early Warning? To Better Famine Protection? *Internet J. African Studies* 2 (1997). http://www.brad.ac.uk/research/ijas/ijasno2/tapscott.html. Accessed September 2002.
24. Walker, P.: *Famine Early Warning Systems.* Earthscan, London, 1989.
25. Richet, H.M., Mohammed, J., and McDonald, L.C.: Building Communication Networks: International Network for the Study and Prevention of Emerging Antimicrobial Resistance. *Emerg. Infect. Dis.* 7 (2001), pp.319–322.
26. WHO: *Malaria Early Warning Systems: Concepts, Indicators, and Partners. A Framework for Field Research in Africa.* World Health Organization, Geneva, 2001.
27. McMichael, A.J., Martens, P., Kovats, R.S., and Lele, S.: Climate Change and Human Health: Mapping and Modelling Potential Impacts. In: P. Elliott, J.C. Wakefield, N.G. Best, and D.J. Briggs (eds): *Spatial Epidemiology: Methods and Applications.* Oxford University Press, Oxford, UK, 2000, pp.444–461.
28. Greenland, S. and Rothman, K.J.: Fundamentals of Epidemiologic Data Analysis. In: K.J. Rothman and S. Greenland (eds): *Modern Epidemiology.* Lippincott-Raven, Philadelphia, PA, USA, 1998, pp.201–229.
29. HIMAL and MARA: Developing New Approaches for the Surveillance and Control of Malaria Epidemics in the Highlands of East Africa. Summary Workshop Report, 20–21 March. Highland Malaria Project and Mapping Malaria Risk in Africa, Salt Rock, South Africa, 1999.
30. Thomson, M.C. and Connor, S.J.: The Development of Malaria Early Warning Systems for Africa. *Trends Parasitol.* 17 (2001), pp.438–445.
31. Huss-Ashmore, R.: Local-Level Data for Use as Early Warning Indicators. *Internet J. African Studies* 2 (1997). http://www.brad.ac.uk/research/ijas/ijasno2/ashmore.html. Accessed September 2002.
32. Boughton, C.R., Hawkes, R.A., Naim, H.M., Wild, J., and Chapman, B.: Arbovirus Infections in Humans in New South Wales: Seroepidemiology of the Alphavirus Group of Togaviruses. *Med. J. Australia* 141 (1984), pp.700–704.
33. Foran, J.A. and Brosnan, T.M.: Early Warning Systems for Hazardous Biological Agents in Potable Water. *Environ. Health Persp.* 108 (2000), pp.993–995.
34. McNeil, B. and Pauker, S.G.: Decision Analysis for Public Health: Principles and Illustrations. *Ann. Rev. Publ. Health* 5 (1984), pp.135–161.
35. Dilley, M.: Warning and Intervention: What Kind of Information Does the Response Community Need from the Early Warning Community? *Internet J. African Studies* 2 (1997). http://www.brad.ac.uk/research/ijas/ijasno2/dilley.html. Accessed September 2002.
36. Watts, M. and Bohle, H.: The Space of Vulnerability: The Causal Structure of Hunger and Famine. *Prog. Hum. Geog.* 17 (1993), pp.43–67.

37. Hawkes, R.A., Boughton, C.R., Naim, H.M., and Stallman, N.D.: A Major Outbreak of Epidemic Polyarthritis in New South Wales during the Summer of 1983/1984. *Med. J. Australia* 143 (1985), pp.330–333.
38. Field, J.O.: Beyond Relief: Toward Improved Management of Famine. In: H.G. Bohle, T. Cannon, G. Hugo, and F.N. Ibrahim (eds): *Famine and Food Security in Africa and Asia: Indigenous Response and External Intervention to Avoid Hunger.* Naturwissenschlaftliche Gesellschaft Bayreuth, Bayreuth, Germany, 1991, pp.151–166.
39. Hutubessy, R.C., Bendib, L.M., and Evans, D.B.: Critical Issues in the Economic Evaluation of Interventions Against Communicable Diseases. *Acta Trop.* 78 (2001), pp.191–206.
40. Mishan, E.J.: *Cost-Benefit Analysis: An Informal Introduction.* George Allen & Unwin, London, UK, 1971.
41. Martin, S.W., Meek, A.H., and Willeberg, P.: *Veterinary Epidemiology: Principles and Methods.* Iowa State University Press, Ames, USA, 1987.
42. Last, J.M.: *A Dictionary of Epidemiology.* Oxford University Press, New York, 2001.
43. Harley, D., Sleigh, A., and Ritchie, S.: Ross River Virus Transmission, Infection, and Disease: A Cross-Disciplinary Review. *Clin. Microbiol. Rev.* October (2001), pp.909–932.
44. Australian Bureau of Statistics: *Average Weekly Earnings: States and Australia (catalogue number 6302.0).* Australian Bureau of Statistics, Canberra, 2002.
45. Mylonas, A.D., Brown, A.M., Carthew, T.L., McGrath, B., Purdie, D.M., Pandeya, N., Vecchio, P.C., Collins, L.G., Gardner, I.D., De Looze, F.J., Reymond, E.J., and Suhrbier, A.: Natural History of Ross River Virus-Induced Epidemic Polyarthritis. *Med. J. Australia* 177 (2002), pp.356–360.
46. Rose, R.I.: Pesticides and Public Health: Integrated Methods for Mosquito Management. *Emerg. Infect. Dis.* 7 (2001), pp.17–23.
47. NSW Department of Health: NSW Arbovirus Disease Control Strategy: A Green Paper for Public Comment. Better Health Centre, NSW Department of Health, Sydney, 1998.
48. Russell, R.C.: Vectors vs. Humans in Australia – Who Is on Top Down Under? An Update on Vector-Borne Disease and Research on Vectors in Australia. *J. Vector Ecol.* 23 (1998), pp.1–46.
49. Beaglehole, R., Bonita, R., and Kjellstrom, T.: *Basic Epidemiology.* World Health Organization, Geneva, 1993.
50. Tai, K., Whelan, P.I., Patel, M.S., and Currie, B.: An Outbreak of Epidemic Polyarthritis (Ross River Virus Disease) in the Northern Territory during the 1990–1991 Wet Season. *Med. J. Australia* 158 (1993), pp.522–525.
51. Weinstein, P.: An Ecological Approach to Public Health Intervention: Ross River Virus in Australia. *Environ. Health Persp.* 105 (1997), pp.364–366.
52. Queensland Department of Health: Guidelines to Minimise Mosquito and Biting Midge Problems in New Development Areas. Draft Paper. Queensland Department of Health, Brisbane, 2002.
53. Fraser, J.R.E. and Marshall, I.D.: *Epidemic Polyarthritis (Ross River Virus Disease) Handbook.* Health Department, Melbourne, Victoria, December 1989.
54. Murray-Smith, S., Weinstein, P., and Skelly, C.: Field Epidemiology of an Outbreak of Dengue Fever in Charters Towers, Queensland: Are Insect Screens Protective? *Austr. New Zeal. J. Public Health* 20 (1996), pp.87–92.
55. Weinstein, P. and Cameron, A.S.: A Houseboat Outbreak of Epidemic Polyarthritis. *Med. J. Australia* 155 (1991), pp.721–722.
56. Westley-Wise, V.J., Beard, J.R., Sladden, T.J., Dunn, T.M., and Simpson, J.: Ross River Virus Infection on the North Coast of New South Wales. *Austr. New Zeal. J. Public Health* 20 (1996), pp.87–92.
57. Commonwealth Department of Industry, Sport and Tourism: Inbound Visitor Arrivals. 2002. http://www.isr.gov.au/sport%5Ftourism/Forecasts/inbound/in1.html. Accessed November 2002.

58. Tourism Forecasting Council: December 2002 Forecasts. 2002. http://www.isr.gov.au/library/content_library/TFCForecastsFullAnalysisDec2002.pdf. Accessed January 2003.
59. Kelly-Hope, L.A., Kay, B.H., Purdie, D.M., and Williams, G.M.: The Risk of Ross River and Barmah Forest Virus Disease in Queensland: Implications for New Zealand. *Austr. New Zeal. J. Public Health* 26 (2002), pp.69–77.
60. Trewin, D.: *Population Projections, Australia, 1999 to 2101*. Australian Bureau of Statistics, Canberra, 2000.
61. State of the Environment Advisory Council: *Australia: State of the Environment 1996*. CSIRO Publishing, Melbourne, 1996.
62. NSW Department of Health: Report of the New South Wales Arbovirus Forum. University of Sydney, 24 June 1997.
63. Aaskov, J., Williams, L., and Yu, S.: A Candidate Ross River Virus Vaccine: Preclinical Evaluation. *Vaccine* 15 (1997), pp.1396–1404.
64. Yu, S. and Aaskov, J.G.: Development of a Candidate Vaccine against Ross River Virus Infection. *Vaccine* 12 (1994), pp.1118–1124.
65. Climate Impacts Group: Climate Change Scenarios for the Australian Region. CSIRO Division of Atmospheric Research, Melbourne, 1996.
66. Russell, R.C.: Mosquito-Borne Arboviruses in Australia: The Current Scene and Implications of Climate Change for Human Health. *Int. J. Parasitol.* 28 (1998), pp.955–969.

7 The history of malaria control in Africa: lessons learned and future perspectives

Andrew K. Githeko and Clive Shiff

ABSTRACT: The tropical African climate is ideal for the breeding of the most efficient malaria vector in the world, *Anopheles gambiae*, and it is favorable for transmission of *Plasmodium falciparum*, the deadliest type of malaria parasite. This vector-parasite combination results in extremely high transmission intensities. The early motives for malaria control in Africa were associated with commercial and trade development. In the later years of the last century, the level of development in Africa could not support the efforts required to eradicate malaria, particularly in rural areas; control involved the most basic tools for early diagnoses and effective treatment. Toward the end of the 20th century, the mainstay for malaria control, chloroquine, began to fail, and other drugs followed the trend. Newer tools under development, such as vaccines and transgenic mosquitoes, have not yielded practical applications. The introduction of insecticide-treated bed nets brought new hope; however, few people acquire them for reasons ranging from cost to availability. Climate change is likely to worsen the situation, particularly in the eastern and southern African highlands as these areas become warmer and wetter. Some parts of West Africa have become drier and less favorable for malaria transmission. New control targets with time limits have been set; however, the enormity of the problem raises doubt about these expectations unless new tools are found or the economy of the continent improves rapidly. We are faced with a parasite/vector system that has a high capacity to adapt to most current intervention tools. The history of malaria control in Africa suggests that past successes using vector control were obtained in areas of unstable malaria, which are now under the greatest threat of resurging epidemics as a result of more frequent and intense climate variability. Studies should be directed in these areas to determine how to best control malaria. Africa needs to strengthen its capacity to deal with these emerging issues rather than paying attention to new scientific endeavors that have little immediate application.

1. INTRODUCTION

Malaria has been in Africa for thousands and perhaps millions of years. There are few early records of how the disease was managed by African medicine men. Part of the reason is that African medical knowledge was passed on orally; in later years it was repressed by the entry of Christianity, which considered folk medicine to be largely witchcraft. Antigens attributable to this parasite were demonstrated in cranial scrapings taken from Egyptian mummies from the Middle Kingdom period [1]. Clearly, the disease has affected the health and life of African peoples for millennia.

Peruvians used the bark of the Cinchona, the Chinese used extracts of the *Artemisia annua,* and Indians used the neem tree to treat malaria fever and possibly the disease.

Quinine and artemisinin are particularly potent in treating malaria. A recent interest in African traditional medicine has revealed that indeed Africa has a rich pharmacopoeia of herbal remedies for treating fevers and malaria in particular. However, it appears that many of the remedies were able to suppress only the parasitemia and did not have the potency of quinine. In some African communities, malaria was part of a collection of illnesses that appeared in certain unhealthy seasons. The Luo of Kenya referred to these illnesses as *yembe,* which included malaria or *mithusu.* One condition particularly associated with the maturing of maize and millet is *yugni*, and most likely includes malaria, respiratory infections, and allergies caused by maize pollen. This condition coincides with the end of the main rain season and the peak period for malaria transmission. All these diseases were treated as a single syndrome with a concoction of herbal medicines. Exposing the patient to steaming herbs and letting the patient sweat profusely treated *mithusu.*

Disease statistics demonstrate the enormity of the problem in the continent. For example, every year an estimated 320 million cases of malaria are diagnosed and 0.9 million deaths occur. In addition, 40 million disability adjusted life years (DALYs) are lost annually [2]. In recent years, many of the drugs used for treatment of malaria have lost their usefulness because of parasite resistance. And as this situation deteriorates, the economies of many African countries have stagnated or are declining, thus increasing their vulnerability to disease.

In recent years, epidemics are increasingly being reported in highland areas where malaria transmission has been restricted by cool climates. At the same time, it has been demonstrated that both the frequency and intensity of climate variability have been increasing in the last decade of the 20th century [3]. This trend in climate variability is expected to continue as global warming intensifies [3].

Given this scenario, it is pertinent to consider how the African continent will deal with the increased burden of disease arising out of intensified transmission in areas that have had low-level transmission of unstable malaria.

This chapter approaches this issue by examining the history of malaria control in Africa from the perspectives of the availability of control tools and methods, their development, application, and sustainability. From this analysis of past experience, we can define the future outlook and prospects for dealing with a fast deteriorating situation with reference to climate change and variability.

The story of malaria control in Africa starts with the discovery of the life cycle and transmission of the disease and proceeds through various periods that saw the rapid development of control tools and doctrines, economic and political changes, drug resistance, and climate change.

2. AFRICAN CLIMATE

The African climate is influenced by the surrounding oceans, the Mediterranean Sea, various land cover types, lakes, and varied topography. The Intertropical Convergence Zone (ITCZ) primarily controls the main rainy seasons. While the tropical regions of the continent have two rain seasons, the extreme poleward regions have only one distinct rain season. The mean annual rainfall ranges from as low as 10 mm in the middle of the Sahara to more than 2,000 mm in the tropical regions and other parts of West Africa. Coefficients of rainfall variability are highest in the deserts and lowest in the wet tropical regions [4]. The amount of rain in many parts of Africa depends on the prevailing sea surface temperatures (SST), atmospheric winds, the El Niño/Southern Oscillation (ENSO), and regional climate fluctuations in the Indian and Atlantic Oceans [5].

The African climate is generally warm (mean daily temperature >25°C); however, the extreme northern and southern regions experience some cooler weather (mean daily temperature <20°C). And while the low altitudes are warm, the higher altitudes in the highlands are cool.

One-quarter of Africa is hyper-arid desert, one-third is in the humid climate zone, and the remainder of the continent is dry land consisting of arid, semiarid, and dry subhumid areas [6]. All parts of the continent, except the Republic of South Africa, Namibia, Lesotho, and the Mediterranean countries north of the Sahara, have tropical climates. These tropical climates can be divided into three distinct climatic zones: wet tropical climates, dry tropical climates, and alternating wet-dry climates [4]. The tropical African climate is a natural home for the many insects that are vectors of human diseases such as malaria.

Hulme et al. [7] showed that Africa is now warmer than it was 100 years ago and that the mean rate of warming has been 0.5°C per century. The six warmest years in Africa occurred since 1986, and 1998 was the warmest year on record. Glaciers on Mount Kilimanjaro and Mount Kenya are retreating and shrinking. These observations suggest that warming is occurring in high altitude regions, and this could make the areas suitable for malaria transmission.

3. DISCOVERY OF THE BIOLOGY OF MALARIA TRANSMISSION AND ITS IMPLICATIONS FOR MALARIA CONTROL

Malaria parasites were first seen in human blood in 1888 in Algeria by Charles-Louis-Alphonse Laveran; however, his contemporaries rejected his observation. The 18 December 1897 issue of the *British Medical Journal* reported that Dr. Ronald Ross discovered malaria cysts in the stomach wall of anopheline mosquitoes that had fed on a malaria patient [8]. By July 1898, the Italian scientist Giovanni Batista Grassi traced the course of the parasite through the mosquito, and proved that human malarias were transmitted by species of Anopheles. On 28 November 1898, he reported his findings to the *Accademia dei Lincei*, in Rome [9].

The basic malaria transmission cycle was believed to be fairly simple. The discovery of the cycle led many to believe that malaria control would be fairly straightforward. However, it became clear that the ecology, biology, and stability of transmission, vector behavior, human genetics, and immunity interact in a complex way. For example, 28 anophelines were identified in Nigeria and, of these, several were infected with oocysts or sporozoites [10]. These were in the *Anopheles gambiae* group, *An. funestus, An. hargreavesi, An. pharoensis, An. nili, An. moucheti var. nigeriansis, An. hancocki*, and *An. rufipes*. These vectors have different seasonal distributions, habitats, biting, resting behaviors, and vectorial capacities.

The efficiency of *An. gambiae* and *An. funestus* in transmitting malaria in Nigeria is unmatched, with annual average sporozoite infected mosquitoes being 6% for the latter and 5% for the former [10]. (In Asia and South America, this figure is about 0.1–1%.) This scenario is quite common in the rest of equatorial Africa [11–13].

The earliest and the most sustained attempts at malaria eradication were made in the early 1950s by Bagster Wilson in East and West Africa [14] and in Zimbabwe Alves and Blair [47], in South Africa Nethercott, (1974) [77] and in Swaziland, Mastbaum [41]. Soon after, it was recognized that despite a high reduction of vectors, transmission of malaria continued because of the exophilic behavior of *An. arabiensis*.

4. EUROPEAN SETTLEMENT AND IMPLICATIONS FOR MALARIA

The early motivation for malaria control in Africa was to protect settlers and laborers involved in agriculture, trade, communication, and law enforcement. Thus, malaria control was deeply intertwined with trade and commerce. Europeans first arrived in Africa in the 15th century [15]. Many of the early Europeans were explorers, but in later years were followed by traders along coastal towns. It was not until the 19th century that a few Europeans ventured inland in search of natural resources. However, in most cases tropical diseases, and hostile warriors such as the Maasai in Kenya and the Zulus of South Africa, discouraged these expeditions. In the last quarter of the 19th century, following the Congress of Berlin, 1884–1885, most of Africa was colonized by European powers bent on obtaining resources using cheap labor. Most of the commercial activities included mining, while a few other European settlers operated plantations. These settlers had little interest in the health of the Africans apart from when diseases interrupted labor. For example, in South Africa during the 1920s and 1930s, malaria had a marked effect on agricultural development along the coastal belt. In 1928–1929, many employees in the sugar mills and plantations were afflicted by malaria: 6 whites out of 6,000 died; of 20,000 Asians, 151 died; and among 215,000 Africans, 2,600 died [16]. Larval control was initiated in this area; however, the districts of Ingwavuma, Ubombo, and Hlabisa were excluded because it was postulated that control measures would interfere with the natural immunity of the native populations.

The end of World War I imposed some changes in the remote African lands. Germany handed over all its colonies, and Tanganyika fell under British control. After the conflict, a British government program designed to stimulate the settling of veterans in East Africa multiplied the number of settlers, which at the end of the Great War was well over 10,000.

The Boers who migrated from the Cape into the hinterland of Southern Africa in the mid-19th century were mainly successful because there was little malaria in the highveld. However, in one trek to Delagoa Bay (the present Maputo area), almost the entire population of settlers succumbed. The few survivors were eventually moved to Natal, and their tales affected expansion eastward [17]. In West Africa, exploration and settlement of areas of interest to exploitation for ivory, gold, and slaves became infamous to European and American ventures because of the ravages of malaria (e.g., the Bight of Benin). However, quinine was becoming readily available in most parts of the world in the late 19th century and early 20th century. Efforts were introduced to prevent the disease by reducing transmission [18].

The health infrastructure was poorly developed for the mainly rural African populations; in addition, health budgets were small. Garnham and Harper [19], working in western Kenya, concluded that the usual methods of malaria control applied to towns and estates or in more temperate climates were, for various reasons, chiefly financial, inapplicable to rural areas.

For most of the colonial period, malaria control activities in East Africa were conducted by European experts with the assistance of African technical staff (personal communication, W.O. Obudho, Tropical Pesticide Research Institute Amani, Tanzania and Kenya Medical Research Institute, 2001). As a result, there were few if any African experts on malaria, and this situation remained for many decades after independence.

5. THE EARLY ROLE OF THE WORLD HEALTH ORGANIZATION IN MALARIA CONTROL IN AFRICA

The victorious Allied Powers of World War I established the League of Nations. The League's charter, known as the Covenant, was approved as part of the Treaty of Versailles at the Paris Peace Conference in 1919. The mission, as stated in the Covenant, was "to promote international co-operation and to achieve international peace and security". The Charter of the Health Organization was contained in the Covenant of the league [20]. Article 23 endeavored the member States to "undertake steps in matters of international concern, prevention and control of diseases". Following the collapse of the League of Nations, the United Nations was created in 1945. The World Health Organization (WHO) was formed in 1946 under the United Nations structure at a time when there was much work on the development of tools for malaria control.

The WHO led the world in the eradication and control of malaria. One of the great landmarks in the history of malaria in Africa was the First African Malaria Conference in Africa held in Kampala, Uganda in 1950 and organized by the WHO and the Commission for Technical Co-operation in Africa south of the Sahara. This conference recommended that whatever the original degree of endemicity, malaria should be controlled by modern methods a soon as feasible and without awaiting the outcomes of further experiments [21]. Whereas before malaria control in Africa was driven by the need to protect the settlers and their interests, the WHO was driven by the ambitious concept of health for all. Major decisions about malaria control were made more at the international level than at the national level.

6. DEVELOPMENT OF MALARIA ARMAMENT

Following the discovery of the life cycle of the malaria parasites in the late 19th century, it was not until the mid-1940s that the epidemiology of the diseases was unraveled and the first effective synthetic insecticides and drugs for transmission, morbidity, and mortality control were discovered (see Table 7.1). The most significant epidemiological discovery was that of the existence of immunity in populations that had been exposed to the diseases over several years. It was appreciated that malaria control programs would have to be sustainable to avoid epidemics following loss of immunity after reduced exposure to transmission. It is important to note that all the major tools for malaria control have been developed outside Africa. Although the initial concept of indoor application of insecticide was tested in South Africa [17].

6.1 Insecticides

The discovery of DDT in 1942 was perhaps the greatest step toward the control of malaria transmission. In 1936, De Meillon [17] showed that apart from being more

Table 7.1 Chronology of development of tool for malaria control

	1910s	*1920s*	*1930s*	*1940s*	*1950s*	*1960s*	*1970s*	*1980s*	*1990s*	*2000s*
Larvicide	Oil	Oil Paris Green	Oil, Paris Green. Used in S. Africa from 1932–1946.				Abate			
Organo-chlorides				DDT discovered in 1942. Used in S. Africa from 1948.	Dieldrin tested in Taveta 1956–1959					
Organo-phosphates					Malathion discovered in 1954	Malathion tested in Uganda 1963–1964	Fenitrothion tested in Kenya 1972–1976			
Carbamate					Propoxur 1959		Abate tested in Nigeria			
Pyrethroids			Pyrethrum spray used 1931 in S. Africa	Pyrethrum dust (Kenya)			Permethrin and Deltermethrin licensed 1976	Lambda-cyhalothr		
Drugs	Quinine	Quinine	Atebrine	Chloroqine discovered in 1945 Amodiaquine 1946 Proguanil 1945	Pyrimetha-mine discovered in 1952		Artemisinin extracted in 1972	QC resistance appears Fansidar available in 1982 Halofantrin available 1987	SP resistance *Malarone* (atovaquone + proguanil) 1995	Artenusate Artimether and combinations
Bednets								Insecticide Impregnated bednets (1985)		
Vaccines								First synthetic vaccine 1986		

effective, the control of adult mosquitoes cost only about a third that of larval control, and this finding established the foundation of residual insecticides [16]. This being the case, a number of insecticides such as dieldrin, HBC, malathion, fenitrothion, and propoxur were tested and proven effective for residual spraying. However, the newer insecticides were several times more expensive than DDT: for example, malathion is 7-fold, fenitrothion is 18-fold, and propoxur is 27-fold more expensive than DDT [22]. Larviciding using petroleum oils and Paris Green (copper acetoacenate) was restricted to urban areas where transmission is focal.

Development of second generation, synthetic, photo-stable pyrethroids was delayed by concerns of their cross resistance to DDT. However, in 1976, permethrin and deltermethrin were licensed for public health use. These two pyrethroids are now used for impregnating mosquito nets. Later, a third generation synthetic pyrethroid, lambdacyhalothrin, was cleared by the WHO Pesticide Evaluation Scheme (WHOPES) for use as an ultra low volume spray and contact insecticide [23].

6.2 Drugs

The discovery of chloroquine was a major milestone in the control of morbidity and mortality. Following chloroquine, other drugs such as amodiaquine, proguanil, and pyrimethamine were discovered in the mid-1940s and early 1950s; however, chloroquine remained the drug of choice because of its safety, efficacy, and cost. In the early 1980s, resistance to chloroquine was reported in East Africa, and it spread to West Africa. A number of countries in East Africa, such as Malawi, Kenya, Tanzania, Botswana, Uganda, and South Africa, have switched their first line of treatment from chloroquine to sulfadoxine-pyrimethamine (SP) [24,25]. However, SP resistance has appeared in eastern and southern Africa, and the drug will lose its efficacy in the next few years [25,26].

Since the discovery of artemisinin in the early 1980s, the WHO program on tropical diseases (TDR) has made the development of derivatives of artemisinin for the treatment of severe and multidrug resistant malaria a priority [27]. However, these derivatives are too expensive for most of the affected populations in Africa.

6.3 Vaccines

Many viral and bacterial infectious diseases in Africa have been controlled by vaccination. Given the persistent problem of drug resistance by *P. falciparum* and the cost of developing new drugs, it is of strategic importance to develop malaria vaccines. However, in the world of parasites there exist few effective vaccines.

Despite research and development work on malaria vaccines, only one, SPf66, has undergone full efficacy and delivery trials in Africa. When the vaccine was delivered through the Expanded Programme of Immunization (EPI) in Tanzania, the efficacy was only 2%. This has raised concerns about potential difficulties in inducing protective immune response against malaria through immunization of infants [28]. Several other vaccines are under development, including anti-disease vaccines that target the pathology and not the parasite.

6.4 Bed nets

Bed nets have been used to prevent night biting insects for a long time in Africa. In The Gambia, where the use of mosquito nets is a cultural practice, 95% of some communities owned untreated bed nets [29]. Following impregnation of these nets with permethrin, Alonso et al. [30] demonstrated a 63% reduction on mortality in children 1 to 4 years old. As a consequence of this remarkable result, the WHO/TDR initiated four multicenter trials of large-scale bed net projects comprising 100,000 individuals in different epidemiological settings [31]. The trials were carried out in Burkina Faso, The Gambia, Ghana, and Kenya. The pooled overall efficacy on child mortality from the four trials was 18% [32]. One study demonstrated a reduction of 45% in the frequency of severe disease. These reductions in morbidity and mortality are substantial, considering the large numbers of people at risk [33]. Despite these encouraging results, acceptability and affordability of impregnated bed nets have remained a barrier to their use. Currently only about 5% of the populations under threat from malaria are using insecticide impregnated bed nets.

7. TRANSMISSION CONTROL OR MORBIDITY/MORTALITY CONTROL?

At the beginning of the 20th century, Ronald Ross [34] postulated that there was a critical vector density below which transmission was not sustainable, and this led to the concept of source reduction using larvicides. However, this method was applicable to very few situations in tropical Africa because of the extensive availability of larval breeding habitats in Africa.

Macdonald [35], using a mathematical model, showed that reducing the malaria vectors' adult life expectancy was critical in malaria transmission and control. This gave impetus to the application of residual indoor spraying using insecticides. A major assumption in this hypothesis was that there would be a uniform exposure of all female adult vectors to insecticides [36]. In a number of pilot trials, despite achieving good vector control, malaria transmission continued, and further research indicated that some mosquitoes were either biting people outdoors and resting outdoors or biting people indoors and resting outdoors. Some insecticides also caused exicto-repellency, which drove fed mosquitoes out of sprayed houses, thus avoiding contact with the insecticide.

Coluzzi [37] postulated that the house leaving and human biting behaviors were genetic traits of some vectors. It was later confirmed that *An. arabiensis,* a sibling species of the *An. gambiae* group, had a predisposition to bite indoors and rest outdoors, thus escaping the effects of the insecticide.

In Ethiopia, a small proportion of *An. arabiensis* rests outdoors, while most of the individuals rest indoors and avoid surfaces sprayed with DDT – an indication of behavioral resistance. Furthermore, bioassays on female *An. arabiensis* indicated only partial susceptibility – an indication of physiological resistance [38]. However, in South Africa, DDT remained effective where intradomiciliary spraying has been carried out in houses in malarious areas since 1958 [39].

Within the doctrine of malaria control many advocated the use of chemoprophylaxis and mass drug administration for prevention and control. Integrated malaria control methods using indoor residual spraying and mass drug administration were shown to dramatically reduce the prevalence of the disease in pilot studies. However, this was not sustainable as a regular operation because of financial and technical difficulties [24].

From the early days, local as well as expatriate clinicians, public health specialists, and entomologists dominated the malaria treatment world in Africa. More recently, there has been entry of other professionals such as epidemiologists, health economists, molecular biologists, and immunologists. Although all agree that the end point in malaria control is a reduction of malaria morbidity and mortality, there are different opinions on which tools should be used to achieve this goal. Arguments continue about the use of different tools and their cost, sustainability, and efficacy. As a result, allocation of resources for the application of different control methods and, particularly, for the development of new tools has been problematic.

8. MALARIA CONTROL CASE STUDIES

While the malaria control community has been deeply frustrated about the level of transmission in the sub-Saharan region, it has been accepted that in theory malaria could be controlled using insecticides in the mid-altitudes in southern Africa, Zimbabwe, and Ethiopia [40]. The following case studies confirm these beliefs.

8.1 Swaziland

Swaziland is a small, independent nation that borders South Africa and southern Mozambique. According to Mastbaum [41], who was the Chief of the Malaria Section, the territory has three major climate zones (termed bushveld, between 150 and 300 m, middleveld, from 300 to 1000 m, and highveld, from 1,000 to 1,800 m), with malaria confined to the lower altitudes, although seasonal epidemics occurred in the middleveld and even highveld areas. The bushveld areas converged into Mozambique, and there was continual back and forth migration of people. In 1946, training and survey preparations began for a project on the indoor application of HCH (hexachlorohexane). In 1954, the project was extended to the whole country. In irrigated areas, control was also directed against mosquito larvae because of the extensive breeding areas. Later HCH was replaced with dieldrin, which had a greater residual effect [42].

The application of insecticide alone reduced the parasite rate in children under 16 years old from 53.6% in 1955/1956 to 1.8% in 1956/1957 at a per capita cost over the entire nation of 13 pence (US$0.15 in 1956 rates). In irrigated areas, despite additional larviciding activities, the parasite rate was not so greatly reduced, and was still at 6% in 1955/1956. Ramsdale [42] found that the indoor resting variety of *An. gambiae s.l.* was eliminated, whereas the outdoor form was still detectable. In Mozambique, where no spraying was being carried out, both forms were collected. Ramsdale reports that this is different from the situation reported by Gillies [43] in the Taveta-Pare area of Tanzania, where indoor spraying with dieldrin for over 3 years had little effect on the "behavior" of *An. gambiae s.l.* The change of behavior that Ramsdale reported was actually a reduction of one of the members of the species complex, not a behavior change. Subsequent work by Davidson [44] laid the groundwork for unraveling this complicated situation. In fact, what was thought to be a single species of a vector was shown to be a species complex of six members, all morphologically identical, but with different habitats, feeding preferences, and vectorial capacity [37]. This latter research, carried out primarily by African entomologists, helped explain important inconsistencies in the epidemiology and control of malaria.

The situation in Swaziland was complicated by the proximity of Mozambique on the eastern border and the unrestricted movement of people from holoendemic regions into an area where transmission was under control. When the program in Swaziland moved to the consolidation and maintenance phase, eradication efforts declined and there was a resurgence of malaria. By this time, the WHO policies had moved away from eradication mode, and control strategies relied on delivery of drugs (mainly chloroquine) at the periphery. A resurgence of malaria in Swaziland followed as the efficacy of chloroquine declined. Following a serious deterioration of the malaria situation due to Cyclone Domina in 1994, the absence of vector control, large-scale agricultural development, and the influx of refugees from Mozambique, the National Malaria control program was overhauled with assistance from USAID and the South Africa Trade Mission. Vector control was reintroduced using DDT [45], and malaria declined to levels seen in the 1950s.

8.2 Zimbabwe

The development of Rhodesia (now Zimbabwe) was severely hindered by malaria, which affected both European settlers and local African populations. Apart from the high transmission areas of the lowveld (below about 800 m), malaria was very rare on the highveld (above 1,500 m) and epidemic in the middleveld, with severe periodic impacts. Work by Leeson [46] suggested that the vector mosquitoes over-wintered in the lowveld, and during the spring began to invade the middleveld via river conduits. At this season, there was little flow in the rivers, which were more like a series of residual pools with small runnels between them. These were ideal habitats for *An. gambiae, An. arabiensis*, and *An. funestus,* which were responsible for the transmission. Using this knowledge, malaria control programs were initiated in the northeast of the country in 1949 based on the application of residual insecticide indoors in a systematic manner similar to that laid out by Mastbaum in Swaziland [47]. The program was very successful, with parasite rates declining precipitously [48,49]. In 1958, a WHO supported program was initiated that included spraying initially with HCH, then with DDT and dieldrin. These insecticides were alternated to limit the spread of insecticide resistance in mosquitoes that was being observed in many parts of Africa. Eradication was achieved, particularly in the southeast part of the country [42].

The spraying programs were effective, particularly in the southern part of the country. In 1961, the parasite rates were less than 0.05% and the country moved to the surveillance phase [50]. This was to coincide with malaria control activities in neighboring Botswana. During 1962–1963, only 138 cases of malaria were diagnosed in the entire surveillance area (of a population estimated at over half a million). All cases with fever were given a presumptive treatment consisting of a single dose of chloroquine and primaquine combined. The presumptive treatment was also given to all residents living within a 5 mile radius of each case. Blood film diagnosis was made for each case, and if positive, the patient was revisited and given a full course of chloroquine. Those infected with *P. malariae* were hospitalized and treated with 14 mg/kg primaquine over 14 days. The surveillance program continued for several years, but terminated with the breakup of the Central African Federation and the reorganization of the Malaria Control program under the Director of the Blair Research Laboratory in Salisbury (now Harare). Malaria control in the southern areas of the country remained focused on passive case detection, with curative doses of chloroquine administered through the excellent system of rural clinics operating in the country. When the southeastern areas were opened to large-scale irrigation, control devolved to the private sector (commercial farmers and irrigation estates). In the north and northeast of the country, vector control operations proceeded, as did passive case detection, through the state health services. Vector control operations relied on indoor spraying with DDT and later with malathion. It is of interest that after over 40 years of use, there is no evidence of resistance to DDT by local vector species of mosquitoes.

Malaria surveillance was systematically carried out using blood film surveys as indicators, and barrier spraying was implemented to protect the high-density population areas and the agricultural productive regions of the country. In some instances where transmission was low (with parasite rates less than 0.05% in school children), spraying was done every second year. Negative records are hard to substantiate; particularly in the areas that were affected by the liberation war in the 1970s, but until the early 1990s, malaria remained under control, and almost an inconsequential health issue in most of the country. Its breakdown has occurred mainly through lack of supervision and direction due to decentralization of malaria control operations and a pervasive breakdown of the entire health services system [51]. Epidemics of malaria now occur regularly in a non-immune population, with disastrous circumstances.

Notwithstanding the current situation, Zimbabwe did support a health system operated by local scientists who succeeded in controlling malaria by modifying the WHO eradication concepts to fit its own ecological and geographical conditions. One hopes that it will return to this system in the future.

8.3 Zanzibar

The islands of Zanzibar are located off the east coast of Africa; it is part of the United Republic of Tanzania. Ecologically, islands represent an easy target for the control of insect populations because they are isolated and largely protected against re-invasion by the surrounding oceans.

Malaria has long been rife in the islands of Zanzibar and was the subject of a comprehensive report to the Colonial Development Fund in 1937 [52]. This report detailed the high morbidity and mortality that prevailed in Zanzibar and on the island of Pemba. In 1933–1934, the prevalence of malaria by microscopy was 43.3% in urban Zanzibar, while in rural Ingunja in the south, it was 50%; in a hospital in Pemba, it was 53% [52].

This situation was addressed in 1957 when the WHO supported a malaria control scheme on the two main islands. Initially, malariometric surveys were carried out in Zanzibar with geographical reconnaissance, numbering of houses, and training of staff. In 1958, this was extended to Pemba, and houses were sprayed with dieldrin in Zanzibar. Although this was planned to be 2 g/m^2, calculations established that the mean application level was about 1.3 g/m^2. Spraying was carried out twice annually because bioassays showed poor mosquito mortality about 14 weeks after application. After the fourth round of spraying, in 1961, the insecticide was changed to DDT, also on a two cycle per annum basis. Spraying and updating of the houses, and additional recruiting and training of personnel, continued to follow the plan for eradication as proposed by the WHO. Spraying continued until 1965. From 1963 to 1968, active case detection and followup were implemented throughout the two islands. The program was terminated in June 1968 following numerous difficulties such as aging vehicles and often broken-down bicycles. In Zanzibar, the number of houses sprayed per round was over 52,000, including nearly 20,000 temporary huts and "other structures". The number of structures sprayed in Pemba was unrecorded, but coverage was extensive, with only 13% in Pemba and 17% in Zanzibar not sprayed. At the time of census (1967), the population of Zanzibar was 190,117 and that of Pemba was 164,243.

The results were phenomenal. After only the first two rounds of spraying, the infant parasite rate dropped from 47% to 1.0% in Zanzibar and from 52.5% to 1.7% in Pemba [53]. Although the goal of eradication was not achieved, malaria in both islands was successfully controlled to the point where it was no longer considered a public health problem. From 1963 to 1967, malaria case detection revealed an annual parasite incidence ranging from 5.8 to 11.6 per thousand in Zanzibar and from 0.4 to 1.9 per thousand in Pemba.

The program was closed as a result of local political decisions, and the existing organization was dissolved. However, by 1973 passive case detection revealed a resurgence of malaria, to a prevalence of 49.3% in Zanzibar and 9.9% in Pemba. Once again, malaria was rapidly asserting itself as the major cause of morbidity and mortality on the islands.

These case studies demonstrate that vector control using indoor spraying was successful in some particular ecosystems of Africa. Success in malaria control reduced the public health threat of the disease, thus lowering its priority on the health agenda. However, in all the cases, the disease returned to previous levels. Malaria control must be a continuous effort.

9. ENVIRONMENTAL CONSCIOUSNESS AND MALARIA CONTROL

Millions of tons of DDT were used in agriculture world-wide in the 1950s and 1960s. After constant pressure by environmentalists, its use was banned in the United States in 1972. DDT was put on the list of "the dirty dozen", a group of banned persistent organic pollutants (see Box 7.1).

A few epidemic prone countries, including South Africa, Zimbabwe, and Ethiopia, used DDT for malaria control. Following the ban on DDT in South Africa and South America, the number of malaria cases increased rapidly [54,55].

While environmentalists have been quick to protect the environment, they seem to be less willing to protect human life. Pressure groups appear to be more concerned with dying birds and fish than with loss of human life in the tropics. Furthermore, the main source of DDT pollution is use in agriculture, not health related activities where the insecticide is confined to indoor use only.

Box 7.1 Pesticides that fall under the banned persistent organic pollutants (POPs)

DDT, chlordane, toxaphene, mirex, aldrin, deildrin, endrin, and heptachlor are now banned under the Stockholm Convention on persistence organic pollutants. However, DDT has been retained for vector control until an alternative safe, effective, and affordable insecticide is found. Discussions are under way to allow the use of DDT as an interim method until an alternative insecticide is found.

10. FROM MALARIA ERADICATION TO MALARIA CONTROL

With 80% of all global cases and 90% of all mortality in Africa [24,56], the continent is undoubtedly the home of malaria. George Davidson, an eminent entomologist from the London School of Tropical Medicine, wrote that "despite successes and significant reduction in many other parts of the world, including the tropical areas, the World Health Assembly of the United Nations at its 22nd meeting in 1969 decided to abandon the concept of the eradication, considering it as practically unattainable. It urged countries to revert to containing the disease at levels that their own general health services, rudimentary as many of them were, could cope with" [22]. This statement summarizes a major milestone of malaria control in Africa. Following this declaration, the WHO recommended that the only practical policy in malaria control was that of quick diagnosis and effective treatment. This worked fairly well as long as chloroquine was effective. The policy is now under threat because of the cost of treatment following parasite resistance to most of the safe and cheap drugs. In Africa, the situation is exacerbated by increased poverty and the cost-sharing policy introduced by the World Bank's structural adjustment policies. In some countries, there has been a 30% reduction in people going to government hospitals where treatment had previously been free [57].

10.1 Drug resistance and its implications

The first confirmed records of chloroquine resistance in Africa emerged from Kenya in 1977 [58], and by the early 1990s resistance levels had reached 70%. The rate of development of resistance has been slower in other African countries such as Mali, where in the late 1990s parasitological resistance level was 14%. The current distribution of resistance to chloroquine in Africa appears to be related to the level of transmission intensity [24].

The emergence of resistance to anti-malaria drugs has several implications in terms of treatment policy and efficacy. Chloroquine resistance in Kenya serves as an example. Following several studies in the 1980s that demonstrated that the drug had lost its efficacy, the ministry of health had difficulties changing the first line of treatment from chloroquine to SP until 1995, when it was demonstrated that the mortality rate in patients treated with chloroquine alone was 20% and those treated with SP was only 4% [59]. The failure to change drug policy had a particularly severe implication during epidemics in western Kenya, where up to 80% [60,61] of self-diagnosed malaria cases are treated at home. Many people continued to use chloroquine despite its failure to clear parasites in the blood, partly because the drug has an excellent property of clearing clinical symptoms of malaria very quickly. However, the patient goes into a chronic phase and can easily develop life-threatening complications. And now widespread resistance to SP is emerging at a time when there seems to be an increasing number of epidemics in the African highlands.

10.2 The deteriorating malaria situation and renewed control efforts

In the 1980s, it was recognized that the global malaria situation was deteriorating and renewed action was required to contain this state of affairs. Sachs and Malaney [62] observed that malaria impedes development through its effects on fertility, population growth, savings and investment, worker productivity, absenteeism, premature mortality, and medical costs. For example, the direct and indirect losses due to malaria in Africa rose from US$800 million in 1987 to more that US$2 billion in 1997. The WHO Expert Committee on malaria noted that most of Africa south of the Sahara continued to face an increased public health crisis as a result of the disease [63]. The situation was aggravated by lack of financial support, shortage of expertise, and the emergence of resistance to chloroquine. The WHO then proposed a Ministerial Conference on Malaria, which was held in Amsterdam in 1992, to formulate a Global Malaria Control Strategy.

The health community recognized that to implement effective malaria control programs, there was a need for political support by African governments at the highest level. Consequently African heads of state and government at their meeting in Harare in June 1997 issued a declaration on the need to control malaria to ensure greater social development in the continent. This became known as the Harare Declaration, whose basic thrust was early treatment of malaria cases; promotion of the use of insecticide-impregnated nets; prevention, detection, or containment of epidemics; and strengthening of the capacities of countries and communities to combat the disease.

To put the global malaria plan into action required a concerted effort of UN agencies, governments, and the private sector. Led by the WHO, the effort evolved into the Roll Back Malaria (RBM) initiative, which is based on intersectoral partnership, evidence-based interventions, political mobilization, and participation of civil society. African heads of state and governments met in Abuja, Nigeria, on 25 April 2000 and endorsed the RBM initiative; they set time limits and targets to reverse the malaria situation in the continent [2]. The initiative aims to halve the global burden of the disease by 2010. In a concerted effort, the Multilateral Initiative on Malaria (MIM) was launched in Senegal in 1997 with the aim of increasing the research capacity of scientists from the endemic areas. Within the WHO framework for malaria control, member countries are required to institute National Malaria Control Programs where resources can be directed. Although national and international political will is now in place to control malaria, Africa is still faced with a shortage of tools and technical and financial resources.

10.3 Increasing malaria epidemics

The climatology of the African continent in the past century indicates two recent periods of warming: the mid-1930s to the 1940s and from the beginning of the 1980s to date [4]. Widespread malaria epidemics in highland areas have been associated with these periods. For example, in 1941, a community living at 2,360 m in western Kenya was severely affected by a malaria epidemic. This was a period of abnormal warming and unusually heavy rainfall [19].

Between 1994 and 1996, malaria epidemics in 14 countries of sub-Saharan Africa caused an unacceptably high number of deaths, many in areas previously free of the disease. Similarly, during the 1997/1998 El Niño event, several countries were adversely affected by malaria epidemics. Adolescents and young adults are now dying of severe forms of the disease [64]. In Kenya, malaria spread from 3 districts in 1988 to 15 in 1999 [65]. In Tanzania, the disease has moved into the Western Usambara Mountains, which were malaria free until 1997 [66]. In general, these epidemics occur in highland areas, which are generally cool and wet. In the East African region, they have been associated with El Niño. However, during the 1997/1998 El Niño event, epidemics occurred in the arid areas of northeastern Kenya and Somalia because of extensive flooding. A number of factors have been associated with the increase of malaria epidemics:

- climate variability
- land use change
- parasite resistance to drugs
- cost of drugs and hospitalization
- lack of effective vector control activities
- policy, e.g., cost sharing and drug options
- inadequate capacity of medical facilities
- uncertainty in predictability of the disease epidemics.

It is well established that climate is an important determinant of the spatial and temporal distribution of vectors and pathogens. In theory, a change in climate could be expected to cause changes in the geographical range and seasonality (intra-annual variability), and an increase in the incidence rate (with or without changes in the geographical or seasonal distribution) [67].

Climate change includes a change in the mean meteorological values and the departure from the mean. While a change in the mean values has been a slow process, the seasonal and annual departure from normal weather has been more pronounced.

Figure 7.1 shows the trend in mean monthly maximum temperature anomalies in the Lake Victoria basin. The thick line indicates normal conditions. An anomaly above 3°C combined with suitable rainfall can precipitate a malaria epidemic in the

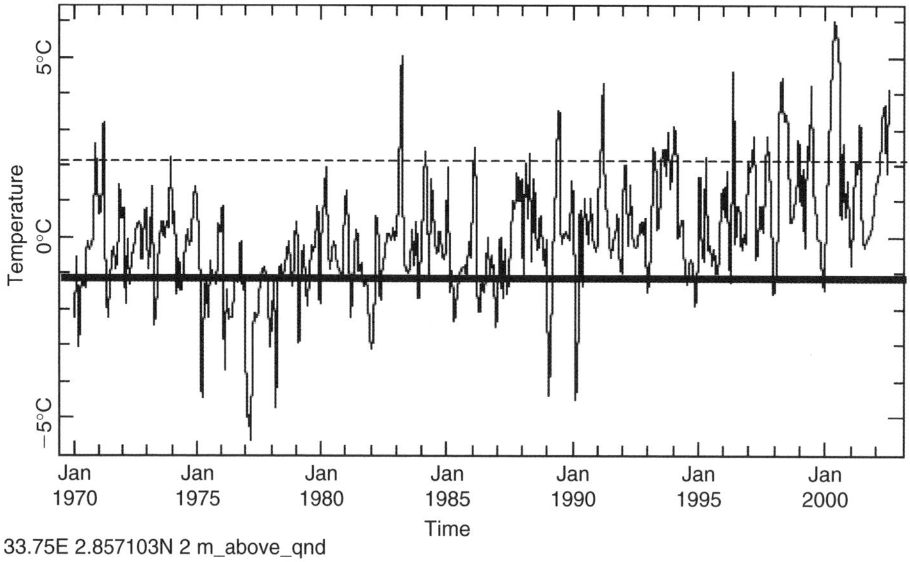

Figure 7.1 Trends in mean monthly maximum temperature anomalies in an area covering the western Kenya highlands. An anomaly above 3°C combined with suitable rainfall can precipitate a malaria epidemic in the western Kenya highlands. The frequency and intensity of such events are increasing, as can be seen by the points above the dashed line

Source: Updated version of data in Ref. 65

western Kenya highlands. The frequency and intensity of such events are increasing, as can be seen by the points above the dashed line.

Epidemic malaria in sub-Saharan Africa remains a major threat: it places a heavy burden on overstretched health services and poses a major constraint to economic development [68]. The ability to forecast epidemics is a major step in their prevention and management. Progress in developing forecasting tools has been slow; however, recent research indicates that it is possible to forecast climate driven epidemics with sufficient lead time for effective action to be taken [65,69].

Land use changes that have been associated with an increase in malaria transmission include deforestation in Tanzania [66,70] and cultivation of swamps in Uganda [71,72]. Recently it was demonstrated in western Kenya highlands that cultivation of swamps can substantially increase the water temperature, thus increasing the potential for prolific breeding of malaria vectors, particularly *An. gambiae*. While deforestation on the slopes of the highlands primarily affects the rate of parasite development in adult mosquitoes, cultivation of swamps increases the rate of production of adult mosquitoes [73].

Drug resistance results in failure to cure the disease, leading to multiple treatments. Consequently the number of in- and outpatient cases increases, leading to an unexpected load on the medical infrastructure. Furthermore, failure to treat the infections increases the risk of the more severe forms of the disease, which require expensive hospital management of patients.

The Roll Back Malaria program's initial focus was on sub-Saharan Africa, where the greatest burden on health is malaria. To launch effective interventions, studies were carried out to determine the cost-effectiveness of various options [74]. The study concluded that cost-effective interventions are available. However, a package of interventions to decrease the bulk of the malaria burden is not affordable in very low-income countries. Coverage of the most vulnerable groups in Africa will require substantial assistance from external donors. Detailed cost estimates prepared by the WHO Commission on Macro Economics and Health find that effective malaria prevention and treatment will require an additional US$2.5 billion per year by 2007, increasing to US$4 billion per year by 2015 [62].

10.4 Future climate change and its implications for malaria in Africa

According to the Third Assessment Report of the Intergovernmental Panel on Climate Change (IPCC), Working Group I, the global surface temperatures increased by 0.6 ± 0.2°C over the 20th century. Models have projected that the global averaged surface air temperature will warm 1.4 to 5.8°C by 2100 relative to 1990 [3].

Malaria transmission requires suitable rainfall and temperature. While it is projected that under climate change regimes some parts of Africa will become warmer and wetter, others will become warmer and drier. In some parts of the Sahel, malaria transmission has decreased by about 30% over the last decades because of a decrease in rainfall [71]. Elsewhere in eastern and southern Africa, the main driving force for increased malaria transmission seems to be the warming of the highlands, which have been free of malaria or have had very low and seasonal transmission. Under these conditions, most of the resident population is immunologically naïve and very vulnerable to severe disease.

The great malaria epidemic that occurred in Ethiopia in 1958 serves as a lesson. This epidemic, which occurred during a period of abnormally warm, wet weather and an early shift of the rain season [75], affected 3 million and caused the death of 150,000 in a span of 6 months [76].

11. PROSPECTS FOR ADAPTATION

A major concern in the process of adaptation is adaptive capacity. According to the IPCC, adaptive capacity is the ability of a system to adjust to *climate change* (including climate variability and extremes), to moderate potential damages, to take advantage and opportunities, and to cope with the consequences [3].

Malaria transmission in tropical Africa is intense, and few countries have effective control programs. After many years of pilot control projects, it is clear that Africa does not have the human and financial resources required for sustainable control programs. More recently, some of the tools that have been effective have lost their efficacy, mainly due to parasite resistance to drugs. It is clear that we are dealing with a parasite that may have a higher adaptive capacity than the human population.

The economies of many African countries are declining because of low technological development, arising out of poor management of human and natural resources. The impacts of globalization will have further negative impacts in Africa because of its inherent low priority for local social development. In many countries there is a shift toward private health, which only a few can afford. The cost sharing policy has resulted, in some cases, in a 30% reduction in hospital use.

Africa can hardly cope with the current burden of malaria, and it seems that the adaptive capacity for increased malaria transmission is very low. The history of public health indicates that the rate of decline of infectious diseases was strongly associated with economic development.

Because the prospects for mitigation against climate change seem to be diminishing, the people of Africa must develop indigenous strategies to fight against malaria. New results arising from studies on the ecology of highland malaria in Kenya [73] suggest that there are nonchemical, affordable, and sustainable ways to prevent mosquito breeding in the highlands. These methods would have multiple benefits to the environment and to the local climates.

There has been a very high expectation of a magic bullet coming out of hi-tech laboratories (e.g., vaccines, genetically modified mosquitoes), and as a result, African malariologists have failed to optimize the use of existing tools such as selective indoor spraying of houses at the foci of transmission in the highlands. Sufficient knowledge exists to modify houses to make them less accessible to mosquito vectors, yet this knowledge has been unused. While it is attractive to apply new tools such as GIS and GPS for mapping transmission intensities, it should not be forgotten that many health facilities do not even have microscopes, slides, and stains required for malaria diagnoses. It is our view that Africa can develop some simple homegrown and scientifically sound solutions for malaria transmission.

While trendy or fashionable science (transgenic vectors and remote sensing technologies) continues to be intellectually fascinating, African scientists must, however, have the wisdom to determine what will be of benefit to the suffering populations.

Table 7.2 Epidemic malaria in Africa[a]

Determinants of adaptive capacity	*Requirements for public health prevention*
Range of available technological options (4)	Awareness that problem exists (5)
Availability and distribution of resources (5)	Sense that the problem matters (5)
Structure of critical institutions (3)	Understanding of the causes (3)
Human capital (3)	Capability to deal with problem (2)
Social capital (2)	Political will to influence (3)
Access to risk spreading (4)	
Ability to manage information (3)	
Public's perceived attribution (3)	

Note
a Scale = 1 (low) to 5 (high).

Table 7.2 presents rankings of determinants of adaptive capacity and requirements for public health prevention. The determinants of adaptive capacity and the requirements for public health prevention are measures of the ability of a community, nation, or region to cope with and respond effectively to an external stress. Elements with the lowest ranking indicate where the greatest efforts are required.

12. LESSONS LEARNED

Malaria epidemics are becoming more frequent in the African highlands. Effective malaria control was once carried out in some of these areas using residual insecticides as the main tool. One of the unintended outcomes with this method is that due to its high efficacy, malaria stopped being a major public health problem and many of the control programs were closed down. In recent years, a popular view has emerged that DDT should continue to be used as a residual insecticide for vector control. Vector ecology studies in the highlands indicate that transmission is focal, and this raises the possibility of limited application of residual insecticides around the foci of transmission. Such an approach would keep the cost of the operations low and sustainable. Judicious application of remote sensing and GPS/GIS technologies and epidemiological data would be critical in mapping these foci. It is thus essential to start building capacity for the application of indoor residual spraying in countries that are affected.

ACKNOWLEDGEMENTS

We are most grateful to Dr. Melanie Renshaw and Mr. Francis Atieli for their assistance in retrieving historical literature. Thanks are due to Frank Hansford in Tzaneen, South Africa, for providing many records on the situation in South Africa, to Robert Bos of the WHO, and to Pepe Najera, Norman Gratz, and Socrates Litsios for providing numerous references as well as a wealth of documents from WHO archives pertaining to Zanzibar malaria control operations. Finally, grateful recognition is given to the late David F. Clyde for his wisdom and advice, as well as for his records of the history of malaria research and control in Tanzania.

REFERENCES

1. Miller, R.L., Ikram, S., Armelagos, G.J., Walker, R., Harer, W.B., Shiff, C.J., Baggett, D., Carrigan, M., and Maret, S.M.: Diagnosis of Plasmodium Falciparum Infection in Mummies Using the Rapid Manual *Para*Sight-F Test. *T. Roy. Soc. Trop. Med. Hyg.* 88 (1994), pp.31–32.
2. RBM: Roll Back Malaria. http://www.who.int/rbm/Presentations/MIP-RBM-final/sld020.htm.
3. IPCC: Summary for Policy Makers. Working Group 1, 2001. Intergovernmental Panel on Climate Change. http://www.ipcc.ch/pub/spm22-01.pdf.
4. IPCC: *Regional Impacts of Climate Change: An Assessment of Vulnerability. Special Report:* R.T. Watson, M.C. Zinyowera, R.H. Moss and D.J. Dokken (eds). Intergovernmental Panel on Climate Change. University Press, Cambridge, UK, 1998.
5. CLIVAR: An International Research Programme and Climate Variability and Predictability. http://www.clivar.org/.
6. UNEP: *World Atlas of Desertification:* N.J. Middleton and D.S.G. Thomas (eds). Edward Arnold Publishers, Sevenoaks, UK, 1992.
7. Hulme, M., Doherty, R.M., Ngara, T., New, M.G., and Lister, D.: African Climate Change: 1900–2100. *Climate Res.* 17 (2001), pp.145–168.
8. Ross, R.: On Some Peculiar Pigmented Cells Found in Two Mosquitoes Fed on Malarial Blood. *BMJ* 2 (1897), pp.1786–1788.
9. Dobson, M.J.: The Malariology Centenary. *Parassitologia* 41 (1999), pp.21–32.
10. Bruce-Chwatt, L.J.: Malaria in Nigeria (Part II). *W. Afr. Med. J.* 1 (1952), pp.2–7.
11. Beier, J.C., Perkins, P.V., Koros, J.K., Onyango, F.K., Gargan T.P., Wirtz, R.A., Koech, D.K., and Roberts, C.R.: Malaria Sporozoite Detection by Dissection and ELISA to Assess Infectivity of Afrotropical Anopheles (Diptera: Culicidae). *J. Med. Entomol.* 27 (1990), pp.377–384.
12. Githeko, A.K., Service, M.W., Mbogo, C.M., Atieli, F.K., and Juma, F.O.: *Plasmodium falciparum* Sporozoite and Entomological Inoculation Rates at the Ahero Rice Irrigation Scheme and the Miwani Sugar-Belt in Western Kenya. *Ann. Trop. Med. Parasit.* 87 (1993), pp.379–391.
13. White, G.B.: The Anopheles Gambiae Complex and Malaria Transmission around Kisumu, Kenya. *T. Roy. Soc. Trop. Med. Hyg.* 66 (1972), pp.572–581.
14. Bruce-Chwatt, L.J.: Man Against Malaria: Conquest or Defeat. *T. Roy. Soc. Trop. Med. Hyg.* 73 (1979), pp.605–617.
15. Greenwood, B.: Traditional Medicine to DNA Vaccines: The Advance of Medical Research in West Africa. *Trop. Med. Int. Health* 3 (1998), pp.166–176.
16. Sharp, B.L., Ngxongo, S., Botha, M.J., and le Sueur, D.: An Analysis of 10 Years of Retrospective Malaria Data from the KwaZulu Area of Natal. *S. Afr. J. Sci.* 84 (1988), pp.102–106.
17. De Meillon, B.: The Control of Malaria with Special Reference to the Contributions Made by the Staff of the South African Institute for Medical Research. *S. Afr. Med. J.* 11 (1986), pp.67–69.
18. Jamot, E.: Note sur des Essais de Quininisation Preventative et Curative au Cameroun. *B. Soc. Pathol. Exotique* 22 (1929), pp.555–568.
19. Garnham, P.C.C. and Harper, J.O.: The Control of Rural Malaria by Pyrethrum Dusting. *E. Afr. Med. J.* 21 (1944), pp.310–320.
20. League of Nations: Health Organization. Information Section, Geneva, Switzerland, 1931.
21. WHO: Technical Report Series Vol. 38. World Health Organization, Geneva, Switzerland, 1955.
22. Davidson G.: Who Doesn't Want to Eradicate Malaria? *New Sci.* December 16 (1982).
23. Miller T.A.: Mechanisms of Resistance to Pyrethroid Insecticides. *Parasitol. Today* 4 (1988), pp.S8–S12.

24. D'Alessandro, U. and Buttiens, H.: History and Importance of Antimalarial Drug Resistance. *Trop. Med. Int. Health* 11 (2001), pp.845–848.
25. EANMAT: Monitoring Antimalarial Drug Resistance within National Malaria Control Programmes: EANMATA Experience. East African Network for Monitoring Antimalarial Treatment. *Trop. Int. Health* 6 (2001), pp.891–898.
26. Brendenkamp, B.L., Sharp, B.L., Mthembu, S.D., Durrheim, D.N., and Barnes, K.I.: Failure of Sulphadoxine-Pyrimethamine in Treating *Plasmodium Flciparum* Malaria in KwaZulu Natal Province. *S. Afr. Med. J.* 91 (2001), pp.970–971.
27. WHO: UNDP/World Bank/WHO Special Programme for Training and Research in Tropical Disease. Tropical Diseases Research; Progress 1995–96: Thirteenth Report of the UNDP/World Bank/World Health Organization Special Programme for Training and Research in Tropical 1997, 1997.
28. Acosta, C.J, Galindo, C.M., Schellenberg, D., Aponte, J.J., Kahigwa, E., Urassa, H., Armstrong Schellenberg, J.R., Masanja, H., Hayes, R., and Kitua, A.Y.: Evaluation of the SPf66 Vaccine for Malaria Control when Delivered through the EPI Scheme in Tanzania. *Trop. Med. Int. Health* 5 (1999), pp.368–376.
29. MacCormack, C.P. and Snow, R.W.: Gambian Cultural Preferences in the Use of Insecticide-Impregnated Bed Bets. *J. Trop. Med. Hyg.* 89 (1986), pp.295–302.
30. Alonso, P.L, Lindsay, S.W, Armstrong, J.R, Conteh, M., Hill, A.G, David, P.H, Fegan, G., de Francisco, A., Hall, A.J, Shenton, F.C. et al.: The Effect of Insecticide-Treated Bed Nets on Mortality of Gambian Children. *Lancet* 337 (1991), pp.499–502.
31. WHO: UNDP/World Bank/WHO Special Programme for Training and Research in Tropical Disease. Tropical Diseases Research; Progress 1991–92: Thirteenth report of the UNDP/World Bank/World Health Organization Special Programme for Training and Research in Tropical 1993, 1993.
32. Lengeler, C.: Insecticide Treated Bednets and Curtains for Malaria Control (Cochrane Review). In: *The Cochrane Library*. Issue 3. Update Software, Oxford, UK, 1998.
33. Shiff, C.J., Checkley, W., Winch, P., Premji, Z., Minjas, J., and Lubega, P.: Changes in Weight Gain and Anaemia Attributable to Malaria in Tanzanian Children Living under Holoendemic Conditions. *T. Roy. Soc. Trop. Med. Hyg.* 90 (1996), pp.262–265.
34. Ross, R.: Logical Basis of the Sanitary Policy of Mosquito Reduction. *BMJ* 1 (1905), pp.1025–1029.
35. MacDonald, G.: *The Epidemiology and Control of Malaria*. Oxford University Press, Oxford, UK, 1957.
36. Githeko, A.K., Service, M.W., Mbogo, C.M., and Atieli, F.K.: Resting Behavior, Ecology and Genetics of Malaria Vectors in Large Scale Agricultural Areas of Western Kenya. *Parassitologia* 38 (1996), pp.481–490.
37. Coluzzi, M.: Heterogeneities of the Malaria Vectorial System in Tropical Africa and their Significance in Malaria Epidemiology and Control. *Bull. WHO* 62 (1984), pp.107–113.
38. Ameneshewa, B. and Service, M.W.: Resting Habits of *Anopheles Arabiensis* in the Awash River Valley of Ethiopia. *Ann. Trop. Med. Parasitol.* 5 (1996), pp.515–521.
39. Sharp, B.L., Le Sueur, D., Wilken, G.B., Bredenkamp, B.L., Ngxongo, S., and Gouws, E.: Assessment of the Residual Efficacy of Lambda-Cyhalothrin 2. A Comparison with DDT for the Intra-Domiciliary Control of Anopheles Arabiensis in South Africa. *J. Am. Mosq. Control Assoc.* 9 (1993), pp.414–420.
40. WHO: 50 Years of International Public Health. The World Health Organization, Geneva, 1998.
41. Mastbaum, O.: Organization and Administration of Malaria Control in Swaziland. WHO/MAL/133. Report to Lagos Conference on Malaria in Africa, 1955, September.
42. Ramsdale, C. and Rivola, E.: A Note on the Response of *Anopheles gambiae* to Spraying with HCH in Swaziland. WHO/MAL/481-64. 1964.

43. Gillies, M.T.: The Problem of Exophily in *Anopheles gambiae*. *Bull. WHO* 15 (1956), pp.437–449.
44. Davidson, G.: A Distribution Map of the Member Species of the *Anopheles gambiae* Complex. *T. Roy. Soc. Trop. Med. Hyg.* 61 (1967), p.454.
45. Packard, R.M.: Agricultural Development, Migrant Labour and the Resurgence of Malaria in Swaziland. *Soc. Sci. Med.* 22 (1986), pp.861–867.
46. Leeson, H.S.: Anopheline Mosquitoes in Southern Rhodesia. 1926–28. *Mem. London School Hyg. Trop. Med.* 4 (1931), pp.1–55.
47. Alves, W. and Blair, D.M.: Malaria Control in Southern Rhodesia. *J. Trop. Med. Hyg.* 58 (1955), pp.273–280.
48. Alves, W.: Malaria Parasite Rates in Southern Rhodesia: May–September 1956. *Bull. WHO* 19 (1958), pp.69–74.
49. Reid, E.T. and Woods, R.W.: Anopheline Mosquitoes of Southern Rhodesia: A General Survey. *Proc. Trans. Rhodesia Scientific Assoc.* 45 (1957), pp.47–72.
50. Wolfe, H.L.: Epidemiological Data Concerning One Year of a Malaria Surveillance Pilot Project in Southern Rhodesia. *Bull. WHO* 31 (1964), pp.707–720.
51. Lancet Editorial February 09: Focusing on Health in the Crisis in Zimbabwe. *Lancet* 359 (2002), p.455.
52. McCarthy, D.D.: Report on the Zanzibar Research Unit, June 1934 to September 1937. Colonial Development Fund (Malaria Research Scheme) Government Printer, Zanzibar, 1937.
53. Chayabejara, S. and Payne, D.: The Malaria Situation in Zanzibar/Pemba: A Report on a Mission 03 March-13 April, 1974. WHO/AFR/MAL/141. 1974.
54. Kleinschmit, I., Sharp, B., Mueller, I., and Vounatsou, P.: Rise in Malaria Incidence Rates in South Africa: A Small-Area Spatial Analysis of Variation in Time Trends. *Am. J. Epidemiol.* 155 (2002), pp.257–264.
55. Roberts, D.R., Laughlin, L.L., Hsheih, P., and Legters, L.J.: DDT, Global Strategies and a Malaria Control Crisis in South America. *Emerg. Infect. Dis.* 3 (1997), pp.295–302.
56. Breman, J.: The Ears of a Hippopotamus: Manifestation, Determinants and Estimates of the Burden. *Am. J. Trop. Med. Hyg.* 64 (2001), pp.1–11.
57. Waddington, E.G., Catriona, J., and Enmayew, K.A.: A Price to Pay: The Impact of User Charges in Ashanti-Akim District, Ghana. *Int. J. Health Plan. Manag.* 4 (1989), pp.17–47.
58. Fogh, S., Jepsen, S., and Effersoe, P.: Chloroquine-Resistant *Plasmodium falciparum* Malaria in Kenya. *T. Roy. Soc. Trop. Med. Hyg.* 73 (1979), pp.228–229.
59. Shretta, R., Omumbo, J., Rapuoda, B., and Snow, R.W.: Using Evidence to Change Antimalarial Drug Policy in Kenya. *Trop. Med. Int. Health* 5 (2000), pp.755–764.
60. Mwenesi, H.A.: The Role of Drug Delivery Systems in Health Care: The Case of Self-Medication. *Afr. J. Health Sci.* 1 (1994), pp.42–48.
61. Ruebush, T.K., Kern, M.K., Campbell, C.C., and Oloo, A.J.: Self-Treatment of Malaria in a Rural Area of Western Kenya. *Bull. WHO* 73 (1995), pp.229–236.
62. Sachs, J. and Malaney, P.: The Economic and Social Burden of Malaria. *Nature* 415 (2002), pp.680–685.
63. WHO: Technical Report Series vol. 892. World Health Organization, Geneva, Switzerland, 2000.
64. IFRC: Malaria in Africa. International Federation of the Red Cross and the Red Crescent. http://www.ifrc.org/what/health/archi/index.htm.
65. Githeko, A.K. and Ndegwa, W.: Predicting Malaria Epidemics Using Climate Data in Kenyan Highlands: A Tool for Decision Makers. *Global Change Hum. Health* 2 (2001), pp.54–63.
66. Mboera, L.E.G. and Kitua, A.Y.: Malaria Epidemics in Tanzania: An Overview. *African J. Health Sci.* 8 (2001), pp.17–23.
67. Kovats, R.S., Campbell-Lendrum, D.H., McMichael, A.J., Woodward, A., and Cox, J.S.: Early Effects of Climate Change: Do They Include Changes in Vector-Borne Disease? *Philos. Trans. Roy. Soc. Series B* 356 (2001), pp.1057–1068.

68. OAU: Harare Declaration of 4th June 1997 on Malaria Prevention and Control in the Context of African Economic Recovery and Development. Organization of African Unity. http://www.massiveeffort.org/html/abuja_declaration.html.
69. Connor, S.J., Thomson, M.C., and Molyneux, D.H.: Forecasting and Prevention of Epidemic Malaria: New Perspectives on an Old Problem. *Parassitologia* 41 (1999), pp.439–448.
70. Matola, Y.G., White, G.B., and Magakuya, S.A.: The Changed Pattern of Malaria Endemicity and Transmission at Amani in the Eastern Usambara Mountains, North-Eastern Tanzania. *J. Trop. Med. Hyg.* 90 (1987), pp.127–134.
71. Mouchet, J., Manguin, S., Sircoulon, J., Laventure, S., Faye, O., Onapa, A.W., Carnevale, P., Julvez, J., and Fontenille, D.: Evolution of Malaria in Africa for the Past 40 Years: Impact of Climatic and Human Factors. *J. Am. Mosq. Control Assoc.* 14 (1998), pp.121–130.
72. Lindblade, K.A., Walker, E.D., Onapa, A.W., Katungu, J., and Wilson, M.L.: Land Use Change Alters Malaria Transmission Parameters by Modifying Temperature in the Highland Areas of Uganda. *Trop. Med. Int. Health* 5 (2001), pp.263–273.
73. Guiyun and Githeko: Ecology of African Highland Malaria; Western Kenya 2002–2006: Project Report. 2003. Available from http://wings.buffalo.edu/academic/department/fnsm/bio-sci/faculty/yan.html.
74. Goodman, G.A., Coleman, P.G., and Mills, A.: Cost-Effectiveness of Malaria Control in Sub-Saharan Africa. *Lancet* 354 (1999), pp.378–385.
75. Fontaine, R.S., Najjar, A., and Prince, J.S.: The 1958 Malaria Epidemic in Ethiopia. *Am. J. Trop. Med. Hyg.* 10 (1961), pp.795–803.
76. Ministry of Health of Federal Democratic Republic of Ethiopia, Malaria and Other Vector-Borne Diseases Control Team. Epidemiology and AIDS Department: Malaria Control Profile. February 2000.
77. Nethercott, A.S.: Forty years of malaria control in Natal and Zululand. *South African Medical Journal* (8 June, 1974), pp.1168–1170.

8 Heat waves: past and future impacts on health

R. Sari Kovats and Christina Koppe

ABSTRACT: Although heat waves are rare events, they are associated with significant mortality impacts. A range of measures are available that potentially reduce the impact of extreme temperature episodes on mortality and morbidity, including health promotion, building design, urban planning, and heat wave early warning systems. The effectiveness of health interventions in response to heat waves has not yet been formally evaluated, although there is evidence that they have been effective during major events in the United States. Populations are likely to acclimatize to the changes in mean climate forecast by global climate models. However, it is not known how the frequency or intensity of heat waves will change in the future. Heat deaths are of concern because many populations are aging and the elderly are the most vulnerable to heat-related mortality. The impact of climate change on the exposure to heat stress in urban areas may be exacerbated by increases in the urban heat island effect and other environmental changes.

1. INTRODUCTION

Heat waves kill. In countries such as Australia and the United States, heat waves in the 20th century caused more deaths than any other natural hazard [1]. Heat waves are natural hazards that also affect property as well as people. They can cause expensive livestock and crop losses and can damage roads, electrical equipment, railways, bridges, and other types of infrastructure. In some countries heat waves are accompanied by bush fires.

In August 2003, a major heat wave affected France, Spain, Portugal, and other countries in Western Europe. France was most affected; at the time of writing, more than 10,000 excess deaths occurred across the country during the heatwave, with approximately 2,000 in Paris [2]. The effect of the high thermal load was exacerbated by a lack of staff in hospitals because the heat wave occurred during the fixed holiday season. In Portugal, 1,316 excess deaths were attributed to the heatwave [3], and forest fires caused extensive damage.

The midwestern United States has experienced several major heat wave events, most recently in 1995 and 1999. The heat wave in Chicago in 1995 caused 514 heat-related deaths (12 per 100,000 population) and shocked the nation [4]. All heat wave deaths are considered to be preventable: the 1999 heat wave of similar magnitude caused 114 heat deaths in Chicago [5], indicating that "adaptation" was relatively effective in reducing mortality, but not sufficient to prevent all the excess mortality.

Table 8.1 Mortality and morbidity attributed to selected heat waves in Europe, United States, and Australia

Heatwave event	*Attributable (all cause) mortality and references*
1966-United States	32.9% during first heat wave and 36.4% during second heat wave [6]
1976-London, UK	9.7% increase England and Wales and 15.4% Greater London [7] Almost twofold increase in mortality rate in geriatric hospital inpatients (but not other inpatients) [34]
1980-St Louis, United States	62 heat stroke deaths, 57% increase in all cause mortality in July [8], 226 admitted for heat-related illness 14.3% increase in ER visits 5.1% increase in hospital admissions
1980-Kansas City, United States	17 heat stroke deaths and 64% increase in all cause mortality in July [8] 276 patients admitted for heat-related illness 7.6% increase in ER visits
1980-Memphis, United States	483 emergency admissions: 58% heat exhaustion, 17% heat stroke, 6% heat cramps [9]
1981-Portugal	1906 excess deaths (all cause, all ages) in Portugal, 406 in Lisbon [July] including 63 heat deaths [10]
1983-Rome, Italy	65 heat-stroke deaths during heatwave in Latium region. 35% increase in deaths in July 83 compared to July 82 in 65+ age group in Rome [11,12]
1987-Athens, Greece	2690 heat-related hospital admissions and 926 heat-related deaths, estimated excess mortality > 2000 [13]
1988-Allegheny County, United States	105 excess deaths, all aged 65+ [14]
1993-Southeastern Australia	17 heat deaths and 100 people admitted to hospital in New South Wales, South Australia and Victoria [1]
1995-Chicago, United States	514 heat-related deaths (defined by Cook County criteria) and estimated 696 excess deaths [4] 11% increase in emergency admissions, with 35% increase in emergency admissions in 65+ age group [15]
1995-London, UK	8.9% increase in all-cause mortality (768) in England and Wales, and 15.4% (184) increase in Greater London [16]
1997-southern Australia	10 heat deaths, and 15% increase in hospital attendance in Adelaide region [1]
2000-Queensland, Australia	22 heat deaths and 280 cases of nonfatal heat stroke in Brisbane [1]

Source: Data from Ref. 17.

Note

Heat deaths are assumed to be deaths due to heat stroke.

The majority of heat waves that have been studied have been in the United States and Europe (Table 8.1). In Athens in 1987, a 10-day heat wave resulted in 926 deaths that were classified as heat related. However, the attributable excess mortality was estimated to be more than 2,000 [13]. In all urban areas except Athens, a 32.5% increase in mortality was observed in July 1987 compared to the average July mortality for

Table 8.2 Total killed and number of heat wave events by region, reported in EM-DAT disaster events database (1975–2001)

Region	*Total killed*	*Number of events*
East Asia	160	3
Europe	117	2
European Union	1,124	8
North Africa	54	3
North America	1,872	13
Oceania	23	4
Rest of Europe	203	12
Russian Federation	328	3
South America	73	2
South Asia	5,434	21
West Asia	(Unknown)	2

Source: Data from Ref. 20.

1981–1986 [18]. In 2000, a severe heat wave affected many countries in the Balkans (southeastern Europe). For example, in Croatia, 40 people died of heart attacks caused by the heat in its four major cities, while hundreds more were hospitalized with serious health problems, according to local news reports [19].

When defined as the extremes of the daily temperature range, heat waves occur in all countries. Some countries are much more vulnerable than others – such as those in mid-continental areas where seasonal and diurnal ranges of temperatures are greater. The EM-DAT disaster database includes records of disaster events where at least 10 people died or more than 100 people were affected; countries in South Asia are the most vulnerable to heat wave events (Table 8.2). The information on heat waves is derived mainly from the US Office of Foreign Disaster Assistance (OFDA) [20,21]. Further information on attributable mortality is not available. The impacts, however, are likely to be indicative and highlight the need for some epidemiological research into the impacts of high temperatures in vulnerable populations in South Asia. Heat waves are a recurrent feature of the climate in India and Pakistan. A heat wave in India in June 1998 was estimated to cause over 10 weeks of high temperatures [22]. In Ores, India, the air temperature rose to 49.5°C and was reported to have caused many deaths. The high temperatures were exacerbated by recurrent power failures that affected cooling systems and hospital services. Power failures are a regular feature of life in many cities in developing countries.

This chapter reviews what is known about the current impacts of heat waves on human health, and identifies those populations that are the most vulnerable to dying in a heat wave. Global climate change is one of many factors that may affect exposure to higher temperatures in the future. We also review some of the policies and measures that have been used to reduce the impact of heat waves in both developed and developing countries.

2. PHYSIOLOGICAL EFFECTS OF HEAT

Heat stress causes some well-described clinical syndromes such as heat stroke, heat exhaustion, heat syncope, and heat cramps [23]. The cause of death most easily attributable to heat is heat stroke, which occurs when the core body temperature reaches 40.6°C, leading to multiple organ dysfunction. However, many causes of death have been observed to increase during heat waves, particularly deaths from cardiovascular and respiratory disease. Even when evidence suggests otherwise, such deaths are not routinely recorded as heat related. Further, underreporting of heat stroke deaths may occur because heat stroke is similar to other, more familiar causes of death, particularly coronary or cerebral thrombosis, once the body is no longer hot itself or in a hot environment [24,25]. Following the Chicago heat wave, a committee of medical examiners recommended that coroners use the following definition of a heat-related death [26]: "a death in which exposure to high ambient temperature either caused the death or significantly contributed to it".

Heat stroke has a high case-fatality ratio and a rapid onset. Complications of heat stroke include adult respiratory distress syndrome, kidney failure, liver failure, and disseminated intravascular coagulation [26]. Nonfatal heat stroke can lead to long-term illness [27]. Severe functional impairment was observed in 33% of 58 patients admitted with heat stroke during the Chicago heat wave, with no improvement after one year in those still alive [28].

Ambient thermal conditions are an important environmental exposure and are responsible for a quantifiable burden of mortality and morbidity. The best epidemiological evidence is provided by time series studies of daily mortality. These methods are considered sufficiently rigorous to assess short-term associations (days, weeks) between environmental exposures and mortality if adjustment is made for longer term patterns in the data series, particularly the seasonal cycle and any long-term trends. Time series methods are used to quantify the relationship between mortality and temperature across the whole temperature range. The methods are not often applied to assess the impact of individual extreme events. A study of mortality in London (1976 to 1996) found that hot days (defined by the 97th percentile of the 3-day moving average) were associated with a 3.34% increase in deaths for every 1° increase in average temperature above 21.5°C [29]. The major heat waves in 1976 and 1990 were modeled separately and were associated with an average increase of 31% and 17% in deaths on days during the respective heat wave periods.

3. HEAT WAVES AND THEIR IMPACTS ON HEALTH

Heat waves are rare events that vary in character and impact. It is has proved difficult to come up with a standardized definition of a heat wave using absolute or relative temperature thresholds, or combined meteorological indices. The essential component of a heat wave is sustained duration of extremely high heat load that cause increased morbidity and mortality in the population under study (Table 8.3). The threshold air-temperature above which mortality is significantly associated with heat varies between populations, and is an air temperature that is not considered extreme. For example, in London, deaths associated with heat occur at air temperatures above 19°C [15]. In Boston, Massachusetts, deaths associated with heat occur from 21°C [30].

Some cities are more vulnerable to heat waves than others. Clearly, climate factors are an important determinant of vulnerability. In the United States, populations in the north-eastern and midwestern cities appear to be the most vulnerable to heat waves, because heat load episodes there occur infrequently or irregularly [31]. During the 1995 Chicago heat wave, air temperatures reached 38°C and this was combined with high humidity, resulting in average apparent temperatures exceeding 36°C over a large area on several

Table 8.3 Concepts relating to heat waves

Factor	*Implications*	*Methodological issues*
Duration	Must be longer than 2 days.	Heat waves are of variable length – need to specify minimum duration.
Intensity	Must be above threshold temperature – although thresholds not often clear from the mortality data series.	Heat waves are of variable magnitude – magnitude and duration should be combined somehow.
Time in season	Impact of heat wave early in season is greater than late in season.	Need enough events to make a comparison – not likely to get two events in same year.

days [32]. The impact of the extreme weather was further exacerbated by the urban heat island effect, which raised night-time temperatures by a further 2°C.

In the United States, approximately 400 deaths directly attributed to "heat" (ICD9 = 992) are reported each year [33]. Only approximately 48% of these deaths are due to extreme weather conditions, the remainder occurring in periods that are not heat waves. From 1979 to 1994, heat-related mortality due to weather conditions in the United States, age adjusted, was 2.7–3.7 per million population in the four highest reporting states (Arizona, Arkansas, Kansas, and Missouri) [34]. A significant proportion of these (38%) was in the <55 age group.

The number of deaths due to heat stroke is often reported, but those deaths do not represent the total impact of the heat wave on mortality. The attributable or "excess" mortality can be estimated by subtracting the "expected" mortality from the observed mortality. The expected mortality is calculated using a variety of measures, including moving averages and averages from similar time periods in previous years. Estimates are therefore very sensitive to the method used to estimate the "expected" mortality [4]. Published studies have used different methods, and this makes comparison difficult. A review of published mortality studies during 14 US heat waves found that the numbers of deaths certified as heat related as a proportion of all "excess deaths" ranged from 0% to 78% [23].

A proportion of the "acute" effect on mortality of heat waves may be due to the hastening of death in already ill persons by a few days or weeks. The mortality displacement effect can sometimes be seen in the lower than expected mortality immediately following a heat wave (Fig. 8.1). The key question is, What proportion of deaths were brought forward by more than a matter of days/weeks? Several methods have been suggested for estimating the contribution of mortality displacement in daily time series models. The extent of mortality displacement due to extreme temperatures has yet to be resolved. Deaths due to heat stroke are considered 100% preventable.

Overall, increases in emergency hospital admissions during heat waves (an indicator of heat-related morbidity) are not comparable with the dramatic increases observed in mortality. During the 1995 Chicago heat wave, emergency hospital admissions increased by 11% in total, and by 35% in the over 65 age group [15]. Most of these admissions were for heat-related illness (dehydration, heat exhaustion, and heat stroke) in persons with underlying chronic disease. Admissions for acute renal failure also increase during hot weather.

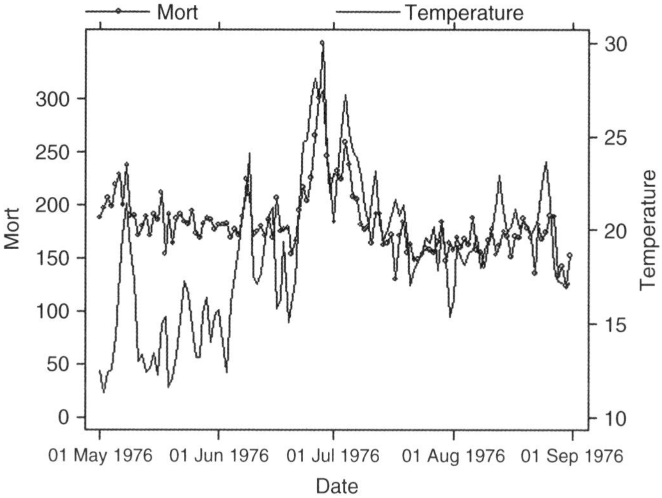

Figure 8.1 Heat wave events: 1976 in London. Source: Data from Ref. 35

Studies of the effect of temperature on emergency room visits and emergency hospital admissions may provide valuable information on the physiological mechanisms by which temperature affects morbidity and mortality. There are also important implications for the response of the health service providers, and in particular the need to ensure that emergency room staff are aware of the coming influx of heat-related cases (see below).

4. WHO IS VULNERABLE TO HEAT-RELATED MORTALITY?

Reviews of mortality have identified the following risk factors for heat stroke or heat-related death and illnesses [36,37]:

- age (see below);
- pre-existing disease – primarily chronic respiratory or cardiovascular disease;
- social factors (e.g., living alone);
- use of certain drugs, e.g., phenothiazines, antidepressants, alcohol, diuretics;
- impaired cognition, e.g., dementia;
- housing (e.g., building type, which floor);
- presence and use of air conditioning in the home or residential institution; and
- physical activity – overexertion or inactivity.

A case control study after the 1995 Chicago heat wave confirmed that those at increased risk were people who were already ill, confined to bed, unable to care for themselves, isolated, and without air conditioning [38].

In developed countries, studies indicate that urban populations experience higher rates of heat-attributable mortality than rural populations. One reason is that temperatures are often higher in urban environments because of the urban heat island

Table 8.4 Heat wave events and deaths (heat stroke) in the Indian subcontinent

Year	*No. of wave days*	*States affected*	*No. of heat deaths*
1979	16	Orissa, Andhra Pradesh, Gangetic West Bengal, Uttar Pradesh, Bihar	365
1980	7	Uttar Pradesh, Maharashtra, Tamil Nadu, Andhra Pradesh	106
1981	7	Uttar Pradesh, Madhya Pradesh, Bihar	63
1982	4	Uttar Pradesh	11
1983*	11	Bihar, Punjab, Uttar Pradesh, Maharashtra	185
1984	11	Uttar Pradesh, Punjab, Maharashtra	58
1985	5	Punjab, Bihar	141
1986	8	Andhra Pradesh, Punjab, Rajasthan, Himachal Pradesh, Bihar	155
1987	6	Orissa, Punjab, Haryana, Uttar Pradesh	90
1988	21	Rajasthan, Gujarat, Saurashtra and Kutch, Uttar Pradesh, Punjab	924
1989	15	Rajasthan, Madhya Pradesh, Maharashtra	43
1990	6	Rajasthan	–
1991	10	Rajasthan, Madhya Pradesh, Gujarat, Maharashtra	250
1992	13	Rajasthan, Madhya Pradesh, Haryana, Bihar, Uttar Pradesh, Maharashtra	114
1993	13	Punjab, Haryana, Rajasthan, Uttar Pradesh	73
1994	25	Punjab, Haryana, Rajasthan, Madhya Pradesh, Uttar Pradesh, Maharashtra	234
1995	29	Haryana, Rajasthan, Punjab, Uttar Pradesh	410
1996	9	Rajasthan, Haryana, Uttar Pradesh	17
1998	27	Punjab, Rajasthan, Gujarat, Uttar Pradesh, Maharashtra, Orissa, Andhra Pradesh, South Tamil Nadu	1,300

Source: Data from Ref. 39.

Note
The heat waves in 1995 and 1998 had the greatest duration.

effect (see below). During the 1999 heat wave in the US Midwest, the highest air temperatures were concentrated in urban areas [5]. The populations of urban and rural areas also differ in their characteristics and vulnerability to heat-related impacts. For example, in the United States, individuals are more vulnerable to heat stress if they are isolated and live in certain types of housing, with no access to functioning air conditioning. These risk factors are more likely to be present in urban areas than in rural areas. The situation may be different in developing countries. Evidence from India indicates that rural populations may also be at risk during heat waves.

India is vulnerable to heat waves, because of both the climate and population characteristics. Table 8.4 lists heat stroke deaths in India in the last few decades. A heat wave there is considered to be severe when maximum air temperature remains 7°C or more above its long-term normal value for a station having a normal maximum temperature $\leqslant$ 40°C, or remains 5°C or more above its long-term normal value for a station having a normal temperature >40°C [39]. India typically experiences very high temperatures in early spring. These high temperatures are moderated, however, by low pressure weather patterns that originate in the Mediterranean and move across Pakistan and into India. Heat waves occur when these western disturbances are fewer in frequency, as occurred in 1999 and 2002.

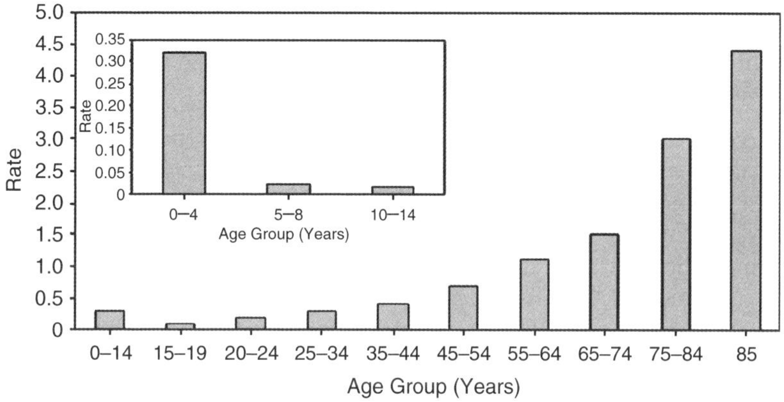

* Per 1 million population.
† Underlying cause of death attributed to excess heat exposure classified according to the *International Classification of Diseases, Ninth Revision* (ICD-9),as code E900.0 "due to weather conditions (deaths)."

Figure 8.2 Average annual rate* of heat-related deaths†, as the result of weather conditions, by age group – United States, 1979–1997. Source: Data from Ref. 46

4.1 Age and aging

There is evidence that the elderly are more vulnerable to heat stress, particularly those in hospitals or long-term care institutions. Vulnerability to heat in old age occurs because of intrinsic changes in the regulatory system or because of the presence of drugs that interfere with normal homeostasis. Few studies of heat tolerance have been undertaken in older people [40,41]. As homeostasis is impaired, an elderly person may not be aware that they are becoming ill because of high temperatures and therefore may not take action to reduce their exposure. Several studies have shown that the elderly in institutions such as residential care homes are vulnerable to heat-related illness and death [42,43,44]. This population is likely to be more frail. In the United States, such institutions are likely to have air conditioning, but this may not be typical in European countries.

Epidemiological studies indicate that there is no significant difference in risk between men and women. Studies, however, vary concerning the age at which the vulnerability is shown to increase. Most population-based time series studies show an effect in adult age groups [45] with the effect larger in those over 65 compared to those of other ages. Since these studies use predetermined groups for the "elderly," there has not been a more detailed examination of the age at which vulnerability is increased between different populations.

Children and infants are also at risk of heat-related death (Fig. 8.2). Overall mortality in this group is very low in developed countries. In the United States, less than 4% of all heat-related deaths due to weather conditions occur in those aged 4 and under [33]. Figure 8.2 indicates the children under 4 years of age are more at risk than adolescents or young adults, and have comparable rates to adults (35+). A study of deaths attributed to heat [ICD9 = 992] in Japan (1968–1994) found a high proportion of deaths in children (4 years and under) [47]. Children are more at risk of dehydration than adults because they have a higher volume of water in their bodies. However, a significant proportion of heat deaths in children is caused by being left in cars on hot days [26].

4.2 Poverty and socioeconomic factors

It has been suggested that people with lower socioeconomic status are more vulnerable to heat-related mortality because of poorer quality housing and a lack of air conditioning. Populations in more deprived areas within a city are also more likely to have other risk factors for heat-related death. Several studies that have investigated heat-related mortality rates in different neighborhoods reveal the importance of socioeconomic factors [38,48,49]. The physical and social isolation of elderly people in the United States further increases their vulnerability to dying in a heat wave [50].

An analysis of heat wave mortality (defined in this study as all deaths of those over 65 years of age during a 1980 heat wave) in St. Louis, Missouri, USA found evidence that socioeconomic factors were associated with heat-related mortality [49]. The deaths were mapped, and those areas with high death rates had a lower prevalence of air conditioning and owner occupied housing, and higher vacancy rates (derived from census information). A study of five cities in Australia, however, found no association between deaths rates on hot days when compared with other days by selected small area indicators of deprivation [51].

Few studies have analyzed heat-related mortality in developing country cities. A study in Sao Paulo, Brazil, found little evidence of modification of heat-mortality relationships by socioeconomic status, but this analysis did not include any heat wave events [52]. In 2002, a heat wave was reported to have killed 622 people in the southern Indian state of Andhra Pradesh. Information from news reports indicated that one of the populations most at risk are the rural daily wage earners, such as laborers and rickshaw pullers, who have no option but to work outdoors under any conditions. National and state governments issued advice during heat waves, such as to stay indoors and drink water. For many farmers, however, conditions indoors are not cooler than those outdoors because their thatched houses offer no protection against the heat. Many people do not have access to clean drinking water. Remote villages have no piped water, and because heat waves are often accompanied by drought, people have to walk farther to gain fresh water supplies.

5. ADAPTING TO A WARMER WORLD

Global climate change is likely to be accompanied by an increase in the frequency and intensity of heat waves, and by warmer summers and milder winters. The impact of extreme summer heat on human health may be exacerbated by increases in humidity [53]. A central question in estimating future heat-related mortality is the rate at which populations will adapt to a warmer climate. Populations are likely to acclimatize to warmer climates via a range of behavioral, physiological, and technological adaptations. The initial physiological acclimatization to hot environments can occur over a few days, but complete acclimatization may take several years. The rate at which changes will take place in infrastructure is likely to be much slower, however.

Extreme weather events are, by definition, rare stochastic events. With climate change, even if the statistical distribution of such events remains the same, a shift in the mean will entail a nonlinear response in the frequency of extreme events. This has been demonstrated using statistical theory [54]. Changes in the future frequency of daily air temperature extremes are of concern for estimating future burdens of thermal environment-related mortality. Current methods of impact assessment of climate change use 30-year averages of monthly climate data. Very few studies of the impacts of climate change have considered climate variability or extremes, and this has been recognized as a severe limitation in climate change impact assessments [55]. Populations and systems are, in general, more vulnerable to changes in extremes than changes in mean conditions.

Most countries have experienced some climate warming during the last three decades [56]. Increases in the frequencies of hot days have already been observed in the United States [53] and the UK [57]. In general, minimum temperatures are rising faster than maximum temperatures [56].

Climate change is anticipated to increase the frequency of days above a air temperature threshold because even small increases in average air temperature can result in big shifts in the frequency of extremes. Several approaches have been used to estimate the future impact of climate change on exposure to "heat stress" that do not directly estimate mortality or other health outcomes. These methods use biometeorological indicators of heat stress that are then mapped spatially or to show changes in temporal risk. For example, current and future frequencies of heat stress days were estimated using a temperature-humidity index for Oxford and Thessaloniki [58,59].

5.1 Heat wave emergency planning and heat health warning systems

Many cities in developed countries that are vulnerable to heat waves have some form of emergency planning that can be deployed for heat waves. According to the US Centers for Disease Control and Prevention (CDC) in Atlanta, spending time in an air-conditioned area is the strongest factor in preventing heat-related deaths [33]. The Greek government implemented its Xenocrates emergency plan for natural disasters during severe heat waves in 1998 and 2002. The plan requires state buildings to provide air-conditioned spaces to the public and for public beaches to stay open late into the night. In addition, hospitals are placed on full alert and all leave is cancelled for ambulance drivers.

In countries with a high coverage of air conditioning, power failures are common during heat waves because of sudden increases in electricity demand. The impact of a power failure is therefore likely to exacerbate the impacts on health. In Australia, power failures occurred in the 1997 heat wave in Adelaide [1]. Also in Athens in 2002, the power system was overwhelmed, leading to failures in supply.

Heat health warning systems (HHWS) use meteorological forecasts to reduce the impact of heat waves on human health. The challenge lies in determining at which point heat stress conditions become sufficiently hazardous to human health in a given population to warrant intervention [60]. A range of methods has been developed to identify dangerous meteorological conditions for a given population. Simple methods are based on thresholds of air temperature (mean, maximum, or minimum) or apparent temperature (a combination of air temperature and relative humidity). More complex indices are derived from a combination of temperature and duration indicators as well as of temperature and relative humidity thresholds. In the United States, the

National Weather Service (NWS) issues heat warnings based on the Mean Heat Index. More complex methods rely on complete heat budget models or on a temporal synoptic index. The synoptic approaches are based on the identification of air masses that have been associated with adverse effects on mortality in that population [61]. The synoptic classification includes additional meteorological variables such as cloud cover and wind speed, in addition to temperature and humidity.

As part of the warning system, public health messages should be disseminated to all age groups to increase awareness of symptoms of heat-related illness. There are different warning procedures in place, but most involve the issue of a warning through the mass media (TV, radio). A one-tiered system will issue a warning or "heat advisory" when the threshold is expected to be exceeded. Cities in the United States also have many two- or three-tiered warning procedures, including a "watch" or an "alert" when a particular level of heat stress occurs or is forecast, and an "emergency" ("warning") stage when the indicator becomes dangerous and action is needed [60]. An example of such a three-step warning procedure is shown in Fig. 8.3. The US NWS initiates advisories and warnings based on the threshold value of the Mean Heat Index, for each city.

In North America and Europe, media warnings are typically issued for extreme weather, high pollen counts, high UV levels, and poor air quality. Warnings issued to the general public aim to modify the behavior of the people and to raise short-term awareness of the dangers that are connected with "heat waves" to reduce health impacts. The US CDC has issued guidelines in reducing heat-related illness (Box 8.1). Heat advisories are linked to general advice about behavior, and they often ask people to check on relatives and neighbors, especially the elderly. Box 8.2 lists some of the active interventions associated with HHWS in the United States. The inclusion of specific interventions varies from city to city and over time. Intervention plans should be best suited for local needs, through coordination between the local health agencies and meteorological agencies.

HHWS when accompanied by specific health interventions are generally considered to be effective in reducing deaths during a heat wave. Heat waves vary in magnitude and in the time that they occur during the summer season. Direct comparison between events in the same city can provide limited information about the effectiveness of interventions. Comparison between events in different populations is unlikely to provide useful insights because population responses to heat waves are sufficiently different.

The 1995 heat wave in the midwestern United States was followed by a similar event in 1999. This has afforded the opportunity for some evaluation of the heat health interventions implemented in Chicago in 1999. The 1995 heat wave in Chicago caused more than 500 deaths, whereas 119 heat deaths were recorded during the 1999 event. After the 1995 heat wave, response plans were significantly improved. A formal meteorological comparison of the two heat waves found that the 1995 heat wave was a bit hotter and more sudden [5]. During the 1999 event, the level of awareness was very high, with warnings issued throughout the region and emergency plans implemented, such as the opening of cooling centers (see Box 8.2). The fewer heat-related deaths during the 1999 heat wave were attributed by Palecki et al. [5] to the characteristics of the heat wave and the successful implementation of the emergency activities. An additional factor in preventing deaths may have been the maintenance of the electrical supply system, which failed during the heat wave in 1995.

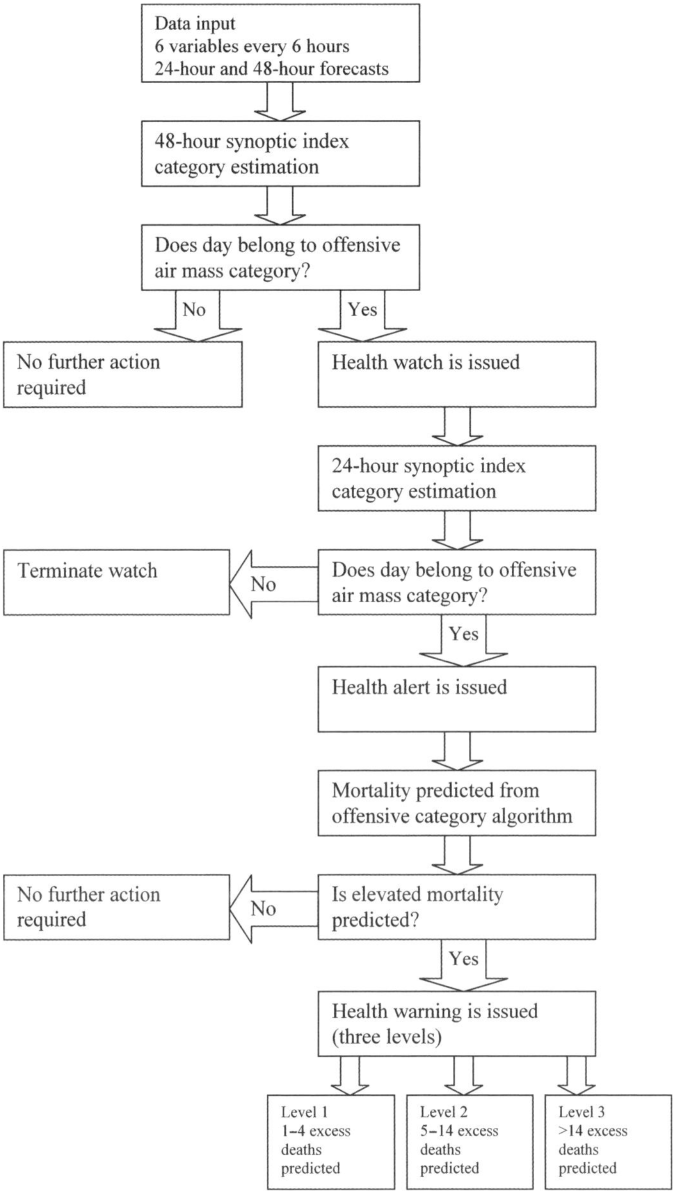

Figure 8.3 Heat watching warning system (Philadelphia model). Source: Data from Ref. 62

The Philadelphia HHWS was evaluated during its first summer, in 1995. The model on which the system is based projects the number of excess deaths (all ages, all causes) on a given day that is associated with a particular air mass type. Usually only one or two air mass types are associated with significant increases in summer deaths. For example, in Philadelphia, the most "offensive" air mass was identified as the "moist tropical", characterized by the highest air and dewpoint temperatures, southwesterly

Box 8.1 CDC Guidelines on preventing and managing heat

- Drink more fluids (nonalcoholic), regardless of your activity level. Don't wait until you're thirsty to drink. Don't drink liquids that contain caffeine, alcohol, or large amounts of sugar – these actually cause you to lose more body fluid.
- Stay indoors and, if at all possible, stay in an air-conditioned place. If your home does not have air conditioning, go to the shopping mall or public library – even a few hours spent in air conditioning can help your body stay cooler when you go back into the heat. Call your local health department to see if there are any heat-relief shelters in your area.
- Electric fans may provide comfort, but when the temperature is in the high 90s, fans will not prevent heat-related illness. Taking a cool shower or bath, or move to an air-conditioned place.
- Wear lightweight, light-colored, loose-fitting clothing.
- NEVER leave anyone in a closed, parked vehicle.
- Although any one at any time can suffer from heat-related illness, some people are at greater risk than others. Check regularly on: infants and young children, people aged 65 or older, people who have a mental illness, those who are physically ill, especially with heart, disease or high blood pressure
- Visit adults at risk at least twice a day and closely watch them for signs of heat exhaustion or heat stroke. Infants and young children need more frequent watching.
- If you must be out in the heat:
  - Limit your outdoor activity to morning and evening hours.
  - Cut down on exercise.
  - Try to rest often in shady areas.
  - Protect yourself from the sun by wearing a wide-brimmed hat, and sunglasses and by putting on sunscreen

Source: Data from Ref. 63.

Box 8.2 Interventions associated with heat early warning systems in the United States

- Opening of cooling centers in local communities, where the elderly can go to experience an air-conditioned environment.
- Alert to hospitals and increased staffing in hospital emergency rooms.
- Activation of a telephone help-line which provides information and health education to the general public on avoidance from heat stress.
- Home visits to persons requiring more attention than can be provided over the hotline, but still those not requiring emergency intervention.
- Interventions in nursing homes and residential homes for the elderly.
- The electricity supply companies do not cut off domestic supply during the heatwave period for non-payment, in order to maintain air conditioning.
- Promotion of a "buddy" system, media announcements encourage friends, relatives, neighbors, and other volunteers to make daily visits to elderly persons during hot weather.
- Daytime outreach to the homeless.

winds, and partly cloudy conditions [62]. Additional factors needed to be identified to forecast increased mortality:

- the number of consecutive days the air mass is present,
- maximum temperature, and
- the time of season (i.e., early or late in summer season).

An evaluation of the system concluded that there was some evidence that the system was effective in reducing mortality during the hot summer of 1995 in Philadelphia [62]: 72 heat-related deaths were recorded within the city, and 32 of these occurred during two identifiable heat wave periods. Weather conditions indicated that Level 2 heat warnings should be issued on 12 days and Level 3 heat warnings should be issued on 3 days (Fig. 8.3). Thus the model indicated an excess mortality (all ages, all causes) for the summer of between 100 and 200 deaths. In practice, actual warnings were issued on only 9 days, and health alerts were issued on 6 days. As previously discussed, however, certified heat deaths do not account for all excess mortality during a heat wave. Therefore, this model will always project more deaths than would be certified as heat related. Since interventions were in place and warnings were issued, it is difficult to validate the predictive model because the number of "avoided deaths" cannot be independently observed.

A study on the cost-effectiveness of the Philadelphia HHWS [64] calculated the benefits of the system for those aged 65 and over. A heat mortality regression model estimated that each warning during a heat wave "prevented" about 2.6 deaths per day (including the 3 days after the warning was issued). This model assumes that the system was 100% effective in preventing heat deaths. From 1995 to 1998, warnings were issued on 45 days. The model, therefore, estimates that the system "saved" 117 deaths. Assuming a statistical value of US$4 million per life saved, the gross benefits of the Philadelphia warning system were US$468 million. Most of the monetary costs of the Philadelphia system are indirect costs, which were not included in this assessment. These indirect costs arise from actions taken by city employees as a normal part of their jobs and actions taken by volunteers. The direct costs emerge from the "Heatline" and additional emergency medical services and are estimated to be about US$300,000 for the 1995–1998 period (see Box 8.2). In general, heat health warning systems are likely to be cost-effective because they are relatively inexpensive to set up and maintain. It has proved difficult, however, to evaluate either the systems as a whole or the effectiveness of specific interventions.

5.2 Reducing heat stress in the indoor thermal environment

The capacity of humans to adapt to varied climates and environments is considerable. Physiological and behavioral differences between cultures have developed over many millennia as a consequence of exposure to vastly different climatic regimes. Most homes have an indoor air temperature of 17°C to 31°C. Humans cannot comfortably live in temperatures outside this range. The tolerance range of an individual is usually less than this and tends to narrow with age or infirmity. The temperature of the surrounding air is the most significant factor for human comfort. Comfort is also dependent on other factors such as humidity, wind, sunshine (short-wave radiation), and long-wave (infrared) radiation. Humidity has a pronounced effect on our sensation of temperature when conditions are hot; wind has a significant effect on our well-being when conditions are cold or very hot.

In developed countries, people spend the vast majority of their time indoors, at home and at work. The indoor environment has been investigated in relation to indices of thermal comfort. Perceptual scales have been developed to evaluate thermal comfort in an individual (e.g., the ASHRAE scale). In temperate climates, the optimum indoor temperature for health is between 18°C and 24°C [65]. In general, recommendations have focused on maintaining a minimum indoor temperature and reducing the impact of cold on health rather the potential impact of heat [66].

The indoor comfort temperature depends on the outdoor temperature. A study in cities in a range of climates found a linear relationship between the comfort indoor temperature and the mean monthly outdoor air temperature [67]. This relationship was valid for an outdoor air temperature range between 5°C and 32°C. The lower outdoor temperature was associated with an indoor "comfort temperature" of 19°C and the upper with 32°C.

Buildings today are designed to have a long lifetime. Most will last more than 100 years. Climate change (such as the increasing numbers of extreme hot days), changing lifestyles, and new technologies all have implications for building design. With respect to climate change, the design of comfortable, energy-efficient, and safe buildings is a priority. In particular, the design should aim to limit both the frequency of occurrence of high air temperature episodes inside the building and their intensity and duration. Conventional building designs have evolved in harmony with the environment and usually provide adequate protection against the heat. In recent decades, it seems that rapid urbanization has led to an increase in poor building design in relation to weather extremes in many cities. Thus, populations in these dwellings are less adapted and perhaps more vulnerable to heat episodes [68]. People living in informal dwellings in large cities may be very vulnerable to weather extremes because these dwellings are often constructed from corrugated iron, which provides no thermal insulation. Further, these populations are already marginalized and in poor health [69].

In developed countries, air conditioning or "comfort cooling" has a direct role in reducing temperature exposures in an individual when they are in the cooled environment. Evidence from the United States indicates that air conditioning is an effective intervention to prevent heat stroke and heat-related illness [38,70,71,72]. Half the cases of heat stroke admitted during the 1995 heat wave in Madison, Wisconsin, USA, were attributed to indoor activity with no functioning air conditioning [27]. The risk of mortality was lower for those with central air conditioning (defined as compared to no air conditioning) in a US cohort study [73]. A study of mortality in nursing homes during four heat waves in New York City in 1972 and 1973 found that air conditioning was significantly protective against mortality [74].

The population groups most vulnerable to dying in a heat wave (the very old and the chronically ill) are also those persons least likely to have air conditioning in place and functioning during a heat wave [38]. Therefore, recommendations to use more air conditioning should address the utility costs needed to support air conditioners. Interviews with Chicago residents after the 1995 heat wave indicated that a significant proportion of elderly people were unable to afford basic utilities, and even restricted the use of lighting in their home. Federal funding to a program that supported utility (electricity costs) in low income households had been steadily reduced during the

1990s. Further, fear of crime deterred individuals from leaving their home and seeking cooling shelters [50].

Technological advancements in the last century permitted the design of structures that emphasize engineered approaches to interior climate control. As a result, modern structures often do not reflect local climates, needlessly consume large amounts of energy, and assume a standard level of comfort for the building occupants. Architectural design can prevent buildings from warming up and so ensure a comfortable indoor environment without the use of energy-intensive air conditioning. Table 8.5 describes techniques that can reduce indoor heat stress in hot and dry environments. Natural cooling techniques should be favored in all future building or retrofitting of old buildings. The effective use of simple natural cooling strategies in hot climate building design can reduce internal temperatures both day and night [66], such as shading all windows and using light colors on outside walls.

5.3 Reducing heat stress in the outdoor environment

Half of the world's population is living in urban areas. It is estimated that at least 60% of the world population will be living in large conurbations by 2030 [76]. Populations in cities are having to deal with a range of current environmental problems, and global climate change is likely to exacerbate many of these problems [77]. Urban environmental problems such as outdoor air pollution have, in general, been decreasing steadily in developed countries because of active control measures. Although concentrations of sulfur dioxide and suspended particulate matter are decreasing in developed countries, those of NO_x and ozone are either constant or increasing. In developing countries, increasing traffic and its exhaust as well as industrial emissions are raising concentrations of SO_2, NO_x, O_3, and suspended particulate matter [78]. Global climate change may increase the frequency of days with high levels of ozone [79]. Adaptation to climate change will require a range of diverse and complex adaptation strategies that are likely to include physical modification to the built environment and changes in decision-making practices [80].

Regional climate is modified at the local scale by human activities through changes in the land surface, such as building a city [68]. The urban heat island (UHI) is defined as the difference in temperature in the city compared to the temperature of the surrounding area. This effect is greatest during the night and in winter. The impact of heat waves in cities is thus exacerbated by the heat island effect, which can raise air temperatures 3° to 12°C.

The most important factors that affect the urban microclimate are [81]:

- *Anthropogenic heat production.* Buildings are permanent heating appliances and discharge heat all day and year round because of space heating and cooling, artificial lighting, and the use of domestic and office appliances. In Northern Europe during winter, the amount of heat dissipated within the urban canopy layer by buildings and traffic can easily exceed that from solar radiation.
- *Airflow.* Wind velocities in cities are generally lower than those in the open country. This results in a reduced rate of heat dissipation by convective cooling. In addition, very tall buildings and the channeling effect of urban canyons lead to complex air flow patterns and produce unpleasant turbulence.

Table 8.5 Building designs that reduce indoor heat stress

Architectural feature	*Effect/building strategies*	*Negative effects on objective no.*	*Positive effects on objective no.*
Building layout	The more compact a house, the smaller the surface area of the walls for a given floor area.	3	1, 2, 5
	A passive solar building can use its southern wall and windows as solar collection elements. Buildings elongated along the east-west axis will be more energy efficient than square buildings inspite of the larger wall surface areas.	1, 2, 5	3, 6
Building orientation and shading objectives	Highest intensities of the impinging solar radiation: summer: eastern and western walls winter: southern wall north-south orientation for the main facades and windows.		1, 2, 5, 6
Window size and location	Natural ventilation. Small windows in hot dry regions.	3, 6	1, 2, 5
Colors of the building	Difference in maximum surface temperature between a white and a black roof in a desert can be 40°C. The heat gain depends on the insulation. White roofs.	6	1, 2
Thermal properties of building materials	High thermal insulation coupled with effective shading. High heat capacity.	3	1, 2, 5
		3	1, 2, 5
Building height	Only for humid regions: Construction of high buildings because windspeed increases with elevation while temperature and humidity tend to increase.	5	3, 4
Kitchens, water heaters, and bathrooms on the leeward side	Heat can be rapidly removed by draft.	5	2, 3

Source: Data from Ref. 75.

Objective no.:
1 Minimize solar heating of the building during the hot season.
2 Minimize the rate of indoor temperature elevation in summer during the daytime hours.
3 Maximize the rate of cooling of the indoor temperature in summer during the evening hours, and ensure indoor comfort during night.
4 Use natural energies for passive cooling in summer.
5 Minimize the heat loss of the building in winter.
6 Use solar energy, by passive systems, for heating in winter.

- *Built environment.* Thermal capacity, surface reflectance, and exposed external surface area are all important determinants of a building's net contribution to urban warming.

Heat waves present special problems in urban areas. Buildings retain the heat during the day and then re-radiate that energy during the night. The inhabitants of urban areas therefore experience sustained high temperatures both day and night [82]. In Chicago, a mean UHI of 2.8°C was observed during the night in summer, which decreased to 1.8°C during the daytime [83]. In Athens, Greece, a mean maximum UHI of 4.6°C in summer (June-September) has been observed [84].

The magnitude of the heat island is related to the size of the city and the density of buildings. A comparative study in the United States found that the maximum urban-rural difference was approximately 2.5°C for towns of 1,000 inhabitants, reaching 12°C for towns of 1 million or more under ideal (clear, calm) conditions [85]. These values are less for European cities, where per capita energy use and hence anthropogenic heat production are lower than in North America. An Australian study found that the maximum UHI was proportional to population size, but values were less than those found in Europe [86]. A study in Tokyo found that the near surface air temperature could be reduced by 0.5°C by reducing energy consumption (assuming a 50% reduction in demand for hot water and an unrealistic 100% reduction in space cooling) [87]. Urban heat islands are greater in higher latitudes. An analysis of 150 cities found that the relationship between UHI and latitude can be explained by anthropogenic heat production, the radiation balance, and annual climate variability [88].

Within a given city, the hottest areas are those with the tallest buildings, with the greatest density of buildings, and without green spaces [89]. Therefore, measures to reduce the urban heat islands focus on:

- increasing green spaces and planting trees in streets (trees provide shade but can also improve air quality),
- increasing ventilation and air flow (which also improves air quality),
- increasing the number of courtyards and other open spaces,
- increasing the albedo of a city (e.g., painting roofs white),
- decreasing anthropogenic heat production (e.g., natural space cooling, see above).

The thermal situation of Berlin was assessed by using the urban bioclimate model UBIKLIM [90]. UBIKLIM enables an assessment of the urban climate at a high spatial resolution (5 km grid) and thus facilitates the consideration of urban climate issues in planning and decision making. A high resolution bioclimatic map enables a relative evaluation and comparison of the bioclimate of different urban areas [91]. The number of days with heat stress during a typical summer is fewer in the outskirts of Berlin and greatest in the city center.

The urban heat island is an inevitable consequence of urban development. Appropriate and climate friendly urban planning, however, may help reduce the magnitude of the urban heat island. Planting trees and increasing green spaces is one method of reducing heat stress within a city. The benefits of tree planting projects are shading, cooling due to evapotranspiration, dust control, runoff control, consumption of carbon dioxide, and water conservation. A study in Munich compared street canyons with and without trees [92]. Trees had little effect on air temperature at human height level, but were effective in reducing heat stress by reducing radiative temperature. An analysis in four US cities of the potential of vegetation to reduce summer cooling loads in residential buildings found that an additional 25% increase in the urban tree cover can save 25–40% of the annual cooling energy [93]. The costs of urban trees in Tucson, Arizona, were estimated at $9.61 per tree, based on over 40 years of planting 50,000 trees, while the total benefits are $25.09 per tree [94]. Further, trees were only half of the cost of metal bus shelters, but provided better shade.

Table 8.6 Estimated maximum decrease in the urban heat island for two days (14–15 July 1995) compared to the base case in selected cities in northeastern United States

Measures	*14/07/1995*	*15/07/1995*
Albedo + vegetation 100%	1.2°K	0.9°K
Only albedo	1.1°K	0.6°K
Only vegetation	0.4°K	0.2°K
Albedo + vegetation 50%	0.7°K	0.4°K

Source: Data from Ref. 95.

Implementing measures to reduce the magnitude of the UHI has been shown to also reduce ozone air pollution in cities in the northeastern United States [95]. The EPA Heat Island Reduction Initiative (HIRI) program found that increasing the albedo and the vegetation cover was effective in reducing the urban heat island. The simulation model was run for the heat episode period of 9–15 July 1995 under four scenarios (Table 8.6). Increasing only the albedo of a city is nearly as effective as implementing both measures. However, the effect of the vegetation cover is greater on radiant temperature than on air temperature, and radiant temperature is the meteorological parameter with the greatest effect on the sensation of thermal stress in humans [92].

There are many competing priorities for urban planning. In practice, climate issues often have a low impact on urban design. Although urban planners are interested in climatic aspects, the use of climate information is unsystematic [96]. Good building designs can provide effective measures to reduce heat stress of individuals living in cities. A UK report on adaptation concluded that many planning processes and systems had the capacity for integrating climate change considerations, but required more robust and reliable information on the potential climate change impacts and on better understanding of the importance of adaptation by stakeholders [97].

6. FUTURE ADAPTATION AND ADAPTIVE CAPACITY

A main uncertainty in estimates of the impact of climate change on heat-related mortality is the extent to which, even without specific adaptation strategies, physiological adaptation and factors such as behavioral changes in hot weather will reduce impacts in the general population. Physiological acclimatization to hot environments can occur over a few days, and can explain why the impact of the first heat wave on mortality is often greater than that of subsequent heat waves in a single summer. The rate at which infrastructural changes will take place without specific advice is likely to be much slower. Neither the magnitude nor the time course of the various modifying factors can be predicted with any confidence. It is clear that preventive measures will be needed to counter the substantial initial adverse effects of heat, and long-term changes are required to housing and urban infrastructure.

As with many environmental health problems, the technology (air conditioning, natural ventilation, drinking fluids) is widely and readily available. However, many barriers to their effective use remain. Table 8.7 outlines an assessment of adaptive capacity with respect to the future impacts of heat stress on health in the United States. The availability and distribution of resources may limit adaptive capacity in

Table 8.7 Adaptive capacity: Heat-related deaths in the United States[a]

Determinants of adaptive capacity	*Requirements for public health prevention*
Range of available technological options (5)	Awareness that problem exists (4)
Availability and distribution of resources (2)	Sense that the problem matters (4)
Structure of critical institutions (2)	Understanding of the causes (4)
Human capital (3)	Capability to deal with problem [public agencies] (3)
Social capital (3)	Political will to influence (2)
Access to risk spreading (n/a)	
Ability to manage information (3)	
Public's perceived attribution (4)	

Note
a Scale = 1 (low) to 5 (high).

this population. The elderly poor have been identified as the population most vulnerable to heat-related mortality. In all developed countries the population is aging, and unless inequities in the distribution of resources are reduced, adaptive capacity will not be improved.

Adaptation strategies require further research to clarify those that are truly effective. The impact of heat waves and climate change in developing countries needs to be investigated further, and in particular, the interactions between global and local environmental changes on health in developing country cities. Measures and strategies for reducing the impact of heat stress in such populations need to be identified. Clearly, we should be building cities that are more sustainable and energy efficient. An important component of this is the use of optimum methods and materials for space cooling. Reliance on energy-intensive technologies such as air conditioning is unsustainable and can be considered a maladaptation. The construction of poorly insulated housing is also a form of maladaptation to the impacts of more extreme summer weather. Heat health warning systems are an important strategy to reduce heat-related deaths, providing they are accompanied by the active detection and care of vulnerable individuals.

ACKNOWLEDGMENTS

This work was funded in part by the European Commission as part of the following project: Climate Change and Adaptation Strategies for Human Health (cCASHh) (EVK2-2000-00070).

REFERENCES

1. EMA: EMA Disaster Events Data Tracking System (EMATrack). Emergency Management Australia (EMA) Disaster Events Data Tracking System, Dickson, 2002.
2. Impact sanitaire de la vague de chaleur en France survenue en août 2003. Département des maladies chroniques et traumatismes, Départment santé environment, Paris, 2003.

3. Falcao, J.M., Nogueira, P.J., Contreiras, M.T., Paixao, E., Brandao, J., and Batista, I.: Onda de calor do Agosto de 2003: Repercussoes sobre a saude da populacao [The heat-wave of August 2003: implications for health]. Instituto Nacional de Saude Dr Ricardo Jorge, Lisbon, 2003.
4. Whitman, S., Good, G., Donoghue, E.R., Benbow, N., Shou, W., and Mou, S.: Mortality in Chicago Attributed to the July 1995 Heat Wave. *Am. J. Public Health* 87 (1997), pp.1515–1518.
5. Palecki, M.A., Changnon, S.A., and Kunkel, K.E.: The Nature and Impacts of the July 1999 Heatwave in the Midwestern United States: Learning from the Lessons of 1995. *B. Am. Meteorol. Soc.* (2001), pp.1353–1367.
6. Bridger, C.A. and Helfand, L.A.: Mortality from Heat During July 1966 in Illinois. *Int. J. Biometeorol.* 12 (1968), pp.51–70.
7. McMichael, A.J. and Kovats, R.S.: Assessment of the Impact on Mortality in England and Wales of the Heatwave and Associated Air Pollution Episode of 1976. LSHTM, London, UK, 1998.
8. Jones, T.S., Liang, A.P., Kilbourne, E.M., Griffin, M.R., Patriarca, P.A., Wassilak, S.G.F., Mullan, R.J., Herrick, R.F., Donnell, H.D., Choi, D., and Thacker, S.B.: Morbidity and Mortality Associated with the July 1980 Heat Wave in St Louis and Kansas City, Mo. *JAMA* 247 (1982), pp.3327–3331.
9. Applegate, W.B., Runyan, J.W. Jr., Brasfield, L., Williams, M.L., Konigsberg, C., and Fouche, C.: Analysis of the 1980 Heat Wave in Memphis. *J. Am. Geriatr. Soc.* 29 (1981), pp.337–342.
10. Garcia, A.C., Nogueira, P.J., and Falcao, J.M.: Onda de calor de Junho de 1981 em Portugal: efeitos na mortalidade. *Revista Portuguesa de Saude Publica* 1 (1981), pp.67–77.
11. MMWR: International Notes Heat Related Mortality – Latium Region, Italy, Summer 1983. *MMWR* 33 (1984), pp.518–521.
12. Todisco, G.: Indagine biometeorologica sui colpi di calore verificatisi a Roma nell'estate del 1983 [Biometeorological Study of Heat Stroke in Rome during Summer of 1983]. *Rivista di Meteorologica Aeronautica* XLVII (1987), pp.189–197.
13. Katsouyanni, K., Trichopoulos, D., Zavitsanos, X., and Touloumi, G.: The 1987 Athens Heatwave [letter]. *Lancet* 2 (1988), p.573.
14. Ramlow, J.M. and Kuller, L.H.: Effects of the Summer Heat Wave of 1988 on Daily Mortality in Allegheny County, PA. *Public Health Rep.* 105 (1990), pp.283–289.
15. Semenza, J.C., McCullough, J.E., Flanders, W.D., McGeehin, M.A., and Lumpkin, J.R.: Excess Hospital Admissions During July 1995 Heat Wave in Chicago. *Am. J. Prev. Med.* 16 (1999), pp.269–277.
16. Rooney, C., McMichael, A.J., Kovats, R.S., and Coleman, M.: Excess Mortality in England and Wales, and in Greater London, during the 1995 Heatwave. *J. Epidemiol. Comm. Health* 52 (1998), pp.482–486.
17. Kovats, R.S.: Global Climate Change and Temperature-Related Mortality in Low Income Populations. London School of Hygiene and Tropical Medicine, London, UK, 2002.
18. Katsouyanni, K., Pantazopoulu, A., Touloumi, G., Tselepidaki, I., Moustris, K., and Asimakopoulos, D., Poulopoulou, G., and Trichopoulos, D.: Evidence of Interaction Between Air Pollution and High Temperatures in the Causation of Excess Mortality. *Arch. Environ. Health* 48 (1993), pp.235–242.
19. BBC News: Heatwave in Croatia. 2000.
20. CRED and OFDA: EM-DAT: The OFDA/CRED International Disasters Data Base. Center for Research for the Epidemiology of Disasters, Brussels, Belgium, 2002. http://www.cred.be/emdat/intro.html.
21. Sapir, D.G. and Misson, C.: Development of a Database on Disasters. *Disasters* 16 (1992), pp.74–80.
22. Kumar, S.: India's Heatwave and Rains Result in Massive Death Toll. *Lancet* 351 (1998), p.1869.

23. Kilbourne, E.M.: Heat Waves and Hot Environments. In: E. Noji (ed.): *The Public Health Consequences of Disasters*. Oxford University Press, New York, 1997, pp.245–269.
24. Keatinge, W.R., Coleshaw, S.R., Easton, J.C., Cotter, F., Mattock, M.B., and Chelliah, R.: Increased Platelet and Red Cell Counts, Blood Viscosity, and Plasma Cholesterol Levels During Heat Stress, and Mortality from Coronary and Cerebral Thrombosis. *Am. J. Med.* 81 (1986), pp.795–800.
25. Mirchandani, H.G., McDonald, G., Hood, I.C., and Fonseca, C.: Heat-Related Deaths in Philadelphia – 1993. *Am. J. Foren. Med. Pathol.* 17 (1996), pp.106–108.
26. Donoghue, E.R., Graham, M.A., Jentzen, J.M., Lifschultz, B.D., Luke, J.L., and Mirchandani, H.G.: Criteria for the Diagnosis of Heat-Related Deaths: National Association of Medical Examiners. Position paper. National Association of Medical Examiners Ad Hoc Committee on the Definition of Heat-Related Fatalities. *Am. J. Foren. Med. Pathol.* 18 (1997), pp.11–14.
27. Dixit, S.N., Bushara, K.O., and Brooks, B.R.: Epidemic Heat Stroke in a Midwest Community: Risk Factors, Neurological Complications and Sequelae. *Wis. Med. J.* 96 (1997), pp.39–41.
28. Dematte, J.E., O'Mara, K., Buescher, J., Whitney, C.G., Forsythe, S., McNamee, T., Adiga, R.B., and Ndukwu, M.: Near-Fatal Heat Stroke During the 1995 Heat Wave in Chicago. *Ann. Intern. Med.* 129 (1998), pp.173–181.
29. Hajat, S., Kovats, R.S., Atkinson, R.W., and Haines, A.: Impact of Hot Temperatures on Death in London: A Time Series Approach. *J. Epidemiol. Comm. Health* 56 (2002), pp.367–372.
30. Curriero, F., Heiner, K.S., Samet, J., Zeger, S., Strug, L., and Patz, J.A.: Temperature and Mortality in 11 Cities of the Eastern United States. *Am. J. Epidemiol.* 155 (2002), pp.80–87.
31. McGeehin, M.A. and Mirabelli, M.: The Potential Impacts of Climate Variability and Change on Temperature Related Morbidity and Mortality in the United States. *Environ. Health Persp.* 109 (2001), pp.185–189.
32. Kunkel, K.E., Changnon, S.A., Reinke, B.C., and Arritt, R.W.: The July 1995 Heatwave in the Midwest: A Climatic Perspective of Critical Weather Factors. *B. Am. Meteorol. Soc.* 77 (1996), pp.1507–1518.
33. CDC: Heat-Related Deaths – Four States, July–August 2001, and United States, 1979–1999. *MMWR* 51 (2002), pp.567–570.
34. CDC: Heat-Related Deaths – Dallas, Wichita, and Cooke Counties, Texas, and United States, 1996. *MMWR* 46 (1997), pp.528–531.
35. ONS: Population and Migration. 2002. Office for National Statistics, London, UK. http://www.statistics.gov.uk/.
36. Kilbourne, E.M.: Heat Waves. In: M.B. Gregg (ed.): *The Public Health Consequences of Disasters*. U.S. Dept. of Health and Human Services, Centers for Disease Control, Atlanta, GA, 1989, pp.51–61.
37. Kilbourne, E.M.: Illness Due to Thermal Extremes. In: J.M. Last and R.B. Wallace (eds): *Public Health and Preventative Medicine*. 13 ed. Appleton Lang, Norwalk, CT, 1992, pp.491–501.
38. Semenza, J.C., Rubin, C.H., Falter, K.H., Selanikio, J.D., Flanders, W.D., Howe, H.L., and Wilhelm, J.L.: Heat-Related Deaths During the July 1995 Heat Wave in Chicago. *N. Eng. J. Med.* 335 (1996), pp.84–90.
39. De, U.S. and Mukhopadhyay, R.K.: Severe Heat Wave Over the Indian Subcontinent in 1998, in Perspective of Global Climate. *Curr. Sci.* 75 (1998), pp.1308–1315, (Suppl. 12).
40. Drinkwater, B.L. and Horvath, S.M.: Heat Tolerance and Aging. *Med. Sci. Sports* 11 (1979), pp.49–55.
41. Basu, R. and Samet, J.: An Exposure Assessment Study of Ambient Heat Exposure in an Elderly Population in Baltimore, Maryland. *Environ. Health Persp.* 110 (2002), pp.1213–1224.
42. Bull, G.M. and Morton, J.: Relationships of Temperature with Death Rates from all Causes and from Certain Respiratory and Arteriosclerotic Diseases in Different Age Groups. *Age Ageing* 4 (1975), pp.232–246.

43. Lye, M. and Kamal, A.: Effects of a Heatwave on Mortality-Rates in Elderly Inpatients. *Lancet* 1 (1977), pp.529–531.
44. Faunt, J.D., Wilkinson, T.J., Aplin, P., Henschke, P., Webb, M., and Penhall, R.K.: The Effect in the Heat: Heat-Related Hospital Presentations During a Ten Day Heat Wave. *Aust. New Zealand J. Med.* 25 (1995), pp.117–120.
45. Sierra Pajares, O.M., Diaz, J.J., Montero, R.J.C., Alberdi, J.C., and Miron Perez, I.J.: Daily Mortality in the Madrid Community During 1986–1991 for the Group Between 45 and 64 Years of Age: Its Relationship to Air Temperature. *Rev. Esp. Salud Publica.* 71 (1997), pp.149–160.
46. CDC: Heat-Related Illnesses, Deaths, and Risk Factors – Cincinnati and Dayton, Ohio, 1999, and United States, 1979–1997. *MMWR* 49 (2000), pp.470–473.
47. Nakai, S., Itoh, T., and Morimoto, T.: Deaths from Heat Stroke in Japan: 1968–1994. *Int. J. Biometeorol.* 43 (2001), pp.124–127.
48. Smoyer, K.E.: A Comparative Analysis of Heat Waves and Associated Mortality in St. Louis, Missouri – 1980 and 1995. *Int. J. Biometeorol.* 42 (1998), pp.44–50.
49. Smoyer, K.E.: Putting Risk in its Place: Methodological Considerations for Investigating Extreme Event Health Risks. *Soc. Sci. Med.* 47 (1998), pp.1809–1824.
50. Klinenberg, E.: *Heat Wave: A Social Autopsy of Disaster in Chicago.* University of Chicago Press, Chicago, IL, 2002.
51. Guest, C., Willson, K., Woodward, A.J., Hennessy, K., Kalkstein, L.S., Skinner, C., and McMichael, A.J.: Climate and Mortality in Australia: Retrospective Study, 1970–1990, and Predicted Impacts in Five Major Cities. *Clim. Res.* 13 (1999), pp.1–15.
52. Gouveia, N., Hajat, S., and Armstrong, B.: Socio-Economic Differentials in the Temperature-Mortality Relationship in Sao Paulo, Brazil. *Int. J. Epidemiol.* 32 (2003), pp.390–397.
53. Gaffen, D.J. and Ross, R.J.: Increased Summertime Heat Stress in the US. *Nature* 396 (1998), pp.529–530.
54. Katz, R.W. and Brown, B.G.: Extreme Events in a Changing Climate: Variability is More Important than Averages. *Clim. Change* 21 (1992), pp.289–302.
55. Goodess, C.M., Hanson, C., Hulme, M., and Osborn, T.J.: Representing Climate and Extreme Weather Events in Integrated Assessment Models: A Review of Existing Methods and Options for Development. *Int. Assess.* 4 (2003), pp.145–171.
56. IPCC: *Climate Change 2001: The Scientific Basis. Contribution of Working Group I to the Second Assessment Report of the Intergovernmental Panel on Climate Change.* Cambridge University Press, New York, 2001.
57. Hulme, M., Jenkins, G.J., Lu, X., Turnpenny, J.R., Mitchell, T.D., Jones, R.G., Lowe, J., Murphy, J.M., Hassell, D., Boorman, P., McDonald, R., and Hill, S.: *Climate Change Scenarios for the United Kingdom: The UKCIP02 Scientific Report.* Tyndall Centre for Climate Change Research, School of Environmental Sciences, University of East Anglia, Norwich, UK, 2002.
58. Gawith, M.J., Downing, T.E., and Karacostas, T.S.: Heatwaves in a Changing Climate. In: T.E. Downing, A.A. Olsthoorn, and R.S. Tol (eds): *Climate, Change and Risk.* Routledge, London, UK, 1999, pp.279–307.
59. Karacostas, T.S. and Downing, T.E.: Heat Wave Events in a Changing Climate. In: T.E. Downing, A.A. Olsthoorn and R.S. Tol (eds): *Climate Change and Extreme Events: Altered Risk, Socio-Economic Impacts and Policy Responses.* Institute for Environmental Studies, Amsterdam, 1996, pp.139–156.
60. Smoyer-Tomic, K.E. and Rainham, D.G.C.: Beating the Heat: Development and Evaluation of a Canadian Hot Weather Health Response Plan. *Environ. Health Persp.* 109 (2001), pp.1241–1247.
61. Kalkstein, L.S., Greene, J.S., Nichols, M.C., and Barthel, C.D.: A New Spatial Climatological Procedure. *Proceedings of the Eighth Conference on Applied Climatology* (1993), pp.237–240.

62. Kalkstein, L.S., Jamason, P.F., Greene, J.S., Libby, J., and Robinson, L.: The Philadelphia Hot Weather-Health Watch Warning System: Development and Application, Summer 1995. *B. Am. Meteorol. Soc.* 77 (1996), pp.1519–1528.
63. CDC: Tips on Preventing and Managing Heat. Centers for Disease Control and Prevention, Atlanta, GA, 2003. http://www.cdc.gov/nceh/hsb/extremeheat/heattips.htm.
64. Teisberg, T.J., Ebi, K.L., Kalkstein, L.S., Robinson, L., and Weiher, R.F.: Heat Watch/Warning Systems Save Lives: Estimated Costs and Benefits for Philadelphia 1995–1998. *BAMS* (in press).
65. WHO: Health Impact of Low Indoor Temperatures. World Health Organization Regional Office for Europe, Copenhagen, 1987.
66. WHO: Indoor Environment: Health Aspects of Air Quality, Thermal Environment, Light and Noise. WHO/EHE/RUD/90.2ed. World Health Organization/United Nations Environment Programme, Geneva, 1990.
67. De Dear, R. and Brager, G.S.: The Adaptive Model of Thermal Comfort and Energy Conservation in the Built Environment. *Int. J. Biometeorol.* 45 (2001), pp.100–108.
68. Oke, T.R.: Urban Climates and Global Environmental Change. In: R.D. Thompson and A.H. Perry (eds): *Applied Climatology: Principles and Practice.* Routledge, London, UK, 1997, pp.273–287.
69. WRI: Urban Environment and Human Health. World Resources 1996–97. World Resources Institute, New York, 2000, pp.31–55.
70. Kiernan, V.: If You Can't Stand the Heat, Go Shopping. *New Sci.* 20 (1996), p.10.
71. Marmor, M.: Heat Wave Mortality in New York City, 1949 to 1970. *Arch. Environ. Health* 30 (1975), pp.130–136.
72. Kilbourne, E.M., Choi, K., Jones, T.S., and Thacker, S.B.: Risk Factors for Heatstroke. A Case-Control Study. *JAMA* 247 (1982), pp.3332–3336.
73. Rogot, E., Sorlie, P.D., and Backlund, E.: Air-Conditioning and Mortality in Hot Weather. *Am. J. Epidemiol.* 136 (1992), pp.106–116.
74. Marmor, M.: Heat Wave Mortality in Nursing Homes. *Environ. Res.* 17 (1978), pp.102–115.
75. Givoni, B.: Design for Climate in Hot, Dry Cities. In: T.R. Oke (ed.): *Urban Climatology and its Applications with Special Regard to Tropical Areas.* Proceedings of the Technical Conference, Mexico DF, 26–30 November 1984. WMO, Geneva, 1986, pp.487–513.
76. de Leeuw, E.: Global and Local (Glocal) Health: The WHO Healthy Cities Programme. *Global Change Hum. Health* 2 (2001), pp.34–45.
77. McMichael, A.J.: The Urban Environment and Health in a World of Increasing Globalization: Issues for Developing Countries. *B. WHO* 78 (2000), pp.1117–1126.
78. WHO: *Air Pollution. Factsheet no. 187.* World Health Organization, Geneva, 2000.
79. DOH: Health Effects of Climate Change in the UK. Department of Health, London, UK, 2002.
80. Anon.: Adaptability of Prairie Cities: The Role of Climate – Current and Future Impacts and Adaptation Strategies. Potential City Climate Change Adaptation Strategies. SRC Publications No. 11296–1E01. SRC Publications, Saskatoon, 2001.
81. Yannas, S.: Toward More Sustainable Cities. *Sol. Energy* 70 (2001), pp.281–294, (Suppl. 3).
82. Clarke, J.F.: Some Effects of the Urban Structure on Heat Mortality. *Environ. Res.* 5 (1972), pp.93–104.
83. Ackerman, B.: The March of the Chicago Urban Heat Island. *J. Clim. Appl. Meteorol.* 24 (1985), pp.547–554.
84. Livada, I., Santamouris, M., Niachou, K., Papanikoloau, N., and Mihalakakou, G.: Determination of Places in the Great Athens Area Where the Heat Island Effect is Observed. *Theor. Appl. Climatol.* 71 (2002), pp.2719–2730.
85. Oke, T.R.: City Size and the Urban Heat Island. *Atmos. Environ.* 7 (1973), pp.769–779.

86. Torok, S.J., Morris, G.J.G., Skinner, C., and Plummer, N.: Urban Heat Island Features of Southeast Australian Towns. *Australian Meteor. Mag.* 50 (2001), pp.1–13.
87. Ichinose, T., Shimodozono, K., and Hanaki, K.: Impact of Anthropogenic Heat on Urban Climate in Tokyo. *Atmos. Environ.* 33 (1999), pp.3897–3909.
88. Weinert, U.: Analyses Respect to the Dependency of the Urban Heat Island on Latitude and Climate Zones. Dissertation. Universität Essen, Germany, 2001.
89. Matzarakis, A.: *Die thermische Komponente des Stadtklimas.* 6 ed. Freiburg: Meteorologisches Institut der Universität Freiburg; 2001.
90. Piehl, H.D. and Grätz, A.: Klimakarten für das Land Berlin. Teil 1: Bioklima Berlin. Potsdam, Deutscher Wetterdienst, 1996.
91. Grätz, A. and Jendritzky, G.: Bewertung des Klimas in städtischen Gebieten mit Hilfe von UBIKLIM. UVP-Report 1998, 1:17–19.
92. Mayer, H.: Human-biometeorologische Probleme des Stadtklimas [Human-biometeorological problems of urban climate]. *Geowissenschaften* 14 (1996), pp.223–239.
93. Huang, Y.J., Akbari, H., Taha, H., and Rosenfeld, A.H.: The Potential of Vegetation in Reducing Summer Cooling Loads in Residential Buildings. *J. Clim. Appl. Meteorol.* 26 (1987), pp.1103–1116.
94. McPherson, G.: Environmental Benefits and Costs of the Urban Forest: Two Examples from Tucson, Arizona. Cities and Global Change. Climate Institute of Washington, Washington, DC, 1991.
95. Hudischewskyj, A.B., Douglas, S.G., and Lundgren, J.R.: Meteorological and Air Quality Modeling to Further Examine the Effects of Urban Heat Island Mitigation Measures on Several Cities in the Northeastern U.S. Final Report SYSAPP-01-001. ICF Consulting, Systems Applications International, San Rafael, 2001.
96. Eliasson, I.: The Use of Climate Knowledge in Urban Planning. *Landscape Urban Plan.* 48 (2000), pp.31–44.
97. ERM: Potential UK Adaptation Strategies for Climate Change. Technical Report. Produced for the Department of Environment, Transport and the Regions, by Environmental Resources Management, London, UK, 2000.

9 Extreme weather and climate events – implications for public health

Roger Street, Abdel Maarouf, Heather Jones-Otazo

ABSTRACT: Extreme weather and climate events (e.g., tropical, winter, and wind storms; floods; droughts; tornadoes; extreme heat and cold) are part of the climatic norm of any geographical region. They have threatened, and will continue to threaten, human health and safety in various regions of the world. Extreme air pollution events, enhanced by the prevailing weather conditions, are also negatively impacting on human health and the public health system. The threats posed by extreme weather and climate events include direct impacts of physical and psychological trauma and indirect impacts on the necessities of life (e.g., food, water, shelter, income, access to health care). The impacts depend on population vulnerability, preparedness of the public health system, and society's capacity to respond and recover. The impacts of recent extreme weather and climate events throughout the globe are exemplary of our vulnerabilities. Exposure is increasing as a result of increasing numbers of people at risk. More people are living and working in, and traveling through, areas characterized by high risks of extreme events. Urbanization is concentrating people, and more and more people are living in coastal areas, on flood plains, and on bare mountainsides at risk of mudslides. Weather extremes and disasters in a warmer world are expected to intensify the disturbance to ecological systems and natural resources, thereby exposing more people to vector-borne, food-borne, and water-borne diseases. This chapter examines the implications of extreme weather and climate events on the public health system, and the projected implications of climate change on these events. The public health sector has typically been dealing with disasters in a reactive mode. Two specific case studies – Hurricane Mitch and the Eastern North American Ice Storm – illustrate a range of impacts and vulnerabilities that can be used to draw lessons and to explore effectiveness and viability of adaptive responses. Most public health systems around the world are currently taxed to their maximum, and struggling to deal with emerging health problems. However, the health risks posed by weather and climate extremes can be mitigated through various proactive adaptation strategies. Numerous measures such as early warning systems, disaster preparedness, protective technologies, diversification and redundancy of infrastructural facilities, legislation on land use, public education, planning, coordination among various agencies, training programs to emergency workers and volunteers, and identifying vulnerable groups are highlighted. These can play a significant role in reducing population exposure and vulnerabilities, and health casualties in the wake of extreme weather and climate events.

1. VULNERABILITIES AND IMPACTS OF EXTREME WEATHER AND CLIMATE EVENTS

Extreme weather and climate events[1] have threatened and will continue to threaten human health and security. The toll on individuals and communities from events such as tropical, winter, and wind storms, floods, droughts, tornadoes, extreme rainfall events, dense fog, and lightning continues to grow, with more people now displaced by natural disasters than by conflicts [1]. Extreme heat and cold events are not usually accompanied by visible damage to infrastructure and buildings such as is the case with other extreme events; however, they can account for significant number of morbidity and mortality cases. Periods of poor air quality, primarily the result of human activity, enhanced by the prevailing weather conditions also negatively affect human health and the public health systems. As a result, extreme weather and climate events (hereafter, extreme events) place the public health systems of affected communities, regions, and nations under pressure.

Extreme events will continue to occur; they are part and parcel of climate. The size of the resulting health impacts, however, depends to a large degree on the preparedness of the public health system, its partners (e.g., emergency measures and relief organizations), and its community of responsibility. It is also worth noting that measures not related to public health, for example, legislation on land use, can play a significant part in reducing population exposure and vulnerabilities.

This chapter examines the implications of extreme events for public health systems, the projected implications of climate change on these events, and potential adaptive measures and determinants of adaptive capacity that could play a role in reducing associated vulnerabilities and impacts. Lessons are drawn from specific case studies – Hurricane Mitch and the eastern North American ice storm in 1998 – and the hazards community. These two events have been chosen because they allow for the presentation of a range of impacts and vulnerabilities that can be used to explore the effectiveness and viability of adaptive responses and any lessons learned.

The threats posed by extreme events include those that occur directly from physical or psychological trauma and those that occur indirectly from effects on the necessities of life (e.g., food, water, shelter, income) or on factors that affect human exposure to diseases or unhealthy conditions. Both place increased demands on and requirements for increased access to the public health infrastructure. The ability of a public health system to meet the demands associated with these events needs to be examined in light of the fact that in many areas of the world the health infrastructure is already stretched to its limits dealing with existing health issues and growing populations. The additional demands associated with extreme events further tax and may go beyond that infrastructure's capacity. Should these events become more frequent (more frequent than that which is considered "normal" by the affected community), without introducing further actions to reduce the impacts on human health (i.e., adaptation strategies and measures), the existing public health infrastructure will not be able to meet ongoing or subsequent event-specific demands.

1 Extreme weather and climate events are defined as those that threaten life, public health, and quality of life, or are economically disruptive. These events range from events of small area and short duration (e.g., tornadoes) to more widespread events of longer duration (e.g., floods, droughts, hurricanes).

For direct impacts, the needs are primarily during and immediately after the event, but they can have prolonged implications (especially psychological trauma) requiring continued intervention by the public health system. Examples of direct physical consequences include trauma, heat or cold stress, and impacts on cardiovascular and respiratory health. For extreme climate events (droughts, prolonged periods of rain, or poor air quality), implications for the public health infrastructure grow with the duration and magnitude of the event and can vary depending on the relative adaptive capacity of the affected region (e.g., see Chapter 2 of this volume). Experience with past natural disasters has shown that 15–20% of affected people will experience symptoms of post-traumatic stress disorder [2], with the potential for significant increases in suicide rates many years after the disaster [3].

Indirect impacts, although they can result in demands during the event, primarily place increased burdens on the public health system after the event, and these demands can persist for a prolonged period. Examples of the manner in which extreme events can indirectly lead to health consequences include the following:

Reduced or lack of access:

- Reduced access to, or quality of, food and water supply, resulting in nutritional impairment, famine, dehydration, food poisoning, and water-borne disease.
- Restricted or loss of access to public health facilities – facilities are destroyed, capacity is reduced, transportation routes are disrupted or destroyed.
- Lack of access to or availability of support and relief services because the people responsible for responding, including health care workers, are unable to reach their workplace or are unavailable due to their own injuries or compromised health status.

Increased exposure:

- Loss of shelter, forced human migration, and overcrowding of living facilities – such population displacements can increase the occurrence of infectious diseases [4].
- Exposure to heat and extreme cold.
- Increased incidence of vector-borne diseases such as malaria, dengue fever, encephalitis, hantavirus, and West Nile virus. Extreme events are known to affect the timing and intensity of such disease outbreaks [5]. What is not clear, however, is whether the noted increased incidence after an extreme event is due to increased human exposure or changes in the vector's range [4,6].
- Changes in the geographic distribution of and increased exposure to airborne allergens and water-borne diseases such as cholera, *E. coli*, cryptosporidiosis, shellfish poisoning, and leptospirosis.
- Physical trauma as a result of movement and exposure of landmines and unexploded bombs after floods and associated landslides or erosion.
- Exposure during efforts to clean up after an extreme event (e.g., exposure to molds, chemicals, and human/animal waste, and animals such as snakes and rodents seeking shelter in homes and other buildings).
- Pollution of land and water with toxic chemicals, human and animal wastes, and other pollutants from storage or disposal sites as a result of flooding, heavy rainfalls, and associated landslides/erosion.
- Increased stress after the event as a result of dealing with builders, insurance companies, etc.

Reduced capacity:

- Loss of access to work or loss of a source of income (particularly when dependent on subsidence or natural resources).
- Contamination of land and water by salt-water intrusions from storm surges in coastal and island communities.

The effects of extreme events on an individual or community are a function of exposure, sensitivity, and vulnerability.[2] Disasters occur when the weather or climate hazard converges with a vulnerable population.

Geographic location, climate, and prevailing weather conditions are the primary determinants of exposure. What constitutes an extreme event can vary with location, time of year, and what is considered normal (e.g., snow and cold temperatures can be normal at higher latitudes during winter months but could have a major impact on lower latitude communities). Examples of how these factors can influence exposure include the following:

- Extreme events (e.g., floods in river and delta areas, droughts in midcontinental areas) and the influence of the El Niño cycle are more common in some parts of the world than in others.
- Populations living in low-lying coastal areas are more exposed to the effects of storm surges and waves than those on higher ground.
- Populations living in floodplains and on deforested slopes are more likely to be exposed to mudslides and flooding.
- Dense population areas increase the exposure of larger numbers of people.
- Populations bordering current distributions of vector-borne diseases are more likely to be exposed to changes in distribution.

For human health, however, exposure is also determined by human activities that could contribute to an increased risk of a weather or climate-related disaster with subsequent health consequences:

- Sanitation and chemical facilities, particularly in areas of high risk (e.g., flood plains).
- Intensity of agricultural activities and storage methods for animal wastes.
- Intensity of emissions from industries and forest/biomass burning; populations near and downwind from these emissions are at a greater risk.
- Environmentally degraded and deforested areas and relative location of populations.
- Landmines and unexploded bombs.

A population's sensitivities to extreme weather and climate events are determined by such factors as [7]:

2 Exposure – the nature and degree to which a system is exposed to significant climate variations. Sensitivity – the degree to which a system is affected, either adversely or beneficially, by climate-related stimuli. Vulnerability – the degree to which a system is susceptible to, or unable to cope with, adverse effects and as such a function of the character and magnitude of the climate stimuli to which the system is exposed, its sensitivity, and its adaptive capacity.

- Existing health status: people already in poor health (e.g., those with cardiovascular disease, respiratory impairments) are more sensitive.
- Socioeconomic factors: the poor are more sensitive (and vulnerable) because of a complex mixture of hazards and human actions [4].
- Demographic factors: the elderly and the young are more sensitive (and vulnerable) [8].
- Integrity of water and sanitation systems and their capacity to resist extreme events.
- Infrastructure quality, including housing quality.
- Sensitivity of the local food supplies and distribution system; those dependent on subsidies and nondomestic food supplies are the most sensitive.

In most areas, exposure increases with the number of people at risk. For example, exposure is high where more people are living in, working in, and traveling through areas where there is a high risk of an extreme event. Urbanization is concentrating people and their property. More people are living in coastal areas, on flood plains, and on bare mountainsides at risk of mudslides. More people are traveling, increasing the potential for exposure and the spread of disease. In addition, vulnerabilities are increasing. For example, there is either no or reduced redundancies in our infrastructures (e.g., water, electricity, communications, transportation), enhancing the potential of disruptions that could lead to health consequences. Furthermore, there is complacency or a lack of understanding: many people and communities believe that they are not at risk and that only others will be affected or that the existing safety net will protect them. Together, these changes in demographics, infrastructure, and social beliefs are increasing vulnerability to extreme events.

The term "hydrometeorological" refers to events that include avalanches and landslides, droughts and famines, extreme temperatures, floods, forest and scrub fires, windstorms, insect infestations, and waves and surges, and exclude geological events such as earthquakes and volcanic eruptions [9]. Throughout the globe, the resulting impacts of recent events are exemplary of our vulnerabilities.[3]

- Globally, there were 2,465 reported hydrometeorological disasters from 1992 to 2001, including avalanches, landslides, droughts, famines, extreme temperature events, floods, forest and scrub fires, windstorms, insect infestations, and tidal waves and surges [9].
  - The number of people affected by these disasters was reported at 1.965 billion, with 457,401 deaths. During this period, the year when these disasters affected the largest number of people was 1998 (342.9 million people) [9].
  - The highest number of people reported affected by hydrometeorological disasters was in Asia (1.774 billion people) followed by Africa (136.9 million people) and the Americas (45.2 million people) [9].
  - More people were reported killed or affected by hydrometeorological disasters in countries with low and medium human development indexes versus those with high development indexes. Countries with the highest estimated economic damage due to hydrometeorological disasters were nations with a high human

3 Although our selected case studies focus on ice storms and hurricanes, the numbers here are included to illustrate the impacts of other extreme events not included in the case studies. Impacts of other extreme events can be unique and can require very different responses and adaptations.

development index. However, economic losses, although relatively smaller, were especially devastating to poor countries, where the losses often represent a large share of the national economy [9].
  - Windstorms proved to be the most economically costly hydrometeorological disaster in countries with a high human development index, while floods proved to be the most economically costly in countries with a medium and low human development index.
- In the 1990s, natural disasters like hurricanes, floods, and fires caused over US$608 billion in economic losses (an amount greater than during the previous four decades combined) and affected over 2 billion people world wide [1,10].
- On average, about 50 million people are confronted with the consequences of flooding each year. Between 1971 and 1995, floods affected more than 1.5 billion people around the world; about 318,000 people died and more than 81 million were made homeless [11]. There were 26 “major flood disasters” worldwide in the 1990s, compared with 18 in the 1980s, eight in the 1970s, seven in the 1960s, and six in the 1950s. The economic costs of floods rose to an estimated US$300 billion in the 1990s, up from about US$35 billion in the 1960s. Major floods are those inflicting heavy environmental and economic losses and human casualties.
- Diarrhea, followed by respiratory diseases resulting from overcrowding in shelters, was found to be the most common illness resulting from the catastrophic 1988 floods in Bangladesh. Watery diarrhea was the most common cause of death for all age groups under 45 [12]. In addition, the proportion of severely malnourished children increased after the flooding [13,14].
- Droughts are also growing more severe and widespread as a result of many factors (e.g., changes in land use and climate variability and change), accounting for up to 45% of reported deaths from natural disasters between 1992 and 2001 [11].
- Current air pollution problems are greatest in cities of developing countries. For example, nearly 40,000 people die prematurely every year in India because of outdoor air pollution [15].
- In Canada, from 1995 to 1998, about 7% of road traffic injuries (affecting about 15,000 people) and damage worth about $1 billion Canadian were directly attributable to inclement weather. The resulting health care costs were estimated at over $10 billion Canadian per year.

Because reporting may vary across countries and years, some observed differences may in fact be artifacts (Personal communication, S. Hajat, London School of Hygiene and Tropical Medicine, April 2003).

1.1 Eastern North America ice storm (4–10 January 1998)

Compared with other major natural disasters around the world, the January 1998 ice storm in eastern North America resulted in relatively low traumatic consequences. There is, however, evidence of large-scale implications associated with the public health systems in the affected areas. This ice storm, at the peak of the event, affected 3.6 million people in a line stretching from southwestern Ontario to the Bay of Fundy in Nova Scotia southward into some parts of New York State and New England states [16]. The majority of affected people lost power for one week or longer [16]. One of the reasons it had such an impact was that temperatures plummeted during the days

Table 9.1 Fatalities in Quebec caused by the ice storm between 6 January and 17 March 1998

Cause of fatality	*Number of fatalities*	*Remarks*
Hypothermia	5	
Falling ice from roof	4	De-icing from roof
Burns	10	From candles, overheating from supplementary heating devices
CO poisoning	6	Generator used indoors, propane stoves
Other	5	Hit by snowplow, crushed by falling snow, falling ice during thawing, contact with electrical cable, head injury
Total fatalities	30	

Source: Adapted from Ref. 18.

when power outages affected many people, reaching -20°C or lower in some areas at some points [17], forcing some people into emergency shelters with access to generators, set up to combat the cold.

Health outcomes related to this ice storm included carbon monoxide poisoning; hypothermia; burns; injuries from motor vehicle accidents, slips, and falls; chainsaw accidents; electrocution; head injuries and cuts from falling ice; food poisoning; flu outbreaks; and stress, anxiety, post-traumatic stress disorder, and a variety of other mental health concerns [18]. The ice storm resulted in 30 fatalities in Quebec directly linked to the storm (see Table 9.1).

Groups especially vulnerable for impacts during and after the ice storm were children, women, senior citizens, apartment tenants, families with many children, single parent families, immigrant families, and the infirm [18].

As for the public health infrastructure, hospitals soon became overwhelmed by the storm and its impacts. Hospitals were quickly operating at 100% ward occupancy and emergency departments were overflowing [19], because patients could not be discharged and beds could not be emptied. Ambulances continued to drop off the "normal" number of emergency cases, along with the increased number of storm victims. In addition, without electricity, doctors' offices were forced to close, further burdening and overwhelming the emergency departments of hospitals. These increased pressures on the health care system continued through February in the most affected regions [18].

1.2 Hurricane Mitch (26 October–4 November 1998)

Hurricane Mitch affected most of the population of Honduras and Nicaragua, large tracts of Guatemala and El Salvador, and to a lesser extent Belize and Costa Rica. It caused more than 11,000 deaths and left behind destruction and hundreds of thousands of shattered lives. The poorest were the hardest hit, given their occupancy of high-risk, marginal areas. Further complicating matters was that Mitch struck at a time when the region was still recovering from the effects of El Niño-related droughts, bush fires, and floods.

Further exacerbating impacts on the region, and increasing vulnerability, were then-current government policies and land clearing and agricultural practices. The governments of Honduras and Nicaragua had recently instituted severe cutbacks in public services to fulfill the Structural Adjustment Programs required by international banks in order to receive loans [7]. These cutbacks severely affected spending in health services and transportation, increasing vulnerability to disasters such as Hurricane Mitch. Furthermore, land clearing and agricultural practices in Honduras and Nicaragua led to deforestation and loss of topsoil, rendering the region more vulnerable to the impacts of heavy rains [7]. Large tracts of forested land had been cleared to make way for plantations of cash crops like bananas and coffee, frequently owned by multinational corporations. These corporations pushed the subsistence farmers deeper into the forest, increasing deforestation and loss of topsoil. The loss of forests caused the topsoil to wash out, resulting in the collapse of many hillsides, which destroyed everything in their path [7].

The impacts on the public health infrastructure were devastating and seriously crippled its ability to provide required health services to affected communities. Mitch destroyed or damaged a significant part of that infrastructure (approximately 30%), including health centers and hospitals. Considering the roles in of these facilities in combating communicable diseases, their loss is expected to have long-term adverse health consequences that could cost the affected countries more in loss of human life and economic losses than the immediate impacts of Mitch. In addition, Mitch weakened efforts directed at reform and modernization of the health sector, aggravating inequities in health services.

Further complicating matters and leading to serious health consequences was the destruction of water pipes and sewers, which led to a lack of potable water, contamination of water sources, and the presence of stagnant pools that served as breeding grounds for disease vectors. Access to drinking water and the dramatic deterioration of sanitary conditions and waste control in all affected countries continue to be problematic. Further examples of the impacts on public health include the following [20]:

- Cholera cases jumped from 2,836 in 1998 before Hurricane Mitch to 3,544 in the nine weeks after Mitch, with the average number of weekly cases increasing from 16.4 to 78.7.
- More than 700 leptospirosis cases were reported after Mitch; only one case was reported in 1998 before the hurricane.
- Malaria jumped 48%, from an average of 433 cases a week per country before Mitch to 642 after.
- Dengue rose from 141 cases a week before Mitch to 162 cases per country after the hurricane.

It is unclear whether the increased incidence of these vector-borne diseases resulted from increased exposure to the vector or from changes in the geographic distribution of the vector itself.

2. IMPACTS OF CLIMATE CHANGE ON EXTREME WEATHER AND CLIMATE EVENTS

The climatic norm of a locale is determined by the average, frequency, and extremes of its meteorological observations and weather events, usually taken over a period of 30 years. Not all extreme events end in disasters, and what constitutes an extreme can vary from place to place and season to season. For example, from an impact perspective, a 20-cm snowfall would be an extreme event in Washington, DC, but just an inconvenience in Montreal, Canada [21]. Global climate change is likely to bring not only changes in mean climate conditions but also changes in climate variability and extreme events [23].

Under projected climate change, more hot days and heat waves are very likely over nearly all land areas (as discussed in Chapter 8 of this volume). Precipitation extremes are expected to increase more than the mean, and the intensity of precipitation events is projected to increase. The frequency of extreme precipitation events is projected to increase almost everywhere [23]. A general drying of midcontinental areas is projected during summer. There is little agreement yet among climate models concerning future changes in midlatitude storm intensity, frequency, and variability. The models show no consistent evidence of changes in the frequency of tropical cyclones such as hurricanes and typhoons or in areas of formation. However, some measures of intensities show projected increases, and some theoretical and modeling studies suggest that the upper limit of these intensities could increase. Mean and peak precipitation intensities from tropical cyclones are also likely to increase appreciably [23].

For some other extreme phenomena, there is currently insufficient information to assess recent trends, and confidence in models and understanding are inadequate to make firm projections. In particular, very small-scale phenomena such as thunderstorms, tornadoes, hail, and lightning are not simulated in global models.

Several models show a mean El Niño-like response in the tropical Pacific, with the central and eastern equatorial Pacific sea surface temperatures projected to warm more than the western equatorial Pacific and a corresponding mean eastward shift of precipitation. Current projections show little change or a small increase in amplitude for El Niño events over the 21st century. However, even with little or no change in El Niño amplitude, global warming is likely to lead to greater extremes of drying and heavy rainfall and to increase the risk of droughts and floods that occur with El Niño events in many regions. Also, it is likely that warming associated with increasing greenhouse gas concentrations will increase Asian summer monsoon precipitation variability. The confidence in such projections is limited by how well the climate models simulate the detailed seasonal evolution of the monsoons [23].

2.1 Air quality

Common air pollutants (e.g., ground-level ozone) are known to cause negative health effects. Baseline levels of air pollutants, however, are critical in terms of dose-response, and if they are exceeded as a result of weather and climate extremes are likely to cause greater public health problems. Weather has a major influence on the dispersal and ambient concentrations of air pollutants. Large, high-pressure systems often create an inversion of the normal temperature profile, trapping pollutants in the shallow boundary layer at the Earth's surface. It is difficult to predict the impact of climate change on local urban climatology and, therefore, on average local air pollutant concentrations. Also, the high uncertainty associated with future cloud cover and future human activity (e.g., fossil fuel consumption) makes it even more difficult to project changes in air pollutant concentrations. However, any increase in anticyclonic conditions in summer would tend to increase air pollution in cities. Formation and destruction of ozone are accelerated by increases in temperature and ultraviolet radiation. Existing air quality models used to examine the effect of climate change on ozone concentrations indicate that decreases in stratospheric ozone and elevated temperatures lead to increased concentrations of ground-level ozone [22]. An increase in the frequency of hot days could also increase biogenic and anthropogenic emissions of volatile organic compounds (e.g., from increased evaporative emissions from fuel-injected automobiles). These studies of the impact of climate change on air quality are indicative but by no means definitive. Important local weather factors may not be adequately represented in these models [22].

Climate change is also expected to increase the risk of forest and rangeland fires. Haze-type air pollution therefore is a potential impact of climate change on health.

Radon, an inert radioactive gas, is emitted from the ground at a rate sensitive to temperature. High indoor exposures are associated with an increased risk of lung cancer [24]. There is some evidence from modeling experiments that climate warming may increase radon concentrations in the lower atmosphere [22].

2.2 Floods, tropical storms, and ice storms

The risk of flooding would increase in some regions as a result of a warmer climate [25]. This increase would come mainly from rainstorm floods, with heavier rainfall expected to come from more frequent, and possibly more severe, thunderstorms and from fewer but larger rainstorms associated with large-scale weather systems. Shorter winters, however, may reduce the risk of snowmelt and ice-jam floods in some areas, although heavier snowfalls could add to the risk in others.

The intensity of tropical storms could also be affected by a warmer climate due to the local energy imbalance between the atmosphere and the ocean. There is some evidence that hurricanes could become more intense in some areas [25]. Current general circulation models (GCMs) can simulate some of the characteristics of hurricane behavior; however, these models provide only limited information because hurricanes are too small to simulate in detail.

Hurricane behavior in a warmer world might also depend on what happens to El Niños. Stronger and more frequent El Niños could cause a geographical shift in hurricane activity, with more hurricanes occurring in the Pacific and fewer in the Atlantic. A weakening of El Niño activity, on the other hand, could shift the balance in the other direction.

Regardless of how climate change might affect the frequency or the intensity of hurricanes, the risk of accompanying floods will almost certainly be accentuated by rising sea levels. Since flooding from hurricane storm surges usually causes more damage than wind, the risk of high death tolls and heavy property damage would actually rise even without an increase in hurricane intensity or frequency. The level of damage and loss will also depend on population density, preparedness, and other related response measures.

In a warming climate, milder winter temperatures could possibly cause an increase in freezing rain events in regions where the average daily temperatures begin to fluctuate around the freezing point instead of remaining mostly below it. This could lead to favorable conditions for ice storm development.

2.3 Trends, sensitivity, and vulnerability

The numbers of severe floods, devastating storms, droughts, and other weather disasters that have occurred over recent decades seem to suggest that such events are becoming more common. The likelihood that weather-related disasters are on the rise is also supported by the findings of the International Decade for Natural Disaster Reduction (IDNDR) [10]. It looked at the rates of change for the four largest categories of major natural disasters – floods, tropical storms, droughts, and earthquakes. Between the mid-1960s and the early 1990s, the number of all of these disasters increased, but weather-related disasters increased at a much higher rate [26].

In floods, for example, most of the deaths are due to drowning within the first few hours of the event. More deaths occur later because of complications of initial injuries and disease outbreaks. Mass migrations and temporary living conditions in crowded areas make maintenance of personal hygiene difficult, and can result in widespread diarrhea and respiratory infection. Cholera outbreaks have been associated with precipitation extremes and with both floods and droughts [27–29]. Climate change factors such as rising temperatures, changing precipitation patterns, greater frequency of storms and floods, and sea level rise could all contribute to an increase in cholera incidence in tropical regions.

Flooding may also result in the release of dangerous chemicals from storage sites and waste disposal sites into floodwaters, and the widespread destruction of food and medical supplies. Decline in the nutritional and health status of children is common after floods in poor countries. Extreme rainfall events cause increased runoff from agricultural lands and pastures, and could increase contamination of water with chemicals such as pesticides and deadly bacteria such as *E. coli* O157:H7.

The major impacts of climate change on human health are likely to occur via changes in the magnitude and frequency of extreme events (see Table 9.2) that trigger a natural disaster or emergency. In developed countries, emergency preparedness has greatly reduced the total number of weather-related injuries and deaths, while property damage has been increasing. However, in developing countries, studies indicate an increasing trend in the number of injuries and deaths, as well as other economic losses, due to all types of natural disasters [30,22]. The consequences of climate change will be greater for poor countries. For example, they do not have the means to protect themselves from floods and their economies are based on weather-sensitive sectors (in particular agriculture). Also, within the poorest countries, the poorest people will be the most vulnerable.

By 2025, over half the world's population will be living in areas that are at a heightened risk from storms and other weather extremes. Populations living off the land may be especially vulnerable – the needed financial, technological, institutional, and knowledge-based resources for rapid adaptation are quite limited for these communities [22].

Low-income people are more likely to live in areas with poor housing, have a lower capacity to cope with difficult conditions, have fewer resources to buy adequate food, have a greater reliance on local food sources, and have less access to health care. Poverty often increases with age and could heighten the risk from severe weather. Older people often suffer from chronic illnesses that may increase susceptibility to extreme environmental conditions. Children who live in poor housing conditions, and lack access to education and medical care, are particularly vulnerable to environmental hazards because of their size, behavior, and the fact that they are growing and developing. For example, children breathe more air than adults relative to their body size and thus they tend to be more susceptible to respiratory exposures.

Weather extremes and disasters in a warmer world are expected to intensify disturbances to ecological systems and natural resources, thereby exposing more people to

Table 9.2 Examples of health-related impacts resulting from projected changes in extreme weather and climate events

Examples of projected changes during the 21st century in extreme events	*Examples of projected health-related impacts*
More intense precipitation events	Increased flood, landslide, avalanche, and mudslide. Increased soil erosion. Increased flood runoff could increase recharge of some floodplain aquifers with contaminated water.
Increased summer drying over most midlatitude continental interiors and associated risk of drought	Decreased crop yields. Increased damage to building foundations caused by ground shrinkage, increasing risk of collapse. Decreased water resource quantity and quality. Increased risk of forest fire.
Increase in tropical cyclone peak wind intensities, mean and peak precipitation intensities	Increased risks to human life, risk of infectious disease epidemics, and many other risks. Increased coastal erosion and influx of biohazards such as algal blooms.
Intensified droughts and floods associated with El Niño events in many different regions	Decreased agricultural and rangeland productivity in drought- and flood-prone regions.
Increased Asian summer monsoon precipitation variability	Increase in flood and drought magnitude and damages in temperate and tropical Asia.
Increased intensity of midlatitude storms (little agreement between current models)	Increased risks to human life and health. Increased risk of health infrastructure losses.

Source: Adapted from Ref. 22.

Note
Heat waves excluded. See Chapter 8 of this volume.

vector-borne, food-borne, and water-borne diseases – and the potential for infectious diseases, malnutrition, and hunger. Adequate protection from those stresses is important and depends on access to sanitation, adequate housing conditions, safe drinking water, and proper health care.

3. ADAPTATION AND ADAPTIVE CAPACITY

The health risks posed by weather and climate extremes can be mitigated through various adaptation strategies. The required public health strategies, measures, and policies should provide a balance between effective programs and policies and strong regulatory mechanisms that are reflective of the social, cultural, and economic situation of the target population. They should also reflect the need to facilitate adaptation, particularly for vulnerable groups, and to limit the potential introduction of maladaptation while providing people with the space to pursue their own adaptation strategies.

A balanced adaptation strategy should be based on an understanding of the targeted population's exposure to weather and climate extremes now and as projected under climate change, and their sensitivities and vulnerabilities in the face of these extremes. Such a strategy should also consider past experiences and best practices. Further complicating matters, however, is that all three elements (i.e., exposure, sensitivity, and vulnerability) can vary in both time and space within and across populations and communities, and that future weather and climate extremes may be different in nature from those of the recent past. As such, effective strategies cannot be simply lifted from the past or other locations, or deemed to be effective based on past performance. Effective health adaptation strategies introduced and supported by governments and civil society should include initiatives specifically targeted at vulnerable groups and should be based on a sound understanding of sensitivities and vulnerabilities. In addition, such strategies should be reviewed periodically to ensure that they remain effective and feasible in the context of changes in risks associated with changes in exposure, sensitivities, and vulnerabilities.

3.1 Determinants of adaptive capacity

The ability of a community, nation, or region to cope with and respond effectively to the health impacts resulting from extreme weather and climate events is a measure of its adaptive capacity. Determinants of adaptive capacity include [31]:

- Economic wealth – determined by the level of poverty and access to economic assets, capital resources, and financial means.
- Technology – relative availability and access, current level and ability to develop, openness to development and utilization.
- Information and skills – access to trained and skilled personnel, relative stores of human knowledge, systems for dissemination, and forums for discussion.
- Social infrastructure – availability of and access to support systems and alternatives (e.g., informal support systems).
- Institutions – state of social institutions; existence of constraints, deficiencies, or weaknesses in institutional arrangements and capacities, including the ability to work cooperatively across institutions with similar interests and responsibilities.
- Equity – distribution of resources through social institutions and arrangements governing the allocation of power and access to resources, availability of resources, differential distribution of information, skills, etc.

Recent critics [32] have particularly noted that a public health system's capabilities to respond to emergencies, including those resulting from extreme weather events, are limited by insufficient funding, lack of availability of specially trained emergency health care workers, inequity in the supply of public health promotion and prevention campaigns, and lack of intergovernmental coordination and planning. The fact that most public health systems are currently taxed to their maximum, and struggling to deal with emerging health problems (e.g., AIDS, increased incidence of vector and water-borne diseases), further limits their capacity to deal with the added health impacts of extreme weather and climate events.

For the two case studies (Hurricane Mitch and the ice storm), a rough ranking of dimensions of the adaptive capacity of the affected communities and their public health system is provided in Tables 9.3 and 9.4, respectively. The ranking is on a scale of 1 (the determinant did not play a significant role in enhancing the impacts) to 5 (the determinant played a significant role in determining the impacts). The identified rankings are relative only to that particular case study (i.e., rankings are not relative between the two case studies). The rankings are presented to suggest where efforts could be focused to enhance adaptive capacity.

Table 9.3 Determinants of adaptive capacity – eastern North America ice storm

Determinants of adaptive capacity	*Relative ranking of role in determining impacts*[a]
Range of available technological options[b]	4
Availability and distribution of financial resources[c]	1–2
Structure of critical institutions[d]	2–3
Human capital[e]	1
Social capital[f]	1–2
Access to risk spreading mechanisms[g]	1–2
Access to and ability to manage information[h]	3–4
Perception of attribution	
• Direct physical impacts	1–2
• Psychological and non-direct impacts	3–4

Notes

Scale = 1 (high level of adaptive capacity exists for this determinant and, as a result, this determinant did not play a significant role in enhancing the impacts) to 5 (low-level or no adaptive capacity exists for this determinant and, as a result, this determinant played a significant role in determining the impacts).

a The rankings provided are relative only within the context of this case study and rankings provided for this case study should not be compared numerically with those provided for the other case study. Rankings identify where efforts could be focused to reduce impacts should similar events occur.

b Relative availability and access, current level, ability and openness to development and utilization.

c Level of poverty and access to economic conditions/assets, capital resources and financial means.

d State of social institutions, existence of constraints, deficiencies and/or weaknesses in institutional arrangements and capacities.

e Access to trained and skilled personnel, relative stores of human knowledge.

f Equity, availability of and access to support systems and alternatives and flexibilities such as informal support systems.

g Insurance and availability of relief support.

h Systems for assembling, disseminating and planning, includes coordination and ability to work across institutional structures.

3.2 Lessons learned from the eastern North America ice storm

3.2.1 *Early warning systems*

In general, early warning systems include the following components: a method of warning, the communication of risk, preparedness, and communication to affected populations (Personal communication, R. Pulwarty, University of Colorado, June 2003).

Canada has a well-developed extreme weather alert system to inform the public of the possibility of extreme weather events occurring. Weather alerts range in severity from a weather warning down to a weather advisory and weather watch [33]. Alerts may be issued from 1 to 12 hours in advance, and recommended behaviors vary with the level of alert [33]. For example, in a blizzard warning, people are advised to stay indoors and to stock up on food and heating fuel; in a cold wave advisory, they are advised to dress warmly and exercise caution before going outdoors [33].

Weather alerts are widely broadcast using mass media such as television, Internet, and radio. Broadcasting systems were successful in reaching a wide population during the 1998 ice storm: 79% of surveyed victims remember hearing a weather alert for freezing rain before the storm hit. Environment Canada's website also received twice as many hits during the storm. Public perception of weather forecast accuracy during the ice storm was very good [16]. Alert systems are useful adaptation tools since they

Table 9.4 Determinants of adaptive capacity – Hurricane Mitch

Determinants of adaptive capacity	*Relative ranking of role in determining impacts*[a]
Range of available technological options[b]	2
Availability and distribution of financial resources[c]	3
Structure of critical institutions[d]	4–5
Human capital[e]	2
Social capital[f]	5
Access to risk spreading mechanisms[g]	3
Access to and ability to manage information[h]	5
Perception of attribution	
• Direct physical impacts	1–2
• Psychological and nondirect impacts	4–5

Note

Scale = 1 (high level of adaptive capacity exists for this determinant and as a result this determinant did not play a significant role in enhancing the impacts) to 5 (low-level or no adaptive capacity exists for this determinant and as a result this determinant played a significant role in determining the impacts).

a The rankings provided are relative only within the context of this case study and rankings provided for this case study should not be compared numerically with those provided for the other case study. Rankings identify where efforts could be focused to reduce impacts should similar events occur.

b Relative availability and access, current level, ability and openness to development and utilization.

c Level of poverty and access to economic conditions/assets, capital resources and financial means.

d State of social institutions, existence of constraints, deficiencies and/or weaknesses in institutional arrangements and capacities.

e Access to trained and skilled personnel, relative stores of human knowledge.

f Equity, availability of and access to support systems and alternatives and flexibilities such as informal support systems.

g Insurance and availability of relief support.

h Systems for assembling, disseminating and planning, includes coordination and ability to work across institutional structures.

are designed to lessen the impact of inclement/extreme weather. They also act as a sentinel to emergency medical and disaster mitigation organizations to remain "on call", reducing reaction time and improving response.

3.2.2 *Infrastructural vulnerabilities*

The ice storm highlighted vulnerabilities in water, heating, and electricity infrastructure. In particular, vulnerabilities in the power distribution systems in Ontario and Quebec were identified. Hydro Quebec was attacked for lacking diversification in power sources, and for lacking diversification and redundancy in the electrical grid network [16]. As a result of such heavy reliance on large distances of single transmission lines, when single lines collapsed under the load of accumulated ice, thousands were left without power. The ice storm and the resulting electrical shortage created a situation in which people were more prone to accidents, infections, stress and confusion, and increased health problems [18]. These vulnerabilities served as a strong wakeup call to the government and to utilities, who are addressing these vulnerabilities in present policy.

3.2.3 *Individual and corporate vulnerabilities*

Individuals and businesses were not prepared for an ice storm of such long duration or intensity, and generally failed to take sufficient action to protect themselves. In particular, citizens were not prepared for long periods without electricity [16], which was mostly their sole source of power.

Proposed adaptive measures include a diversification of home power sources to include battery-operated radios, TVs, smoke and carbon monoxide detectors, flashlights, generators, etc. Additional individual measures include stockpiling several days' worth of emergency supplies such as nonperishable food, bottled water, and so on. An especially important adaptive measure is far-reaching public education programs for both the public, for institutions, and for the media, to increase emergency preparedness and coordination and to prevent some of the impacts caused by such an unprecedented event [34].

3.2.4 *Community and corporate response*

Response to the ice storm by government, community and volunteer groups, corporations, and citizens was fast and charitable. Shelters were opened to look after basic needs for affected people and vulnerable groups. Shelters were improvised out of schools, hospitals, community centers, and other public buildings [34]. Some shelters were criticized for being unhygienic, disorganized, and not responsive to the needs of vulnerable populations [18]. In the future, emergency shelters should be better coordinated and planned in order to provide optimal service.

The Red Cross played a significant role in meeting the basic needs of affected people and in fundraising. Several corporations were also responsible for donating cash and items to meet basic needs. The government mobilized 16,000 reserve troops to aid in the response effort [34]. One can postulate that such an effective and timely response had a lot to do with minimizing the number of fatalities due to such a disaster, and that future adaptive measures should include similar rapid mobilizations of resources and capital by government, the corporate sector, and nongovernmental organizations (NGOs).

3.2.5 *Vulnerabilities of the health care sector*

The ice storm highlighted vulnerabilities in affected health care systems, particularly in Quebec, where there was an increased usage of ambulances and increased hospital admissions into emergency, specialized, and trauma departments [18]. Hospitals became instant community focal points, being some of the few places equipped with emergency generators [34]. Generally, hospitals were not prepared for such an event, because ice storm clauses were not explicitly written into their disaster planning frameworks and because the ice storm occurred when the health care systems in Ontario and Quebec were undergoing adjustments due to cutbacks. A lack of coordination existed, and a certain level of improvisation was necessary; however, hospital associations insist that adequate health services were maintained throughout the storm [19,35,36].

Possible adaptive measures for the public health sector include preparing and planning for hospitals, community centers, and schools to be used as emergency shelters in future disasters. Emergency workers and volunteers also require more training in emergency relief.

A lack of coordination between health care systems at the local, municipal, and regional levels was highlighted by the Government of Quebec [18]. Regional emergency measures plans were generally poorly understood. Institutional emergency measures plans were considered out of practice, inadequate for such a disaster, and poorly linked with municipal plans, which were themselves inadequate.

Particular lessons learned and recommended adaptive measures include the following [18]:

- Local level CLSCs (*centres locales de services communautaires,* local community service centers) displayed some confusion in their role as primary health care providers for people affected by disaster due to the overlap of this role with other health care providers. However, effective partnerships with local community groups were formed. Adaptive measures include clarifying responsibilities of CLSCs and community health care providers in disaster response.
- Volunteer services were deemed generally positive and useful during the ice storm. However, volunteer selection by such bodies as municipalities and the Red Cross was deemed questionable. Adaptive measures could include better screening and management of volunteer personnel.
- Difficulties in the public health care system were caused by redundancy in the system, long delays in demands for expertise, and failure to coordinate between CLSCs and municipalities. Adaptive measures include delineating roles and responsibilities and improving coordination between government and health care providers in disaster planning and response.
- There was a lack of coordination between affected territories and regions defined by administrative boundaries. Adaptive measures include improving organizational structure and delineating responsibilities in disaster planning and response.
- Regional decision-makers were criticized as having too much power in decision making during the ice storm. These regional authorities were perceived as too bureaucratic and lacking in knowledge of local realities. Complaints were also made about lack of information, database limitations, and insufficient telecommunications equipment. Adaptive measures include the decentralization of power during disaster planning and response.
- Many pharmacies were forced to close during the first 48 hours of the ice storm because they depended on a functioning network with the Quebec health insurance provider. Pharmacies were forced to adopt older and more time-consuming methods of payment, with resultant considerable delays. Adaptive measures include developing emergency planning by primary services such as pharmacies.
- Because of power constraints, hospitals in affected regions were forced to postpone elective and non-urgent surgeries. They also had diminished diagnostic capacity, medical surveillance, and treatment of complex cases. This prioritization of health services can be deemed an appropriate response to conditions resulting from the ice storm.
- Reliance on supplies of bottled water by hospitals was required during the brief period when water safety was questionable. Bottled water supplies were sufficient and quickly mobilized.

Primary health care providers identified and contacted vulnerable people in affected areas at the start of the ice storm when volunteers, police, and armed forces conducted house checks, which were coordinated using community databases provided by CLSCs. These databases were not complete, and some vulnerable or handicapped individuals were unknown to CLSCs, who failed to contact and aid them during the disaster [18].

Adaptive measures include creating a complete database of community members and identifying vulnerable groups. This registry should also catalogue critical infrastructure, all municipal and community resources, and their locations [34].

Health outcomes resulting from the ice storm were mostly preventable had access to public health warnings been ubiquitous [37]. For example, the large number of carbon monoxide poisonings and house fires could have been prevented had access to the public health department warnings been widespread. Public health department warnings included those against improper use of heating and cooking devices, and the use of generators indoors [18]. Warnings were also issued about roof injuries, falls and fractures, and food poisoning from ingestion of unrefrigerated foods, but incidences of these problems were also elevated [18]. The many health outcomes resulting from these stressors, despite public health warnings through various communication media, calls for improvements in public health communication facilities and strategies during disaster response.

The response of the public health care system was deemed adequate overall, considering the difficult conditions and lack of preparedness of a system forced to improvise during such an unprecedented event. Nevertheless, lessons learned point to the repeatedly glaring lack of coordination and emergency planning in the affected region.

3.2.6 *Psychological impacts*

In a telephone survey, 34% of affected respondents claimed to have had problems with stress during the ice storm, and a survey in the Monteregie region showed that an elevated level of psychological distress was felt by 25% of respondents in heavily affected areas and by 19% of respondents in less affected areas [18].

Despite the failings of emergency plans to account for a large disaster such as an ice storm, communities were still somewhat equipped to respond to psychological impacts of the ice storm thanks to the existing public health infrastructure. Primary health care providers such as CLSCs are responsible for the provision of psychosocial services, and these community-level organizations were heavily used during the disaster. Adaptive measures can include the explicit inclusion of a response to psychological health impacts in emergency health care planning, to further reduce psychological impacts of disasters.

3.3 Lessons learned from Hurricane Mitch

Because of the limited information on this topic in the published literature, this section is based on the conclusions and recommendations from a Pan American Health Organization (PAHO) conference titled "Evaluation of Preparedness and Response to Hurricanes Georges and Mitch", held 16–19 February 1999 in the Dominican Republic [38].

3.3.1 *Early warning systems*

The poor performance of early warning systems in Hurricane Mitch demonstrated that planning for early warning must encompass all participants at local, regional, national, and international levels. Adaptive measures for early warning planning include management by one competent national agency or multiple decentralized regional agencies, and should include participants from vulnerable populations.

Hurricane Mitch highlighted the need for national governments to issue public warnings on all possible media: radio, television, newspapers, and the Internet. Local communities should also be involved in early warning planning to increase awareness and improve response [38].

3.3.2 *Disaster preparedness*

Adaptive measures for disaster preparedness include improved institutional capacity to respond to disasters at the country level. However, disaster preparedness must begin by strengthening at the community level, because this is always the first line of response in a disaster. Communities should become better prepared for disaster through better organization and training.

Communication media play a key role in disaster preparedness in the affected region, and should become more responsible in disaster response by publicizing preventative measures in a straightforward manner. Vulnerable populations and countries working with national and international organizations should develop their own disaster preparedness plans, and countries should foster a "culture" of disaster preparedness and prevention [38]. For example, in the United States, several groups in California have set up earthquake preparedness programs and websites that effectively communicate pertinent earthquake preparedness information to the public. These include the United States Geological Survey (USGS) Earthquake Hazards Program website [39] and the Governor's Office of Emergency Services website [40].

3.3.3 *Coordination*

Efficiency in the collective response of agencies to Hurricane Mitch was often suboptimal because agencies operated in an uncoordinated way. Adaptive measures for improved disaster planning and response include coordination by one designated national institution [38]. It would be interesting to examine what, if any, regional mechanisms for disaster management have emerged since Hurricane Mitch, and whether these would be sustained in the longer term. With regard to the public health sector, such mechanisms could include the sharing of resources, the mobilization of emergency medical care, shared water purification, and better coordination of the emergency response. Accurate information is also an important factor in coordination efforts, and all agencies and institutions must share accurate information systems.

3.3.4 *Public information and mass media*

During Hurricane Mitch, reporters found it very difficult to acquire information from government about the disaster, and national authorities were perceived to have failed at early warning systems. International media published incorrect information on disaster needs and health aspects, emphasizing the need for authorities to maintain contact with meteorological services and mass media. Adaptive measures include training journalists on disaster issues, and training government officials about media relations with regard to disasters. Information transfer mechanisms from official sources to the media must also be established. Access to communication networks must be improved at the local level, and reliable disaster information must be made available at both national and international levels [38].

3.3.5 *Emergency medical care*

Some affected communities were very difficult to reach during Hurricane Mitch, and disaster response was at the community level until other personnel arrived. Adaptive measures include national health and education authorities providing emergency medical training to medical agencies at the community level and at all levels of education. Similar to the ice storm in eastern North America, adaptive measures should also include disaster planning by health authorities at the national and international level in a way that clearly delineates roles and responsibilities.

Foreign medical personnel can be useful in boosting public health infrastructure and disaster response in some developing countries. However, international aid should enter a country only on request, and should include logistical and technical components as well as health components. Field hospitals were not shown to be an effective local response in providing emergency medical care. Local authorities should manage field hospitals, even if the hospital is internationally funded [38].

3.3.6 *Communicable disease surveillance and control*

There was discrepancy amongst information sources on medical surveillance, and data collection instruments for surveillance were not standardized. Baseline surveillance information on disease frequency was not available, preventing the calculation of disease rates from the disaster. Other lessons learned included the realization that coordination was lacking amongst medical teams, different disease case definitions were frequently employed, and there was an inadequate supply of some vaccines.

Adaptive measures include improving access to medical records, collecting and analyzing baseline disease surveillance data, standardizing data collection, effectively using communication media in conveying disease frequency and control method, improving coordination, and requiring that regional PAHO/WHO offices set case definitions and stockpile vaccine reserves [38].

3.3.7 *Food- and water-borne illnesses*

During Hurricane Mitch, drinking water became contaminated with sewage. Cholera was observed in displaced populations, and the incidence of cholera increased after the hurricane [20]. Information was lacking on how to prevent diarrheal illnesses in some affected areas during Mitch, and local trust in information provided by government is lacking. Problems arose also with expired or perishable donated food.

Adaptive measures should include protecting populations at risk and vulnerable populations. Policy and funding decisions should target sanitation and clean drinking water needs. Multidisciplinary health teams should help with emergency and disaster planning, and disaster planning should involve groups at all levels. Surveillance of food and water should be strengthened, as should the infrastructure to construct shelters during disasters. Neighboring countries should share information about diarrheal diseases, and local education on the prevention of diarrheal diseases should be provided [20].

3.3.8 *Vector-borne diseases*

Increases in vector-borne diseases were not observed during Hurricane Mitch, but incidences of malaria, leptospirosis, and dengue fever did increase afterward.

Ongoing epidemiological surveillance systems are crucial in reducing the impact of vector-borne diseases during a disaster.

Adaptive measures include the availability of technical manuals on ways to control vector-borne diseases for all parties who need them. Local health personnel should be trained on endemic manifestations of vector-borne disease; pesticide use should be undertaken based on epidemiological criteria, and should be appropriately managed. More research should be undertaken on the impacts of pesticides on public health and the environment. Entomological surveillance is recommended in the control of vector-borne diseases, and leptospirosis surveillance is recommended in hurricane-prone areas. Adequate solid waste disposal is important in the control of vector populations [38].

3.3.9 *Psychosocial aspects of Hurricane Mitch*

There is little information available on the psychosocial aspects of Hurricane Mitch, because instruments and personnel were not available to assess such impacts. Psychosocial impacts of disasters are generally not addressed in disaster planning, and communication media generally do not convey information on psychosocial impacts.

Adaptive measures could include the development and sharing of technical tools and models to assess the psychosocial impacts from a disaster. Psychosocial impacts can be included into disaster plans and in the health care response to disasters, and communication media should be encouraged to include these impacts. Medical training for disasters should include mental health training, which does not need to be restricted to specialists; local personnel and national teams with experience can be mobilized. Vulnerable groups should be identified and provided with mental health support, and plans should be in place to reduce vulnerability of such groups [38].

3.4 Adaptation strategies for the public health sector

The health risks posed by weather and climate extremes can be mitigated to varying degrees through various proactive adaptation strategies. It has been relatively well documented (e.g., Ref. 22) that improving adaptive capacity can lead to decreased vulnerability to, and subsequent costs from, the impacts of extreme weather and climate events. It should also be kept in mind that, although the costs of adaptation measures can be quite high, the economic, social, and environmental costs associated with a lack of adaptation measures (e.g., greater loss of life and lower quality of life) can be even higher. Thus a proactive approach to adaptation should be integrated into public health decision-making.

Risk management of the health effects associated with extreme weather events requires identifying vulnerable populations and their adaptive capacities, and selecting different strategies appropriate for the local and regional realities. A community's adaptive capacity depends on its technological capability, its institutional arrangements, the availability of financing, and the exchange of information [41].

Although eliminating human health impacts of extreme events may not be possible, it is possible to moderate or minimize the harmful health impacts by addressing the rudimentary causes of these impacts. Further discussion of potential adaptation options is provided in an appendix to this chapter. It is important to remember that after implementation of any of these adaptation options, their effectiveness and sustainability should be reviewed periodically to ensure that they remain effective and feasible in the context of changes in risks associated with changes in exposure, sensitivities, and vulnerabilities.

3.5 A proactive approach to adaptation in public health

It has become evident that it is virtually impossible to prevent weather-related disasters; however, it is possible to minimize their adverse effects. The natural hazards community has demonstrated significant success over the last few decades in dealing with disasters in an increasingly proactive way. In most cases, certain adaptation and mitigation measures are taken to reduce the vulnerability of the system, e.g., by improving and enforcing building codes. Protective technologies such as flood proofing and preventive measures such as relocating communities from floodplains into safer areas are examples of proactive and effective adaptation. Insurance and compensation programs have been widely instituted to share disaster losses and assist in the recovery process.

On the other hand, the public health sector has typically been dealing with disasters in a reactive mode. However, health casualties could be drastically reduced if the health sector were to play a proactive role to ensure the safety of health facilities and public health services, e.g., water supply and sewage systems. When water supplies are interrupted or contaminated in disasters, public health consequences can be severe, in addition to the high economic costs of reconstruction. Similarly, when hospitals and medical supplies are damaged, adverse health impacts rise significantly.

A community's vulnerability to natural disasters increases as a result of one or more of the following factors [42]: 1) lack of social resources (e.g., income, social support, assets); 2) lack of access to health services, information, etc.; 3) exposure to hazards; and 4) reduced capacity to cope and recover. A successful disaster mitigation program begins with identifying the vulnerabilities in the systems and establishes priorities for retrofitting or repair. The public health sector can complement disaster preparedness proactively in these four areas of vulnerability.

3.5.1 *Social resources*

Higher social and economic status is generally known to be associated with better health. Higher income determines living conditions such as safe housing and the ability to buy sufficient food and medicine. The healthiest populations are those in societies that are prosperous and have an equitable distribution of wealth. While the public health sector cannot influence wealth, it can contribute to improved social support.

The public health sector can promote social support networks (families, friends, and communities), which could be very important in helping people cope with the adverse effects of disasters and maintain a sense of control over the situation. Caring and respect in social relationships seem to act as a buffer against health problems. The importance of social support also extends to the broader community. Civic vitality refers to the strength of social networks within a community, region, or country. It is reflected in the institutions, organizations, and informal practices that people create to share resources and build attachments with others. Social stability, good working relationships, and cohesive communities provide a supportive society that reduces many potential health risks.

3.5.2 *Knowledge, information, and health services*

Education gives people knowledge and skills for problem solving, and it improves their ability to access and understand information to help keep them healthy before, during, and after natural disasters. Effective education for children and lifelong learning for adults are key contributors to health and prosperity for individuals and for society. The public health sector should work with other government agencies and private institutions to enhance public education.

Examples of other proactive activities related to knowledge, information, and health services include the following:

- Improve communication between communities at risk before, during, and after a disaster (e.g., coordinating between public health agencies and other key response organizations to streamline communication procedures, exploring technological alternatives for improved data retrieval, and developing databases about natural hazards and information about regional and international resources available for immediate emergency assistance).
- Enhance health services, particularly those designed to maintain and promote health, to prevent disease, and to restore health and function.
- Enhance collaboration between the public health community and the climate/meteorology community on examining the impacts and vulnerabilities associated with extreme events and on developing collaborative adaptive measures (e.g., enhanced warning and preparedness efforts).
- Integrate key emergency preparedness principles and procedures into ongoing public and primary health programs (e.g., vaccination programs, environmental health, and public health surveillance).
- Improve collaboration on preparedness and response (e.g., strengthening relations between health programs and other sectors involved in emergency preparedness).
- Strengthen human resources and build institutional capacity, e.g., incorporate key principles of emergency preparedness and response into the curricula of institutions such as schools of medicine and public health.
- Organize disaster simulation exercises to test the effectiveness of response mechanisms.
- Develop advanced early warning systems that include the following components: a method of warning, the communication of risk, preparedness, and communication to affected populations.
- Improve technology- and information-transfer strategies.

3.5.3 *Exposure to hazards*

The physical environment is an important determinant of health. Certain levels of exposure to weather hazards or contaminants in the air, water, food, and soil can cause a variety of adverse health effects. In the built environment, factors related to housing, indoor air quality, and the design of communities and transportation systems can significantly influence physical and psychological well-being. The public health sector should therefore:

- Work closely with other specialized institutions (e.g., meteorological and hydrological services) to identify areas exposed to natural hazards and determine the vulnerability of key health facilities and water systems.

- Conduct community-based epidemiologic research, e.g., by developing models that predict the public's vulnerability to different types of weather and climate disasters or identifying populations at increased risk from disasters.

3.5.4 *Coping capacity*

Some community groups may be marginalized because of age (e.g., children, elderly), physical disabilities, gender, language, culture, or social status. These groups need increased attention and recognition from the public health sector to reduce their vulnerability.

Gender refers to the array of society-determined roles, personality traits, attitudes, behaviors, values, relative power, and influence that society ascribes to the two sexes on a differential basis. Many health issues are a function of gender-based social status or roles. In developing countries, for example, women are usually more vulnerable to weather disasters than men, and the public health sector can play a proactive role in closing this gap.

Some persons or groups may face additional health risks due to the socioeconomic environment, which is largely determined by dominant cultural values that contribute to the perpetuation of conditions such as marginalization, stigmatization, loss or devaluation of language and culture, and lack of access to culturally appropriate health care and services.

The public health sector should also promote personal health practices and coping skills, which can lead individuals to actions that help them prevent diseases and promote self-care, cope with disasters, develop self-reliance, and make choices that enhance health.

APPENDIX: POTENTIAL ADAPTATIONS TO REDUCE HUMAN HEALTH IMPACTS DIRECTLY AND INDIRECTLY ASSOCIATED WITH EXTREME WEATHER AND CLIMATE EVENTS

A1. Public education and communication

Behavioral adaptation is an important preventative measure that can reduce the risks and health impacts of weather and climate extremes substantially. Some examples are as follows:

- Provide public education based on sound science giving people information on the risks associated with prevalent extreme events [37].
- Provide information on actions that would reduce exposure before, during, and after extreme weather events, e.g., eliminating vector breeding grounds, boiling water, vaccinating.
- Enhance understanding of vulnerability risk factors.
- Publicize actions to take in preparation for and during severe weather events [37], such as stockpiling nonperishable food, using battery-operated equipment, and avoiding hazardous behavior.
- Motivate donor organizations to prepare lists of services and products, and plans for timely delivery.

- Broadcast actions and precautions to take during smog events and other pollution episodes, e.g., reduce exertion levels and energy consumption, stay indoors, or car pool.
- Provide training to medical and emergency staff as well as volunteer groups.
- Increase coordination and information exchange between authorities at all levels (local, regional, national, and intergovernmental).
- Issue timely weather advisories and warnings, complemented with health advisories.

A2. Surveillance and monitoring

- Improve the linkages between the meteorological and health communities. This includes building the capacity of the meteorological and hydrometeorological community and supporting their engagement/outreach activities with the health and emergency measures community.
- Improve extreme weather and climate advisories and early warning capabilities.
- Enhance the quantity and availability of good quantitative data on impacts of extreme events (both direct and indirect) to help in better understanding of the full range of potential health impacts. This raises the question of whether there is need for a centralized and systematic national reporting of outcomes from extreme events, potentially using a standardized method (improved coordination with the emergency measures organizations).
- Develop health early warning systems (HEWSs). HEWSs involve collaboration between meteorological and health sectors, and enable regions to be better prepared for disease outbreaks influenced by weather and climate events. For example, meteorologists are able to predict El Niño events months in advance, and HEWSs would anticipate increased incidence of vector-borne diseases and respond with vector control programs before any outbreak occurred [43]. The effectiveness of early warning systems in general needs to be developed further and evaluated.
- Increase health surveillance and the capability to act on the concerns identified, such as monitoring and control of infectious diseases.
- Prepare registries of individuals and groups that are seen as vulnerable, at high risk, or of special needs, e.g., children, elderly people with pre-existing health conditions, socially isolated and low-income individuals, immunocompromised people.
- Institute early detection of health outcomes from extreme weather events. This entails improving health data collection and analysis.
- Monitor coastal communities and floodplains for impacts of sea level rise.
- Identify critical and hazardous infrastructure.
- Establish smog/air quality monitoring and advisory systems.
- Establish systems for reporting and repair of polluting vehicles and other polluting sources.
- Foster adaptive strategies in municipalities with direct objectives to reduce or eliminate large-scale potential health problems. Urban areas are especially vulnerable to weather-related disasters due to high population density and concentration of infrastructure.

A3. Ecosystem intervention

- Improve land use planning – institute restrictive land use zoning, including limiting development or encroachment in high risk areas such as flood plains and sensitive coastal areas.
- Foster environmental management, e.g., use of defensive structures, tighter zoning laws and building codes in vulnerable areas, and management of flood-containing wetlands.
- Manage and create green space and plant trees in vulnerable areas.
- Reduce spraying of parks and farmland with harmful chemicals.

A4. Infrastructure development

- Develop national and international cost sharing mechanisms, e.g., insurance, for compensation and recovery to reduce post-event mental and economic stresses.
- Enhance infrastructure for effective interventions by relief organizations, including food, water, and medical support.
- Create or enhance the local and national disaster preparedness capabilities that build on a strong and integrated health and emergency measures capacity, that provide for access to health facilities during extreme events, and that recognize and provide for links to national and international capabilities.
- Develop sanitation programs to deal with human and animal wastes, addressing the need to minimize risks of release into the environment.
- Improve housing insulation and placement of housing away from flood-prone areas, and construct coastal barriers against tidal waves and surges.
- Develop emergency response plans.
- Maintain public shelters and evacuation plans.
- Maintain dams, floodplains, and storm runoff capabilities.
- Improve public transit systems and bicycle lanes to reduce the potential of pollution episodes.
- Develop incentive programs for citizens, households, communities, and corporations to reduce emissions and energy consumption.

A5. Technological/engineering

- Develop protective technologies: hard (sea walls, dams, dykes) and soft (marshes, wet lands, natural buffers, etc.) to reduce the potential for floods. Recognize, however, that these protective technologies, in addition to minimizing flood potential for events to which they are designed, can also lead to increased exposure (i.e., give people a false sense of security) and impacts for events that exceed the design limits [44].
- Enhance water resource management, including enhanced efforts directed at water purification and water conservation.
- Improve housing (establish, modify, and enforce building codes).
- Improve water and pollution control, including fortifying sanitation systems and introducing water purification.
- Increase redundancy, efficiency, and resilience of power supply grids.

- Strengthen and enforce building codes and standards.
- Develop alternative (clean) fuels and zero-emission vehicles.
- Reduce emissions from fossil-fuel power generating stations.

A6. Medical intervention

Public health capacity will determine whether or not a region is equipped to effectively respond to a particular extreme weather event. For example,

- Develop new and effective drugs and vaccines. For example, in Hurricane Mitch, effective malaria drugs or vaccines would have been very useful.
- Conduct public immunization campaigns.
- Improve medical care services, including adequate medical supplies for injuries and illness.
- Maintain disaster preparedness programs, including tools for local public-health facilities to provide rapid health needs.
- Enlist and train volunteers to be recruited during an emergency.
- Include climate change projections in health planning.
- Determine new air quality guidelines and standards to protect human health.
- Increase public-health staff and services during pollution episodes and other emergences.

A7. Research

The predicted health impacts of weather extremes are still clouded with various degrees of scientific uncertainty with regard to their nature and magnitude [41]. This uncertainty is caused by knowledge gaps, which may lead to limited effectiveness of adaptation measures. Therefore, further research is needed to improve our understanding of the underlying causes of vulnerability leading to national and subnational indicators of vulnerabilities, including relative exposure, sensitivities, and adaptive capacity [6,45]. For example:

- Identify vulnerabilities (relative exposure, impacts, and adaptive capacity) associated with extremes and how projected changes in climate could affect those vulnerabilities.
- Evaluate the relationships between extremes in temperature, humidity, and other weather variables and air pollution related health vulnerabilities.
- Identify and evaluate the relationships between weather and climate extremes and vector-borne diseases (e.g., malaria, hantavirus, leptospirosis, dengue, St. Louis encephalitis), especially links between climatological and ecological changes that create the necessary conditions for disease spread.
- Examine water and food contamination in flood and high precipitation events and associated water and food-borne illness.
- Evaluate how changes in the hydrological cycle, water temperatures, frequency of extremes, sea level rise, and land-use changes can affect waterborne diseases.
- Identify and analyze psychological and sociological impacts of extremes (e.g., post-traumatic stress disorder) in relationship to potential impacts on public health access.

- Identify and assess the effectiveness of current adaptive strategies to deal with provision of public health services in extreme events.
- Examine the effectiveness of adaptation strategies (vector management strategies and public health preparedness) under projected changes in climate and changes in the frequency and intensity of extreme events.
- Develop the tools and data needs to prepare for and evaluate the effectiveness of adaptation strategies in responding to future events.
- Evaluate the utility and costs and benefits associated with enhancing weather and climate monitoring and surveillance activities and enhancing the extreme weather warning systems.

REFERENCES

1. Abramovitz, J.N.: *Unnatural Disasters*. Worldwatch Paper 158. Worldwatch Institute, Washington, DC, 2001.
2. Steinglass, P. and Gerrity, E.: Natural Disasters and Post-Traumatic Stress Disorder: Short-Term versus Long-Term Recovery in Two Disaster-Affected Communities. *J. Appl. Soc. Psych.* 20 (1990), pp.1746–1765.
3. Krug, E.G., Kresnow, M., Peddicord, J.P., Dahlberg, L.L., Powell, K.E., Crosby, A.E., and Annest, J.L.: 1998 Suicide after Natural Disasters. *N. Engl. J. Med.* 338 (1998), pp.373–378.
4. McMichael, A.J. and Kovats, S.R.: Climate Change and Climate Variability: Adaptations to Reduce Adverse Health Impacts. *Environ. Monit. Assess.* 61 (2000), pp.49–64.
5. Epstein, P.R.: Chapter 7: Detecting the Infectious Disease Consequences of Climate Change and Extreme Weather Events. In: P. Martens and A.J. McMichael (eds): *Environmental Change, Climate and Health: Issues and Research Methods*. Cambridge University Press, Cambridge, UK, 2002, pp.172–196.
6. NRC: *Under the Weather: Climate, Ecosystems, and Infectious Disease*. Committee on Climate, Ecosystems, Infectious Disease, and Human Health. National Research Council. National Academy Press, Washington, DC, 2001.
7. Comfort, L., Wisner, B., Cutter, S.L., Pulwarty, R., Hewitt, K., Oliver-Smith, A., Weiner, J., Fordham, M., Peacock, W., and Krimgold, F.: Reframing Disaster Policy: The Global Evolution of Vulnerable Communities. *Environ. Haz.* 1 (1999), pp.39–44.
8. Blaikie, P., Cannon, T., Davis, I., and Wisner, B.: *At Risk: Natural Hazards, People's Vulnerability and Disasters*. Second edition. Routledge, UK, 2003.
9. International Federation of Red Cross and Red Crescent Societies: *World Disasters Report: Focus on Reducing Risk*. Kumarian Press, Bloomfield, CT, 2002.
10. IDNDR: IDNDR Archives, International Decade for Natural Disaster Reduction. http://www.unisdr.org/unisdr/idndrarchive.htm. Accessed August 5, 2003.
11. World Water Council: Report on the World Water Forum 3, Water and Climate Theme. http://www.world.water-forum3.com. Accessed August 22, 2003.
12. Siddique, A.K., Baqui, A.H., Eusof, A., and Zaman, K.: 1988 Floods in Bangladesh: Pattern of Illness and Cause of Death. *J. Diarrhoeal Dis. Res.* 9 (1991), pp.310–314.
13. Woodruff, B.A., Toole, M.F., Rodrigue, D.C., Brink, E.W., Mahgoub, E., Ahmed, M.M., and Babikar, A.: Disease Surveillance and Control after a Flood: Khartoum, Sudan, 1988. *Disasters* 14 (1990), pp.151–163.
14. Choudhury, A.Y. and Bhuiya, A.: Effects of Biosocial Variables on Changes in Nutritional Status of Rural Bangladeshi Children Pre- and Post-Monsoon Flooding. *J. Biosoc. Sci.* 25 (1993), pp.351–357.
15. World Bank: *Clear Water, Blue Skies*. World Bank, Washington, DC, 1997.

16. Kerry, M., Kelk, G., Etkin, D., Burton, I., and Kalhok, S.: Glazed Over: Canada Copes with the Ice Storm of 1998. *Environment* 41 (1999), pp.6–33.
17. Environment Canada: Weather Office, Climate Data Online. http://climate.weatheroffice.ec.gc.ca/climateData/canada_e.html. Accessed 5 August 2003.
18. Government of Quebec: *Etudes sectorielles du rapport de la Commission scientifique et technique chargee d'analyser les evenements relatifs a la tempete de verglas survenue du 5 au 9 janvier 1998*. Volume 2: Les impacts sociaux, economiques et environnementaux, (Sectoral studies from the scientific and technical commission in charge of analyzing the effects of the January 5–9th 1998 Ice Storm, Volume 2: Social, economic and environmental impacts). Les publications du Quebec, Sainte-Foy, 1999.
19. Fyfe, S. and Purcell, M.: Ice Storm '98: A Case Study in Response. Queen's University, Kingston, Ontario, 1998, CD-ROM.
20. PAHO: Infectious Diseases with the Greatest Epidemiological Risk in the Post-Mitch Period: Central American Countries, 1998. Pan American Health Organization. http://www.paho.org/English/HCP/HCT/EER/hctmitch.htm. Accessed 29 April 2004.
21. Francis, D. and Hengeveld, H.: Extreme Weather and Climate Change. *Climate Change Digest 98-01*. Environment Canada, Downsview, Ontario, 1998.
22. IPCC: *Climate Change 2001: Impacts, Adaptation, and Vulnerability*. Intergovernmental Panel on Climate Change. Cambridge University Press, Cambridge, UK, 2001.
23. IPCC: *Climate Change 2001: The Scientific Basis. Summary for Policy Makers and Technical Summary of the Working Group I Report*. Cambridge University Press, Cambridge, UK, 2001.
24. U.S. EPA: Indoor Air – Radon (Rn). 2003. http://www.epa.gov/iaq/radon/. Accessed April 29, 2004.
25. UNFCCC: Climate Change Information kit. 2001. http://unfccc.int/resource/iuckit/infokit2001.html. Accessed 29 April 2004.
26. McCulloch, J. and Etkin, D. (eds): *Proceedings of a Workshop on Improving Responses to Atmospheric Extremes: The Role of Insurance and Compensation*. Environment Canada, Downsview, Ontario, 1993.
27. Rowland, M.G.: The Gambia and Bangladesh: The Seasons and Diarrhea. *Dialogue Diarrhoea* 26 (1986), p.3.
28. Bradley, M., Shakespeare, R., Ruwende, A., Woolhouse, M.E.J., Mason, E., and Munatsi, A.: Epidemiological Features of Epidemic Cholera (El Tor) in Zimbabwe. *T. Roy. Soc. Trop. Med. Hyg.* 90 (1996), pp.378–382.
29. Sur, D., Dutta, P., Nair, G.B., and Bhattacharya, S.K.: Severe Cholera Outbreak Following Floods in a Northern District of West Bengal. *Indian J. Med. Res.* 112 (2000), pp.178–182.
30. Munich Re: Topics 2000 – Natural Catastrophes, The Current Position. Munich Reinsurance Group, Munich, Germany, Geoscience Research Group, 1999.
31. IPPC TAR WGII: Adaptation to Climate Change in the Context of Sustainable Development and Equity. Chapter 18 in: *Climate Change 2001: Impacts, Adaptation, and Vulnerability*. Cambridge University Press, Cambridge, UK, 2002.
32. CanWest News Service: Canada Unready for Health Emergencies, 10 March 2003, Mark Kennedy, Ottawa, Ontario, Canada. http://www.canada.com/search/story.aspx?id=8af3be02-2223-45a6-aa6c-8ed54c139994. Accessed 11 March 2003.
33. Environment Canada: Weather Watches, Warnings and Advisories. 2002. http://www.msc.ec.gc.ca/cd/brochures/warning_e.cfm. Accessed 29 July 2003.
34. Kalhok, S. and Maarouf, A.: An Overview of the Health Effects of the Great Ice Storm of 1998. Unpublished manuscript. Environment Canada, Downsview, Ontario, 1999.
35. Jelowicki, A.: January Death Rate Soared: Montérégie Health Board Says it did its Job During Disaster. *The Montreal Gazette* 10 September 1998.
36. OHA: Beacons in the Storm: Ice Storm 1998. Ontario Hospital Association, Ottawa, 1998.

37. Mileti, D.S.: *Disasters by Design: A Reassessment of Natural Hazards in the United States*. Joseph Henry Press, Washington, DC, 1999.
38. PAHO: Conclusions and Recommendations of the Meeting on Evaluation of Preparedness and Response to Hurricanes Georges and Mitch. Emergency Preparedness and Disaster Relief Coordination Program, Washington, DC. Pan American Health Organization, 1999. http://www.paho.org/English/PED/concleng.htm. Accessed 29 April 2004.
39. USGS: Earthquake Hazards Program, Earthquake Preparedness Information. United States Geological Survey, 2003. http://earthquake.usgs.gov/hazards/prepare.html. Accessed 29 July 2003.
40. Governor's Office of Emergency Services: Earthquake Preparedness Information-Tip Sheets. 2003. http://www.oes.ca.gov/CEPM2002.nsf/htmlmedia/pdfs.html. Accessed 29 July 2003.
41. Health Canada: Climate Change and Health and Well-Being: A Policy Primer. Climate Change and Health Office, Safe Environments Programme, Healthy Environments and Consumer Safety Branch, Minister of Public Works and Government Services Canada, Ottawa, 2001.
42. McMichael, A.J., Haines, A., Slooff, R., and Kovats, S. (eds): *Climate Change and Human Health*. World Health Organization, Geneva, 1996.
43. WHO: Climate and Health: El Niño Factsheet. World Health Organization, 2003. http://www.who.int/peh/climate/elnino.htm. Accessed 29 July 2003.
44. Etkin, D.: Risk Transference and Related Trends: Driving Forces Towards More Mega-Disasters. *Environ. Haz.* 1 (1999), pp.69–75.
45. Brooks, N. and Adger, W.N.: Country Level Risk Measures of Climate-Related Natural Disasters and Implications for Adaptation to Climate Change. Work Paper 26. Tyndall Centre for Climate Change Research, University of East Anglia, UK, 2003.

10 Disease surveillance in the context of climate stressors: needs and opportunities

Mark L. Wilson and Martha Anker

ABSTRACT: Our capacity to adapt to future health impacts from climate stressors is related to information, knowledge, and understanding of present disease patterns. Surveillance that detects and describes such time-space patterns allows us to identify and respond to changing trends and unusual events. Surveillance data that permit recognition, confirmation, and response are most useful if there is timely reporting, synthesis, analysis, and dissemination to those who can take action. In addition, careful analysis of such data should improve the understanding of climate and disease links, permitting development of new epidemiological models that include climate as a predictive variable, which could improve adaptive capacity. Increasing global integration and sharing of disease data would enhance international health security and epidemic response capacity for existing climate-related diseases as well as those that are newly emerging.

1. INTRODUCTION

Disease and weather surveillance is a critical component of efforts to measure, recognize, evaluate, anticipate, and respond to climate effects on health. Without such activities and resulting information, analyzing disease patterns, identifying unusual events, and determining appropriate interventions would be impossible.

The data required to anticipate climate effects and to recognize that changing patterns of disease are related to climate are generally long time series of both epidemiological and climate data collected in comparable temporal and spatial units. Careful analysis of such data enables the understanding of the climate and disease links, and the development of epidemiological models that include climate variables. Such models are in various stages of development and evaluation. Once completed, these models should be able to describe weather conditions that increase the risk of outbreaks of specific diseases. The ability to accurately forecast weather conditions months in advance may enable prediction of higher disease risk, and will allow steps to be taken to prevent or minimize such outbreaks. These models are still in the beginning phases of development, and considerable work will be required before their potential can be fully realized.

Data useful to recognize and respond to outbreaks already occurring rely on timely reporting, synthesis, analysis, and dissemination to those who can take action. Thus data required to anticipate climate effects are somewhat different from those to detect and monitor outbreaks, and the differences are not always well understood.

Public health surveillance has been defined as "the continuing scrutiny of all aspects of the occurrence and spread of disease that are pertinent to effective control" [1]. Such surveillance is the means by which public health departments, whether local, regional, national, or global, keep informed about the health status of the populations they are designed to serve, and about real and potential problems they may face. This surveillance involves not only systematically collecting health and disease information but also interpreting and distributing the information to others so that informed public health decisions can be made. Additional surveillance of risk factors (behavioral, socioeconomic) and potential exposures (environmental, occupational) is often critical to interpreting disease data. This chapter primarily addresses disease surveillance, while recognizing that data on weather and other exposures are critical to understanding how adaptation to environmental change is best planned and implemented. Attention is focused on gathering accurate information, but modern surveillance systems also attempt to ensure timely information flow to surveillance personnel, policy makers, and health providers, thereby enhancing decision-making. It is also considered critical that the public be fully informed.

The applied goals of health surveillance are guided by the need to make public health decisions in a context of both constant vigilance and pragmatic responses to unusual events. This may lead to innovative and opportunistic data collection that provides the basis for timely decision-making and more effective treatment or prevention responses. Surveillance data that are not systematically gathered sometimes fall short of what is needed for rigorous scientific research. However, surveillance data are used for a large variety of public health purposes such as monitoring levels and trends in disease occurrence, characterizing geographical spread of disease over time, detecting and investigating outbreaks as they occur, recognizing new strains of pathogens, and anticipating outbreaks before they arise. In many cases, however, surveillance is used to recognize an outbreak well after it has begun – there are not as many early warning systems as we would like. Indeed, surveillance data are also important in monitoring the development of antimicrobial resistant strains, evaluating disease control programs, estimating potential heath care needs, contributing to analyses of the natural history of disease or associated risk factors, and identifying future research objectives. In short, surveillance data are critical to public health prevention and response efforts, but only recently have they been used to understand how weather and climate might affect disease patterns.

2. FEATURES OF SURVEILLANCE

Disease surveillance data have many unique features that stem from the role that surveillance activities play in supporting and informing public health decisions. First, surveillance is complex and multidisciplinary, involving the skills of demographers, statisticians, physicians, entomologists, microbiologists, social scientists, etc. The many sources of surveillance data suggest that coordination, communication, and quality control are important. A critical issue, discussed below, is whether the various diverse surveillance activities are undertaken at comparable spatial and temporal resolutions.

Second, rapid reporting of disease surveillance data has become increasingly important, because information may be needed quickly to control an outbreak or an impending epidemic. This differs from other kinds of disease surveillance data, collected for epidemiological research purposes and focused on understanding conditions influencing endemic patterns. Increasingly, it has become important that standard surveillance data are rapidly transmitted to a central agency that combines, analyses, and interprets such information. This has been critical, for example, in many food-borne disease epidemics and the recent outbreak and spread of Severe Acute Respiratory Syndrome (SARS). New computer- and internet-based systems are being developed to rapidly transmit data and analyses to appropriate authorities and practitioners.

Third, disease surveillance is accompanied by a legal and ethical framework that derives from the broad goal of protecting the health of the public. However, this creates constraints on the data gathering process that are intended to protect the privacy and civil liberties of individuals. However, apart from the International Health Regulations (discussed in the appendix to this chapter), there are no globally accepted "rules and regulations" for gathering surveillance data, and the context and practice of surveillance vary greatly from one country to another. Obtaining informed consent while undertaking epidemiological surveillance, determining contacts during an outbreak, and obtaining authority to quarantine people under particular circumstances are but a few examples of how such concern affects surveillance and response.

Fourth, the cost of surveillance, both monetarily and in terms of effects on personnel in health care facilities, plays an important role in determining the type and quality of data that are collected. Complete reporting on all cases of every disease is never possible. Therefore, good surveillance depends partly on setting priorities and making wise, strategic decisions about what data should be collected, under which circumstances, and with what intended accuracy. The intended application of such surveillance knowledge will influence such decisions.

Fifth, a carefully designed system of disease classification is critical for any surveillance activity. Standard case definitions influence data quality and comparability. Case definitions are commonly based on symptoms and signs of disease, and may also require laboratory confirmation – especially for diseases that have few if any distinct symptoms and signs, or for several diseases with similar signs and symptoms. These case definitions may vary from country to country, even within countries. The World Health Organization (WHO) regularly compiles a list of recommended standard case definitions and surveillance standards that should serve as the basis of case definitions for all countries [2]. Case definitions change over time, however, as more is learned about specific diseases and as diagnostic methods develop. When using surveillance data, it is important to know what case definitions were used, and whether there have been any changes over the period of interest. These changes might be reflected in patterns of observed levels and trends. Unfortunately, case definitions may be difficult to obtain, particularly for data collected in the past. In addition, the level to which case definitions have been strictly adhered to is rarely evaluated.

Sixth, the laboratory component of a surveillance system is crucial for verifying the existence of particular pathogens, especially for those diseases that are difficult to diagnose with certainty solely on clinical grounds. Laboratory data are important for identifying particular disease strains as well as susceptibility and resistance to particular treatments. Accordingly, quality assurance programs are key to maintaining accurate determination. Logistical problems often hamper acquisition of specimens in good condition, and inadequate laboratory capacity may obstruct identification of common and uncommon pathogens. As discussed in the appendix, surveillance systems are complex and typically involve many components that must be coordinated to be effective.

3. SURVEILLANCE APPLICATIONS, LIMITATIONS, AND OPPORTUNITIES

As described in the appendix, disease surveillance methods are used to find, observe, measure, count, and summarize cases or sources. The resulting data provide critical information that should be useful in identifying, analyzing, and understanding public health problems, and in responding to disease events. Such surveillance data and analysis assist many population-level responses and represent one of the cornerstones of disease prevention and health promotion. How, when, and where such information is used varies with its completeness, accuracy, analysis, and interpretation. Critical to successful application are the response and research capacity characteristics in each governmental and cultural context. These issues are discussed below in the context of links to climate-related risk and response.

3.1 Timeliness

Timely surveillance of disease, and of climate-linked anomalies, is critical to an infrastructure of prevention that is based on unusual events. However, timely information on both known and suspected risks is not always available, and epidemics in both developed and developing countries may go undetected for long periods of time. The deaths due to heat waves in Europe in 2003 illustrate this point well. It was not until the heat wave was almost over that the unusually large number of deaths to elderly people was noticed. More needs to be done to sensitize the health infrastructure to potential climate-related problems, if they are to recognize them and take action in a timely manner.

3.2 Coverage and quality of surveillance

The coverage and quality of surveillance in developed countries are adequate for many purposes; however, during recent years there has been a renewed understanding of how important good surveillance is to health security. Surveillance in many developing countries is, not surprisingly, poor and of great concern. Even though the health of many people in developing countries may be most affected by climate variability and change, surveillance data would at least be able to anticipate or document such possible effects. The infrastructure in such settings is often inadequate, and the capacity to respond is typically weak, even if early warning of climate-related risk were available. In part this is due to inadequate surveillance, particularly so if the goal is to focus responses on the most vulnerable. Yet even if surveillance were adequate, most models suggest that impacts of climate variability and change will be felt hardest in countries least able to identify and respond to such changes. This points to an even greater need for disease and climate surveillance in developing countries, and to the importance of data dissemination and analysis so as to be cognizant of climate-related risks to health. It could be argued that this is a rational allocation of very limited resources, since most economic analyses of intervention operating through prevention of disease demonstrate that it is more efficacious than treatment after cases have appeared.

3.3 Response capacity given adequate surveillance

Even when surveillance data are quickly and accurately reported, analyzed, and distributed, effective response depends on adequate resources to support appropriate intervention. This includes ample and accessible funding for the efforts and adequately trained personnel who can be quickly deployed to the epidemic area. Health providers and disease control personnel are needed to swiftly treat cases and reduce transmission. It is understood that focused and intensive active surveillance should be initiated to determine whether the epidemic is expanding in space or increasing over time. Equally important, however, are competent and committed epidemiologists who can undertake special investigations aimed at understanding the genesis of outbreaks. Thus, rapid identification of epidemics may be useless if appropriate response capacities are inadequate. An important lesson is that people who identify, analyze, and evaluate outbreaks and those who implement and direct epidemic responses must work together as an integrated team.

3.4 Suggested research opportunities and needs

Surveillance data often provide opportunities to research disease patterns and underlying processes. Such research can help define time- and space-specific risk factors that could be used to reduce likelihood of future epidemics. The knowledge gained must be considered an essential outcome of surveillance efforts, which should be fully exploited to produce new insights into control and prevention efforts. This implies that the quality and amount of passive surveillance data must be extensive if results of such research are to be sufficiently rigorous to advance prevention. The design of future active and passive surveillance efforts should explicitly consider the needs of research hypotheses that eventually are aimed at intervention.

Another example of research applications involves analysis of environmental and climatic associations that may be covarying with disease incidence. Recent research efforts have focused on the historical records of temporal and spatial disease patterns in relation to those of climate and land use [3]. Various investigations have attempted to address whether environmental and climate variability might partly explain disease incidences hypothesized to have a strong environment/climate forcing. Here again, the success of such research efforts depends on the frequency, spatial distribution, and accuracy of disease surveillance data. The meaning of environment in this context may include land use or land cover (determined by biome, elevation, economic development, and human use histories), or it may involve atmospheric factors that affect weather and climate. Many of these data have historically been collected as part of meteorological records or archived satellite images documenting changing land cover. Such data, when combined with carefully collected disease data, permit critical analyses of patterns that should elucidate underlying processes. Their relevance to understanding and forecasting disease risk depends on the quality of data and the analysis of time-space patterns in relation to those of climate.

3.5 Surveillance-based research aimed at forecasts

Analyses of surveillance-based disease patterns and variation in climate, land use, or other behavioral or ecological factors may permit more rigorous forecasts of transmission risk weeks to months before changes occur. This represents another value of surveillance data in studying and predicting future outbreak risks. The opportunities are enormous, and there should be little hesitation to embrace such efforts if the quality of disease surveillance data is adequate for the forecasting goals.

4. DISEASE SURVEILLANCE AND CLIMATE ASSOCIATIONS

Weather and climate surveillance data are mostly compiled as a matter of policy, often without evaluating how observations might be modified or improved to meet specific needs. However, a number of innovative studies have measured weather variables while prospectively undertaking active disease surveillance, as a research endeavor intentionally aimed at understanding possible links. If well designed and undertaken over a few or more years, such studies can provide considerable insights into how climate variability affects disease risk.

4.1 Climate change vs. climate variability

Surveillance data (both climate and disease specific) can be evaluated to determine long-term trends or to recognize shorter-term variation. The long-term trends in climate (decades to centuries) represent climate change, while monthly to interannual fluctuations can be thought of as climate variability. Although weather measurements have been recorded over the past few centuries in some locations, such records are often incomplete and rarely cover large regions. Most disease-specific information is even more limited in regional coverage; accurate surveillance-based estimates are available for only a few decades in most settings. Thus, it is generally more difficult to analyze how climate change may have contributed to shifting patterns in diseases than how climate variability might alter short-term patterns. For this reason, the systematic gathering of weather and disease surveillance data has taken on new importance.

One approach to analysis of existing data involves identifying and comparing anomalies in weather (heat waves, droughts, floods) as they correspond temporally with surveillance reports of excessive disease incidence [3]. However, understanding how anomalous weather or climate may be associated with corresponding variation in health requires accurate data gathered over many years, including data that may be particularly important to specific diseases. Some data are available for a few diseases in a few contexts, and have been analyzed accordingly.

4.2 Examples of disease and climate surveillance-based studies

This volume describes various examples for which past disease events and present knowledge could inform future response capacity in the face of a changing or more variable climate. We offer a few additional examples of diseases that illustrate the importance of surveillance data in understanding normal patterns, recognizing outbreaks, forecasting when unusual events might occur, and proposing coping strategies that may enhance response effectiveness

Age- and cause-specific mortality surveillance offers a baseline for recognizing what is considered "normal" and allows recognition of unusual mortality periods or locations. Such death data have served as the basis for analysis of extreme weather events that occur over days to weeks. Floods, heatwaves, and "cold snaps" are examples. Historical surveillance data permit evaluation of how these extreme weather periods may have resulted in deaths above what would have been expected for that time and place. Heatwaves have been recognized as causing excess mortality in some settings (e.g., Great Britain [4] and Chicago [5]; see also Chapter 8 in this volume) because there was knowledge of the usual pattern that allowed calculation of the magnitude of excess deaths.

Similar passive or active surveillance data could be useful in analyzing events that affect diarrheal diseases because they could serve as baseline to floods that contaminate potable water, or hurricanes that might dramatically increase mosquito vector abundance. Outbreaks of other water-associated diseases may take place over somewhat longer periods of time, such as if cholera were to increase because of monsoon rains or months that were unusually warm [6]. In this case, organized historical surveillance data of diarrheal disease or of diagnosed cholera would serve as the baseline to suggest that prevention measures should rapidly be instituted when excessive rains occur. Cholera surveillance data have been used in this manner to propose a link with El Niño events, allowing for possible early warning of outbreaks [7].

Sentinel surveillance of mosquitoes and active surveillance of antibodies in sheep and cattle to Rift Valley fever (RVF) virus in eastern Kenya have allowed scientists to identify an association of extreme rainfall with increased risk of human and animal disease there [8]. Caged sentinel chickens that are regularly tested for evidence of antibodies against arthropod-borne viruses such as St. Louis encephalitis are used in some areas of the United States to determine impending periods of elevated transmission [9], thus allowing for enhanced mosquito control efforts.

5. FUTURE NEEDS

A recent report, sponsored by the US National Academy of Sciences (NAS) [3] and written by a panel of experts from diverse fields, clearly articulated many of the needs, problems, and opportunities for surveillance efforts that relate weather, climate, and infectious diseases. Among the key findings was the recognition that epidemiological surveillance programs should be strengthened not only in the United States but globally as well. It was stated that the "... lack of high-quality epidemiological data for most diseases is a serious obstacle to improving our understanding of climate and disease linkages". We consider a few needs that are implied by this report and other such documents.

5.1 Implications for how surveillance could better meet present needs

The NAS report [3] argues that "one of the most critical obstacles to improving our understanding of climate-disease linkages is the lack of high-quality epidemiological data on disease incidence for many locations". Inadequate, incomplete, or nonexistent passive surveillance is what is found in many parts of the world, even in regions of many developed countries. Clearly, this problem must be addressed if present needs are to be met.

Active surveillance undertaken by individual researchers can enlighten us about specific problems in particular areas, and is a welcome complement to often less-complete passive surveillance. However, the structure of government agencies and academic research institutions discourages sharing of data gathered in this manner. As a result, others who might want to analyze these same data for other purposes or with a different perspective or method are hampered from potentially advancing knowledge. Incentives and rules that encourage the appropriate open use of actively gathered disease data should be considered. This will permit more analyses that should increase our adaptive response capacity.

5.2 Special needs in climate-sensitive areas

Greater attention should be focused on areas where climate-sensitive diseases are often found, or where outbreaks that appear to be weather-related are more severe and more difficult to predict. Other areas where global warming is expected to have the greatest and most rapid impacts should be high priority sites for increasing passive surveillance capacity. Similarly, those diseases that seem to be strongly influenced by weather might be candidates for special targeted surveillance efforts.

5.3 New approaches to address disease links to climate

Quality data that are both spatially and temporally explicit, and are of high resolution, serve as the basis for statistical and simulation modeling of climate or weather associations that will become predictive. Many new analytical methods are being used to explore these links. A research agenda that encourages both statistical (spatial, time series, etc.) and theoretical (process-based simulation) modeling will surely improve our ability to anticipate and respond to climate variability. Such model-based disease forecasts cannot be linked solely to climate influences, and must be context-dependent to be useful. Analysis of how other socioeconomic and ecological factors affect risk will improve successful forecasts.

The development of appropriate climate/disease simulation models is only the first step. These models need to be incorporated into early warning systems with sensitivity and specificity that are properly calibrated to the needs and capacities of the users (see Chapter 6 of this volume). The tradeoff between sensitivity and specificity needs to be weighed. A warning system that lacks specificity will generate too many false alarms, whereas one that lacks sensitivity may miss important outbreaks.

5.4 Future of response and prevention

At present, most surveillance data are used to direct responses to outbreaks and to intervene with education or aid if there is a perceived excess incidence. Often, responses occur well after many cases have appeared and are too late to prevent cases than if action been taken earlier. Increased understanding of disease-climate linkages should permit public health workers to begin to shift focus from "surveillance and response" to a more proactive approach of "prediction and prevention" [3]. As surveillance data are assembled and carefully analyzed, a predictive capacity should appear that will give public health officials, and the public, confidence in taking actions that will prevent climate-linked diseases before outbreaks occur. Warnings of extreme heat or precipitation have already permitted people to take precautions against hypothermia-related deaths and water-borne diseases.

6. SOME UNRESOLVED ISSUES

Adaptive capacity can be improved through better surveillance if the resulting data are made available in a timely fashion and appropriate manner. The issue of data sharing is a complicated and important obstacle that needs to be addressed at many governmental levels. Most surveillance data are gathered by regional bodies (state, province), combined at the national, and eventually international level. Typically, each level receives increasingly condensed summaries with less and less detail. Depending on the planned use of such data, there may be inadequate information. Access to surveillance data at local regions is typically restricted, sometimes for valid reasons such as maintaining confidentiality. Nevertheless, even summary disease data can be difficult to obtain, particularly if actively collected for a specific outbreak or research endeavor.

The extent of aggregation and categories of grouping surveillance data that are released may not permit desired analysis or association. For example, if disease cases are summarized by district or region, but a hypothesized effect of climate variability is restricted to particular habitats or economic groups, identifying the association would not be possible. If available data are summarized by year, but the climate link is highly seasonal (e.g., during the rainy season or the winter), then analysis would be severely hampered. The temporal and spatial resolution at which surveillance data are made available will influence how effective those data are in the design and execution of adaptive responses.

With increasing recognition of the global interconnection of people and their diseases, and a greater appreciation of the importance of monitoring and understanding climate and disease links, the need for international coordination of and developed country support for enhanced surveillance has become obvious. The weaknesses of surveillance in developed countries are often compounded in developing nations, some of which lack even the most rudimentary capacity to monitor and report morbidity and mortality. Developed countries have a responsibility to assist in the establishment and proper functioning of systems that are critical to adaptive capacity and the well-being of people in developing countries. International benevolence notwithstanding, many of the new emerging diseases that are affecting people in developed countries are the result of transfer from developing countries (e.g., dengue, SARS).

Related to international support for surveillance efforts is the need for international cooperation in sharing data. This sharing is increasing: some kinds of data are being made available through the internet by individual countries, the WHO, and the World Meteorological Association (WMO). Weather data from many meteorological stations throughout the world can be obtained through the Lamont-Dougherty Climate Data Library at Columbia University (http://ingrid.ldeo.columbia.edu/). The IPUMS project described in the appendix is making historical census data available on the web. Support for such monitoring and data distribution in developing countries is sorely needed. Similarly, the INDEPTH project (also described in the appendix) is a good example of disease surveillance methods in developing countries being coordinated to enhance comparability among regions and improve the predictive power that analyses might produce. The international community needs to take more seriously the issue of producing quality and coordinated disease surveillance data.

7. ADAPTIVE CAPACITY AND PUBLIC HEALTH PREVENTION

The ability to respond to needs and opportunities presented by disease and climate surveillance information depends on the context, goals, and places where understanding is sought and action may be planned. Therefore, the capacity to adapt and reduce risk will vary in different arenas, and may be more or less feasible depending on the social and economic resources of countries and regions. The role of technological

Table 10.1 Ranking of determinants of adaptive capacity and requirements for public health prevention[a] (note that these relative values will differ by specific disease)

Determinants of adaptive capacity	*Requirements for public health prevention*
Range of available technological options (2)	Awareness that problem exists (5)
Availability and distribution of resources (5)	Sense that the problem matters (5)
Structure of critical institutions (4)	Understanding of the causes (3)
Human capital (3)	Capability to intervene (4)
Social capital (3)	Political will to influence (5)
Access to risk spreading (1)	
Ability to manage information (5)	
Public's perceived attribution (2)	

Note
a Scale = 1 (low) to 5 (high).

options and the significance of resource distribution, for example, are likely to depend on the current level of economic development of a country or region. Similarly, the importance of political institutions, of human or social capital, and of information or perceived risk will vary with different cultural and economic contexts. Public perception that a problem exists or matters may be more or less important depending on the nature of proposed interventions. Indeed, it could be of little or great consequence that the public understands the causes underlying a climate-disease link, depending on the available response options and cost offset to society that intervention is believed to offer. We briefly address some of these issues (summarized in Table 10.1).

7.1 Determinants of adaptive capacity

Surveillance is demanding of economic resources: there must be funding for staffing of weather stations and a cadre of public health officials who gather, summarize, and disseminate disease data. This is particularly burdensome where active surveillance is undertaken in the context of special studies or outbreaks.

Thus, institutional infrastructures (public health agencies, weather monitoring stations, data collation and reporting offices) are critical to the adaptive capacity of nations to understand and respond to climate variability and change. Such infrastructure is often inadequate in developing countries. Even in wealthy nations, organizations responsible for surveillance vary among regions. In the United States, for example, the quality of disease surveillance data differs among cities, counties, and states, even though most data are eventually reported to a central national agency (CDC). Indeed, the extent of weather surveillance coverage also varies among sites (mostly airports or meteorological stations), because that effort may not be supported by nationally mandated programs (Table 10.1).

The importance of human and social capital depends on the objectives of adaptation, and may be minor if appropriate responses involve ordinary citizens, but considerably more if coordinated and complicated responses are planned.

The ability to spread risk seems of little application to the problem of surveillance needs. More and better information is critical in all geographic areas and disease domains. Any effort to reduce surveillance in one location or of one type, so as to shift resources elsewhere or to other diseases, would require careful analysis of the health benefits that would justify such change. Similarly, the public's perceived attribution of the importance of disease and climate surveillance is relatively less than the population-level successes of applying such efforts. However, the ability to manage information is critical to successful use of surveillance activities. This includes rapid retrieval of data, its synthesis and analysis, and the dissemination of user-relevant information. Public health agencies need to have summary information to know what are "normal" conditions. Citizens can use other information to make informed decisions concerning their individual risk. Without the management and appropriate dissemination of information, enhanced surveillance will be of little benefit.

7.2 Requirements for public health prevention

Applying surveillance tools to public health risk reduction requires that health workers at many facilities and levels recognize that weather and climate may pose serious threats to the public's well-being. For individuals, proactive or preventative responses depend heavily on an awareness of associated risk. However, such awareness may not affect organizational or individual responses if there is a sense that the problem does not matter. Where climate-health links are considered important, it should be easier to argue for surveillance and analysis of patterns than in contexts where there are other competing priorities.

Whether policy makers and the public understand the underlying causes of links also could be important, but this too depends on the context. Ideally, policy makers would decide on how and how many resources should be devoted to various activities based on an appreciation of the mechanism by which actions translate to benefit. All too often, popular opinion, political pressure, and partisanship determine what policies translate to practice.

Successful public health prevention depends heavily on the capacity to intervene, but this is only partly related to the importance of surveillance. Unless there is strong political will to develop adaptive strategies, even the very best systems of surveillance will produce data and knowledge that are not used. For this reason, politicians and policy makers must be informed and motivated to act on climate health links. Both the examples where surveillance has succeeded in reducing disease and the research results suggesting likely associations should serve as reinforcement that actions be proposed and undertaken.

8. IMPLICATIONS AND IMPACTS

8.1 Multiple benefits of enhanced surveillance

Enhanced surveillance not only will benefit analyses of climate-disease links and improve capacity to respond to weather extremes and climate variability but also will advance public health and aid prevention in many other domains. As part of a larger public health infrastructure, surveillance activities provide basic information useful against many different kinds of health threats, most of which are not strongly linked to climate. Knowledge of underlying disease patterns can allow comparisons of risk among different social or ethnic groups, improve recognition of lifestyle or behavioral threats to well-being, and encourage multinational comparisons that highlight weaknesses or areas where greater attention should be focused. These are all additional benefits that good surveillance provides. Indeed, passive surveillance of all causes of morbidity or registration of cause of death reports enhances public health education that produces cost-benefit-appropriate prevention strategies, whether they are focused on accidents, smoking, violence, or nutrition. Improving the quality and accuracy of surveillance represents a win/win or no-regrets strategy for many kinds of health problems.

8.2 New infectious disease threats

Two specific areas where improved surveillance will benefit domains other than climate-related changes involve our ability to respond to emerging diseases and to bioterrorist threats. Both of these are part of the ability of a system to respond to the unknown. Recognizing and responding to the unexpected is an important attribute of surveillance systems, and it is particularly important today because many new diseases have emerged during recent years, including legionnaire disease, Nipah viral disease, SARS, and many others. Furthermore, previously known diseases such as West Nile virus and dengue fever have spread into new regions. The same strategy that can detect new diseases should also help identify deliberate releases of certain biological agents if they represent an epidemiological anomaly with no typical pattern that past processes might predict. The recent deliberate release of anthrax in the United States illustrated the importance of recognizing and promptly reporting suspicious and unusual symptoms. In this context, plans for improved syndromic surveillance and integration of nonconventional data (e.g., emergency or poison center calls, school absenteeism, nonprescription drug sales) into real-time surveillance centers should improve investigation and response capacity. Thus, current concerns over bioterrorism in the United States will hopefully encourage changes in surveillance systems that will improve their ability to recognize and respond to unexpected events, and at the same time enhance climate and health adaptive capacity.

8.3 Increased understanding of unintended environmental change impacts

Concerns over the larger issue of global change involve many important processes that stand to diminish the public's health. Environmental pollution of water, air, and soil, population growth, increasing inequality within and among countries, urbanization, and crowding all have enormous impacts on health. Some are due to planned development projects that displace people, others are due to plans that did not consider indirect effects, still others may not have considered adequate health infrastructure. Enhanced global surveillance, particularly in those areas that are poor or undergoing rapid change, will only improve the ability to recognize where health services should be placed and intervention planned.

8.4 Public perception of risk

If used properly, surveillance data can be important in conveying to the public the nature of various kinds of risk. Other chapters in this volume address the issue, but surveillance data from various sources is what allows us to define what is "normal" versus "excessive". The accuracy and completeness of such data will determine how confident public health officials are in their presentation of risk to the public. Indeed, more solid surveillance data should improve the forecasts that serve as the basis for what people are told about risk. Inadequate surveillance leading to incorrect forecasts will only diminish public confidence in the important role that surveillance data play in public health practice. Enhancing the capacity to better understand climate-disease links will undoubtedly improve the overall health of the public.

APPENDIX: SURVEILLANCE PRIMER

TYPES AND METHODS OF SURVEILLANCE

Common disease and death records derive from passive and active surveillance, sentinel surveillance, disease registries, and population surveys. In addition, civil registration systems, longitudinal demographic surveillance systems, and census data provide critical demographic information needed to calculate incidence rates and prevalence, and to evaluate underlying patterns and long-term trends. The importance and applications of these methods vary among diseases, differ from country to country, and are adjusted over time in response to changing epidemiological situations. Such considerations, as well as the strengths and weaknesses of these methods, are summarized in Table 10.2 and the following discussion. This information is critical to understanding the response capacity to disease changes linked to climate change and climate variability.

Passive surveillance – routine reporting from health facilities

Routine reporting of disease cases and deaths in health facilities, especially those due to communicable diseases, represents the backbone of most surveillance systems. Almost every country has one or more routine reporting systems. Typically, the number of cases and deaths of each notifiable disease seen by local health services is reported regularly (usually weekly or monthly depending on the disease). Reports from the health facilities are sent to the district level, which compiles data from all health facilities, scrutinizes them, takes action where necessary, and sends the compiled data to the next higher level for further compilation, analysis, and action. Immediate notification is expected for suspected cases of diseases that have severe morbidity, have high case-fatality rates, and spread quickly. Local public health authorities may notify the district facility if they detect a serious or unusual health event for which intervention may be required. Such calls for help by astute practitioners are crucial for outbreak detection and rapid response.

Passive surveillance systems, however important, have many weaknesses. Problems faced by passive reporting include limited information on large portions of the population who may not be seen in health facilities, lack of standard case definitions at the peripheral level, logistic difficulties in sending specimens to laboratories, inadequate laboratory support to confirm cases, problems reporting to higher levels in a timely manner, overburdened and poorly paid health facility staff, and inadequate motivation to comply with reporting requirements and burdensome reporting systems. Recognizing

Table 10.2 Types of surveillance, information gathered, and strengths and their weaknesses

Type of surveillance	*Information typically collected*	*Strengths*	*Weaknesses*
Routine disease reporting or passive surveillance	Number of cases and deaths from notifiable diseases seen in health facilities	Allows regular monitoring of a broad range of diseases, by capitalizing on contacts with health services	Not all cases are seen in health facilities; May lack sensitivity; Underreporting, misreporting, reporting delays
Active disease surveillance	Number of cases and deaths from the population at large for diseases of interest	Provides more complete information than passive surveillance	Requires more resources than passive surveillance; Covers fewer diseases than passive surveillance
Sentinel disease or pathogen surveillance	Regularly updated information from selected labs, practitioners, or animal sources	Timely data that can be used to head off impending problems	Data collected from a small number of sites may not be typical of the population as a whole
Disease registries	Longitudinal data on individuals with particular diseases or exposures	Data can be relatively detailed, providing a wealth of information about the course of disease over time	Expensive; Require sophisticated computers to merge files; Coverage may not be complete
Human population surveys	Data on health of individuals	Rich data source; Representative of the population	Expensive; Respondent knowledge and or recall of illness may be faulty; Samples often too small for rare events
Civil registration	Births, deaths marriages	Where complete, provides necessary data on demographic trends	Civil registration systems in some countries not complete
Prospective demographic surveillance	Births, deaths migration, other health related variables	Usually excellent quality data	Field sites are limited geographically
Periodic censuses	Population size, and age/sex distribution	Complete count of population	Expensive; Usually done only once a decade
Weather surveillance	Hourly or daily temperature, precipitation, RH, winds, cloud cover	Usually accurate and complete for most locations	Coverage may be spotty and data may be difficult to obtain
Climate surveillance	Regional trends and indicators (ENSO, sea surface temp. jet stream patterns)	Usually accurate and complete for most locations	Coverage varies and data may be difficult to obtain

new or unsuspected diseases is more difficult than reporting on known risks.For example, the first cases of West Nile virus in the United States were originally thought to be St. Louis encephalitis virus, because West Nile had never before been seen in the Americas. The extent to which these problems are present in any given surveillance system varies tremendously by the ease of disease recognition and the frequency of occurrence. Indeed, once a disease is recognized in an area, the number of passively reported cases often increases simply because the information is more likely to be sought. Therefore, passively collected data from health facilities must be carefully examined and cautiously interpreted. Underreporting, delayed reporting, and misreporting plague many passive surveillance systems, and numerous epidemics are detected late, sometimes never.

Despite the problems in interpreting data from passive surveillance, most attempts to link climate variability with disease incidence rely on such systematic passive surveillance data. These data, if gathered over long periods of time, may allow creation of forecasts that suggest future periods of increased incidence and enhanced prevention.

As an example, Singh and colleagues [10] investigated relationships between climate variability and the incidence of diarrhea in the Pacific Islands, and demonstrated that diarrhea in adults was positively associated with annual average temperature and negatively with water availability. In addition, there were more diarrhea notifications in Fiji for the period 1978–1998 when temperature and rainfall were elevated. This illustrates how passive surveillance, if carefully monitored, can suggest relationships between climate and disease risk.

Active surveillance – disease-directed intentional gathering of data

Routine passive reporting systems are often supplemented by active surveillance for new cases of a specific disease or syndrome. Diverse activities can be included under the rubric of active surveillance, including actively seeking cases from diagnostic laboratories, carrying out more complete investigations following a suspected outbreak, or regularly promoting population screening (as for African trypanosomiasis [11], for which active case identification and subsequent treatment are important). The type of surveillance that is undertaken depends on the goals of the investigation, because surveillance can be used to determine who became ill, the mode of transmission or exposure, or what control measures are likely to reduce risk. Surveillance can also intensify – moving from passive to active – when risk of an epidemic is suggested or during efforts to locally eliminate or eradicate transmission.

Active surveillance, for example, played an important role in the eradication of smallpox. The approach involved surveillance and “containment”; it combined mass vaccination, active case finding, the media and key informants, house-to-house visits, isolation of infected people, and intensified vaccination of the population immediately surrounding affected individuals. In about a decade, smallpox incidence decreased from ~10–15 million cases per year in 1967 to none. The last naturally occurring case of smallpox was recognized on October 26, 1977. Active surveillance was crucial to the success of this effort.

Similarly, the polio eradication program currently involves active surveillance for cases of acute flaccid paralysis coupled with laboratory testing of specimens. In addition, routine immunization and special days for mass national immunization take place where low-level transmission persists. As a result, polio had been eliminated from the Western hemisphere by the early 1990s, and remains greatly reduced in both Africa and Asia. Nevertheless, it is reappearing in places where it was thought to have been locally removed, thus requiring continuation of surveillance programs. Active and passive surveillance will be critical to successful global eradication of polio.

Civil registration – birth and death surveillance

Civil registration systems represent another important source of information on births and deaths that may be critical to understanding disease patterns. Relatively complete vital registration systems of births and deaths are available from most developed countries, and many Latin American, Caribbean, and Asia countries. Sample registration systems, with information for a representative subset of the population, are available for some countries – for example, India and China. However, such data are largely incomplete for many countries in Africa. Medical certification of cause of death is an important component of many civil registration systems. The International Classification of Diseases (ICD) provides international standard definitions for causes of death used in certification systems. The fact that this standard is used at a global scale means that the data collected on cause of death are comparable across countries. In many developing countries, however, death certificate data are weak or lacking, partly because deaths commonly take place at home without the assistance or knowledge of the health services. Therefore, such data may be limited, unavailable, or even inaccurate in many developing countries. Despite these problems, in developing countries, cause-of-death data, compared over long time periods, are an important source of surveillance data that can be linked to climate data, but only if the data are accurate and unbiased. For some diseases, this seems approximately correct, making data on cause-specific death rates an important surveillance tool for understanding temporal changes in risk factors.

Verbal autopsies

Lacking conventional evaluation of cause of death, several countries (including China, Tanzania, and India) are experimenting with "verbal autopsies" as a useful surveillance alternative. This involves actively gathering information during an interview with next-of-kin of anyone who died. Such information may be augmented by other data or observations on the official medical cause of death, thereby increasing the accuracy. In most cases, however, particularly in developing countries or rural settings, health personnel may never see the deceased, making verbal autopsy information the only available data. Although such data inevitably involve some misclassification, studies indicate that they are a potentially important means of measuring mortality due to specific causes [12,13].

Sentinel surveillance

Another kind of surveillance is based on data obtained from sites, particular events, or suspected organisms; these data are likely to provide an early indication of disease events that could affect health in the population at large. Such sentinel surveillance information may provide early warning of future disease risks, and can enhance predictive and response capacities to climate variability. Various forms of sentinel systems provide more in-depth or timely information on health events than information produced by passive or active systems.

Many countries have sentinel surveillance reporting systems involving selected health care facilities that report more frequently or are asked to provide more detailed information. These facilities can be thought of as a special subset that has agreed to exceptionally complete and rapid passive surveillance, sometimes focusing on particular conditions that carry high risk. Such investment in quality and timely information from a smaller number of sentinel facilities or practitioners can produce considerable dividends, particularly if organized into a sentinel network. In France, a network of general practitioners provides weekly passive surveillance information on the number of cases of a group of diseases that they have seen. Such data have been combined with another network of sentinel microbiological laboratories, for example, to provide information on circulating strains of influenza virus since the late 1940s. These surveillance data are then used as a basis for deciding on the composition of the influenza vaccine to be manufactured for the next influenza season. Another innovative use of sentinel surveillance has involved monitoring the prevalence of HIV/AIDS among pregnant women to estimate prevalence in the general population. (Estimates of HIV/AIDS prevalence are largely based on unlinked anonymous testing of blood collected in sentinel antenatal clinics.)

Sentinel surveillance does not have to be limited to humans. Domestic animals such as cattle, pigs, and poultry are frequently used as sentinels to detect patterns of zoonotic diseases. For example, sentinel chickens are used to monitor the activity of certain mosquito-transmitted viruses that are normally transmitted among birds, but may infect people if they are bitten by these vectors. Chickens, when caged outside, serve as a blood meal source for some mosquitoes that transmit such viruses, and the chickens are likely to be bitten by vector mosquitoes where they are abundant. Blood samples from such sentinel animals can be regularly tested for evidence of pathogens of concern, hence anticipating and perhaps averting a human epidemic. Australia, for example, uses many flocks of sentinel chickens strategically positioned throughout the country to provide early warning of Murray Valley encephalitis and Kunjin. Likewise, sentinel chickens are being used in the United States for surveillance of eastern equine encephalitis, western equine encephalitis, St. Louis encephalitis, and most recently for West Nile infection. Sentinel animal surveillance provides important early information for potential outbreaks, particularly of zoonotic diseases. Long-term use of sentinel surveillance provides another basis for studies in climate variability for zoonotic disease agent transmission and for early warning of periods of intense transmission that may have resulted from unusual climatic events.

Disease registries

Information on morbidity and mortality that otherwise may not be available from death certificates can be obtained through disease registries, which systematically gather information on specific diseases. These registries are increasingly being used to obtain particular risk-factor information from people diagnosed with a specific disease. Eventually, such registries will consist of longitudinal data about individuals suffering from particular chronic diseases or groups of individuals with certain exposures. Registry data may be gathered from a variety of sources, including treatment records, hospital discharge records, or laboratory records, and are particularly useful for chronic diseases such as cancers. Data from such registries can be used for various purposes, including monitoring of incidence levels and trends, changes in disease characteristics, development patterns of symptoms, treatment outcomes, and eventually research on these and other concerns. In addition, other data on individual or group risk factors involving possible environmental exposures, socioeconomic variables, life-style patterns, etc., have become extremely valuable input for interpreting patterns from registry data. Because of the resources needed to establish and maintain such region-wide registries, many registries are restricted to developed countries. Their relevance to adaptive capacity to climate influences remains largely unexplored, because most diseases for which registries exist are not thought to have a strong climate signal.

Longitudinal demographic surveillance systems

Geographically defined and research-oriented field sites exist in Asia and Africa where population dynamic events are continuously monitored, including all births, deaths, and migration. One of the first such systems, named "Matlab", began in Bangladesh in 1963. In addition to demographic data, sites such as these also typically collect considerable information on nonfatal health outcomes, because this is critical to research on a variety of demographic, social, economic, and health related topics. Such demographic surveillance sites often are used to test new interventions before introducing them to the larger population, and typically collect high quality data over relatively long time periods (e.g., decades). Bias in routine data collected from health facilities and other field sites is often minimized, because they actively collect information from all social groups, not only people who have access to health facilities. Although these sites are often situated in rural areas (selected partly because they are considered "typical"), the major weakness is that sites are geographically limited and may not represent the larger population.

Sites such as these recently have been linked together into a network named INDEPTH that has described a standardized surveillance system [13] and proposed a standard verbal autopsy. These sites might be ideal for certain types of climate/health related research.

Census data

Most countries periodically carry out censuses aimed at enumerating all people within their borders. Such data provide key demographic information on population abundance by age, sex, ethnic or racial group, and other characteristics that are summarized by geographic subregions. Depending on the country, other variables such as education and income also are documented, including some that may have relevance to health (e.g., sources of water, presence of electricity, house construction, number of toilets). Census data are critical to estimating disease incidence and prevalence, because they represent the denominators for these values and serve as the basis for understanding changes in disease patterns over time and space. However, these data may be aggregated at the local level, and the geographic scale of aggregation may change as data are summarized at the district or national offices. In addition, the extent of aggregation of summary census data can be considerable, and may have changed over the years. This could hamper analyses of how disease incidence and climate variables may covary over time.

The accuracy and completeness of census data vary considerably among countries. Developed countries generally devote considerable resources to acquiring, verifying, and correcting census data. Because these countries also tend to be undergoing less demographic change (relatively lower birth, death, and migration rates), interpolation estimates for years between censuses are likely to be more accurate. But many poor countries are unable to properly or comprehensively enumerate their people, hampering long-term analyses or comparisons across regions.

Census data are increasingly available to public health workers and research scientists. For example, the University of Minnesota (USA) has a major project to make historical census microdata available to researchers via its website at http://www.ipums.org/international. These data are coded in a harmonized way to allow comparisons of the data over time and space. As of May 2003, data from seven countries are available on the website – China, Columbia, France, Kenya, Mexico, United States, and Vietnam. For all but China, data are available for more than one historical census, spanning between 10 and 40 years. This ease of access and comparability is not yet available for many countries, making comparisons of disease patterns among countries difficult.

GLOBAL NETWORKS AND NOVEL METHODS

Disease surveillance is international in nature, in part because of the desire to protect populations from epidemics occurring elsewhere. Some epidemics are linked to trade and commerce, immigration, or wars. The need for concerted action to prevent the introduction and spread of diseases led to the creation of a framework for international surveillance and response, as well as to the first intergovernmental organizations for health. For example, cholera epidemics that spread from Asia to Europe in the early 19th century led to a series of International Sanitary Conferences, the first of which was held in Paris in 1851. These conferences eventually adopted in 1892 the First International Sanitary Convention (forerunner of the current International Health Regulations) for the control of cholera. Indeed, the cholera and yellow fever epidemics in the Americas during this period led to the creation of the first modern day intergovernmental health organization, the International Sanitary Bureau, forerunner of the Pan American Health Organization (PAHO).

Today, intergovernmental organizations play a crucial role in global disease surveillance. The WHO was mandated to lead and coordinate international disease surveillance. Development of global partnerships for surveillance and response represents one of the cornerstones of a strategy for increasing global health security and enhancing epidemic response capacity. The strategy has three pillars, namely, improve preparedness by strengthening national surveillance systems, contain risks from known diseases, and respond to unexpected events by using innovative approaches of monitoring and verifying suspected outbreaks throughout the world. Such efforts are beginning to be linked to monitoring of extreme weather, and active international surveillance during forecasted periods of unusual climate is being undertaken.

Global outbreak alert and response network

For the past several years, WHO has been using an innovative systematic approach to monitoring, verifying, alerting, and responding to outbreaks of international public health concern worldwide. WHO is uniquely positioned to undertake this effort, because their responsibility for public health allows staff and consultants access to member states that may not be afforded to others. Central to their effectiveness is the Global Outbreak Alert and Response Network (GOARN), which links WHO's member states and over 72 networks and institutions worldwide. This interchange includes technical expertise for monitoring and responding to outbreaks, including expertise with dangerous pathogens. Communication about epidemics of potential international concern is coordinated at WHO headquarters, synthesizing a variety of formal and informal sources, including WHO regional and country offices, WHO collaborating centers, national institutes of public health, UN agencies, nongovernmental organizations, electronic discussion groups, media reports, and personal communications. A major source of epidemic intelligence (39% of all events between January 2001 and October 2002) comes from a multilingual software application, Global Public Health Intelligence Network (GPHIN) (http://gphin-rmisp.hc-sc.gc.ca/index.html), which gathers media reports from electronic news sources around the world.

Incoming reports are screened and discussed with disease specialists and other public health officials at daily meetings. If necessary, further verification is sought through the WHO country or regional office, or from national authorities. A decision is made as to whether a potential outbreak is of international concern, and which of various actions may be recommended. These actions range from simply continuing to monitor the event, to providing assistance in investigating and controlling outbreaks, to informing others who may be affected, such as public health officials in other countries and the GOARN network, to issuing travel warnings as was done for the first time during the recent SARS outbreak. In addition, the public is kept informed through the WHO website (http://www.who.int/csr/sars/goarn/en/), Disease Outbreak News, and the Weekly Epidemiological Record (WER; http://www.who.int/wer/en/) as well as through the media. Any combination of these actions may be taken based on the exigencies of the situation.

The GOARN network capitalizes on a number of important technological innovations such as the ubiquitous coverage of events by the media, the speed with which media reports are made available on the internet, the capacity of modern software to screen information, and innovative approaches of some surveillance practices.

International health regulations

The international health regulations (IHR) provide a legal framework for sharing of epidemiological information. They aim to "...ensure maximum security against the international spread of disease with a minimum interference with world traffic"[15]. Not all epidemics would be controlled by restrictions in terms of travel and trade. Yet, the potential economic consequences of epidemics may reach billions of dollars, even if such disruption to travel and trade is unwarranted from the perspective of disease prevention. On the other hand, some epidemics such as the recent epidemic of SARS require the imposition of travel and trade restrictions to prevent spread, even though these restrictions may be costly for countries involved. Global health security requires that sensible measures be put in place to protect against the international spread of disease, and that countries have the capacity for surveillance and response to epidemics, while limiting unnecessary costs to trade and travel.

The current IHR, last revised in 1981, require that countries officially notify WHO within 24 hours of cases and deaths from cholera, plague, and yellow fever that occur within their borders. These regulations also define public health requirements intended for travelers, such as vaccination certificates, as well as other measures applicable to ships and aircraft, such as removing rats or insects and disinfecting at ports of egress and entry.

The current IHR have many limitations. First they cover only three notifiable diseases and do not adequately include all the possible threats of international spread of serious disease at the beginning of the 21st century, and these three diseases do not necessarily pose a threat to international health if they are under control. In fact, no fixed list would have the flexibility required for a truly effective system. Second, the IHR depend on country notification, and often countries are reluctant to notify because they fear the economic consequences and travel restrictions that could be applied by other countries if it became known that they had an ongoing epidemic of cholera, plague, or yellow fever. This is a powerful disincentive for reporting, and many countries do not comply. Third, the current IHR also lack mechanisms for collaboration. Finally, WHO lacks the capacity to proscribe specific measures to prevent international spread of specific diseases.

The IHR are under revision, adapting to the needs of the 21st century. If successful, they should become an effective surveillance tool to prevent spread of serious diseases internationally. A number of core concepts have been proposed for this revision, including reporting all public health emergencies of international concern (this could be done on a confidential, provisional basis) and ensuring that countries have the capacity to report and analyze national disease risks. Most important, WHO would have the capacity to coordinate an international response to outbreaks of international public health concern, and would issue recommendations when needed. Indeed, this would be based on a transparent process and a nonexhaustive list of protective measures that countries could take. Such a change should enhance the role of surveillance and reporting, and would improve global public health preparedness in response to extremely unusual disease events. These revisions are expected to be submitted to the Fifty-Seventh World Health Assembly in May 2005.

International electronic networks

Computers and the internet have dramatically improved the possibility of rapidly transferring and accessing detailed surveillance data. Today, numerous new initiatives have taken advantage of the rapid transmission of detailed information, creating networks of data collectors and data users. International partnerships for disease-specific surveillance include many specialized surveillance networks of collaborators who both use and contribute information to these networks via the internet. For example, the network of influenza collaborating centers (Flunet) is electronically linked among participants with a web page for data entry from each collaborating center. In this way, data are directly sent to a central location, and become visible in the form of maps and tables as soon as they are entered. Similar surveillance networks exist for other diseases such as rabies (both human and animal); the Antimicrobial Resistance (AR) Infobank provides information on anti-infective drug resistance, salmonella, etc. These networks have been quite successful because participation of each collaborating site is automatically recognized by data being included and feedback is immediate and automatic.

Disease mapping and analysis

Modern computing power is also being used to improve surveillance through geographical information systems (GIS) and spatial statistical analyses. More and more, countries are using simple, low-cost GIS for management and mapping of their data. The health map program of WHO helps countries use GIS systems with standardized indicators to enhance comparability. These computer programs allow data entry and analysis at the local level, and can be transmitted electronically to higher levels without loss of detail. Additional variables such as demographic data, location of population concentration, spatial pattern of clean water, locations of health care facilities, and land use patterns may become part of a surveillance plan that is included in a GIS. When available, such data may be relevant to understanding the pattern of various diseases, depending on their transmission pattern and other relationships.

Other kinds of surveillance data, derived from satellite images (remote sensing data), are often included in GIS and spatial statistical analyses that contribute to disease surveillance and pattern analysis. Such data may be important in understanding how land use, land cover, or soil moisture patterns may affect disease risk. The time-space pattern of precipitation, for example, has been suggested as a cofactor in various water-borne or vector-borne diseases. Such surveillance of the land, water, and atmosphere has a major place in studies of how climate may alter risk of infectious diseases. Indeed, satellite image data, combined with those from ground-based sources such as meteorological stations, are critical for monitoring climate variability and climate change. They may also be important in planning and monitoring adaptations to climate change. These data are now easily incorporated into GIS platforms, and are becoming part of the repertoire of disease pattern associations.

Networks of acute care settings

Another important contemporary surveillance method involves use of acute-care networks that employ innovative software for detecting known risks and unexpected events. Theory suggests that unexpected outbreaks should be more easily detected through aggregated data from an electronically connected network of sentinel sites. Hospital and emergency room settings seem particularly suited to be sentinel sites because patient populations are larger and normally consist of more concentrated acutely ill patients. Such networks, established in recent years, have monitored major public events, e.g., the 1999 World Trade Organization meeting in Seattle, Washington, USA, conventions of the US Democratic and Republican parties in 2000, the baseball World Series in 2001, and the soccer World Cup. A similar system named the Early Warning Outbreak Recognition System (EWORS) has been used in Southeast Asia.

These systems initially report on the occurrence of sets of syndromes (classes of symptoms consistent with various etiological agents) or unusual high frequencies of adverse health events, before a definitive diagnosis of the causative agent or specific disease is available. This allows for more rapid recognition of unusual or aberrant clustering of disease in time or space. Disease syndromes are assessed on admission and transferred to a computer-based reporting system in real time. This is obviously more rapid that waiting for confirmed diagnoses that could take days to weeks for verification. The computer will flag for further investigation any sudden increase in the number of cases presenting with a syndrome (e.g., rash, respiratory symptoms, severe headache). Of particular importance will be devising triggering events or trends that will be both sensitive enough to detect potentially important epidemics at the earliest possible stages and specific enough to avoid overburdening the capacity for investigation. The playoff between sensitivity and specificity may be different in different environments. These systems are relatively new; however, interest in them has increased in recent years owing to an increased perception of the threat of a deliberate release of biological agents.

Weather and climate surveillance

Important to our understanding of how climate may affect disease risk is the nature of weather surveillance. Everyone is familiar with records of temperature, precipitation, wind, relative humidity, and other weather data that have been recorded daily in some locations since about 1860. Weather data have been used to characterize typical climate patterns throughout the world. Such information serves as the basis for forecasts of how local conditions might change under different global warming scenarios. Many countries with various ground and air-based sensors and databases are contributing to this kind of global surveillance. These data have become part of various national and international efforts to understand how weather changes locally, and how local weather events are linked globally. Much of the impetus for this undertaking derives from the goal of reporting on short-term (daily) variations that may affect individual or social activities or predicting short-term (daily or weekly) events that affect travel and commerce. Such surveillance data, when carefully analyzed, can be used to forecast longer-term trends within certain limits of error. Thus weather surveillance is crucial to understanding the importance of disease surveillance in the context of weather and climate.

REFERENCES

1. Last, J.M.: *A Dictionary of Epidemiology. 4th ed.* Oxford University Press, New York, 2001.
2. World Health Organization and UNAIDS: *WHO Recommended Surveillance Standards, Second Edition #8211.* World Health Organization, Geneva, 1999.
3. Committee on Climate, Ecosystems, Infectious Disease and Human Health: *Under the Weather: Climate, Ecosystems, and Infectious Disease.* National Academy Press, Washington, DC, 2001.
4. Hajat, S., Kovats, R.S., Atkinson, R.W., and Haines, A.: Impact of Hot Temperatures on Death in London: A Time Series Approach. *J. Epidemiol. Commun. Health* 56 (2002), pp.367–372.
5. Anonymous: Heat-Related Deaths – Chicago, Illinois, 1996–2001, and United States, 1979–1999. *MMWR* 52 (2003), pp.610–623.
6. Lipp, E.K., Huq, A., and Colwell, R.R.: Effects of Global Climate on Infectious Disease: The Cholera Model. *Clin. Microbiol. Rev.* 15 (2002), pp.757–770.
7. Rodo, X., Pascual, M., Fuchs, G., and Faruque, A.S.: ENSO and Cholera: A Nonstationary Link Related to Climate Change? *Proc. Nat. Acad. Sci. U.S.A.* 99 (2002), pp.12901–12906.
8. Linthicum, K.J., Anyamba, A., Tucker, C.J., Kelley, P.W., Myers, M.F., and Peters, C.J.: Climate and Satellite Indicators to Forecast Rift Valley Fever Epidemics in Kenya. *Science* 285 (1999), pp.397–400.

9. Scott, T.W., Wright, S.A., Eldridge, B.F., and Brown, D.A.: Cost Effectiveness of Three Arbovirus Surveillance Methods in Northern California. *J. Am. Mosq. Control Assoc.* 17 (2001), pp.118–123.
10. Singh, R.B., Hales, S., de Wet, N., Raj, R., Hearnden, M., and Weinstein, P.: The Influence of Climate Variation and Change on Diarrheal Disease in the Pacific Islands. *Environ. Health Persp.* 109 (2001), pp.155–159.
11. Anker, M.: *WHO Report on Global Surveillance of Epidemic-Prone Infectious Diseases.* World Health Organization, Geneva, 2000.
12. Anker, M., Coldham, C., Kalter, H.D., Quigley, M.A., Ross, D., and Snow, R.W.: *A Standard Verbal Autopsy Method for Investigating Causes of Death in Infants and Children.* WHO/CDS/CSR/ISR/99.4. World Health Organization, Geneva, 1999.
13. Sankoh, O.A., Ngom, P., Nyarko, P., Mwageni, E., and Kahn, K. (eds): *Population and Health in Developing Countries. Volume 1: Population, Health and Survival at INDEPTH Sites.* International Development and Research Centre, Ottawa, Canada, 2002.
14. WHO: *Global Crisis-Global Solutions.* World Health Organization, Geneva, 2002.
15. WHO: *International Health Regulations (1969), Third Annotated Edition* (1981). http://policy.who.int/cgi-bin/om_isapi.dll?infobase = Ihreg& softpage = Browse_Frame_Pg42.

11 Stratospheric ozone depletion: successful responses to a global environmental insult

Robyn M. Lucas and Anthony J. McMichael

ABSTRACT: Ultraviolet radiation from the sun is filtered by stratospheric ozone, such that most of the harmful UVB radiation does not reach the Earth. Excessive exposure to UVB causes some skin cancers and some eye disorders, and may have effects on the immune system. Over the last part of the 20th century, stratospheric ozone has been depleted as a result of human activities, namely the production and consumption of halocarbons. With the recognition of first the likelihood and then the reality of ozone depletion, there has been a relatively rapid international response to decrease the production and use of the causative chemicals and to find substitutes that are less destructive of stratospheric ozone. The history of the policy response to stratospheric ozone depletion is explored, and this story is compared to the response to global climate change. The similarities and differences between stratospheric ozone depletion and climate change are highlighted.

1. INTRODUCTION

The role of stratospheric ozone as a filter of harmful ultraviolet radiation (UVR) is well known. The development of ozone in the stratosphere (see Fig. 11.1) occurred originally as a byproduct of other biogeochemical changes. Water-based plants began producing oxygen via photosynthesis – and eventually this waste gas spilled over into the atmosphere. Incoming UVR from the sun chemically converted some of the oxygen in the upper atmosphere to ozone. Oxygen in the lower atmosphere allowed the development of aerobic organisms, and ozone in the upper atmosphere acted as a filter to provide protection from harmful UVR [2]. The scene was set for the development of land-based life.

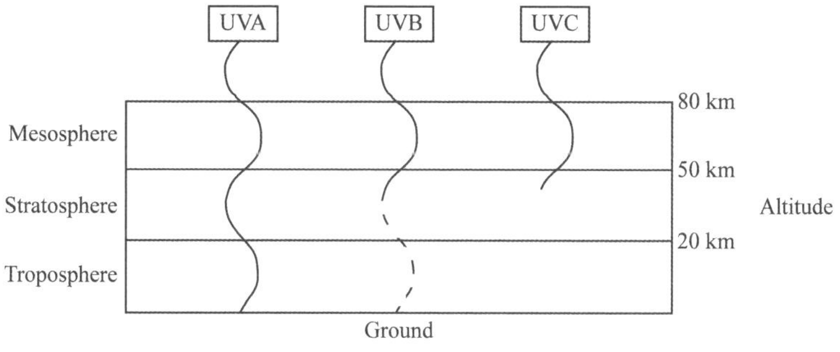

Figure 11.1 Stratosphere and troposphere. Source: Redrawn from Ref. 1

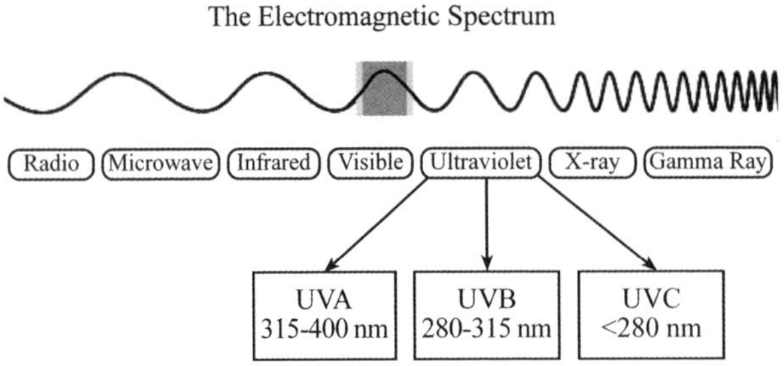

Figure 11.2 Ultraviolet radiation wavelengths in nanometers (nm). Source: Adapted from Ref. 4

The ozone layer in the stratosphere has functioned to limit the transmission to Earth of harmful UVR, and land-based life has evolved within the framework of these limited and varying levels of UVR. Throughout the past half-billion years approximately, ozone production from oxygen and its destruction by UVR have been in equilibrium. Much of the variation in human skin pigmentation by geographic location of traditional habitat, over the past 70,000 years since *Homo sapiens* radiated out of Africa, is likely to have reflected human biological evolution in response to the latitudinal variation in ambient levels of UVR [3].

UVR forms that part of the electromagnetic spectrum with wavelengths that are just shorter than the violet component of visible light. We divide UVR into three bands, based on wavelength, from UVA with the longest wavelength to UVC with the shortest (see Fig. 11.2).

Of these, UVC is completely absorbed in the atmosphere and UVA penetrates to the Earth's surface, virtually unchanged. Around 90% of UVB is "absorbed" by stratospheric ozone, so that only a small proportion reaches the Earth's surface. It is, however, likely that it is the UVB component of UVR that is most damaging to human health, and it is the level of UVB reaching Earth's surface that increases with decreases in stratospheric ozone.

2. REVIEW OF THE OZONE DEPLETION STORY

Concentrations of atmospheric ozone have been measured with ground-based devices since the mid-1920s [5] (see Fig. 11.3).

Most of Earth's ozone is found in the stratosphere (90%), with the remaining 10% in the troposphere [6]. Stratospheric ozone is usually measured in terms of the "total column ozone" (measured in Dobson Units, DU) [7]. The total column ozone is least at the equator (less than 300 DU) and increases with increasing latitude, with greater overall column amounts in the northern hemisphere high latitudes (400–450 DU in some areas) than in the southern hemisphere high latitudes (350–375 DU in some areas). Seasonal and interannual fluctuations in total column ozone result from wind transport and the stratospheric circulation of ozone [7].

Ozone in the stratosphere is continually broken down and re-created. Ozone "absorbs" UVR in the UVB band when UVB breaks an ozone molecule into an oxygen molecule and an oxygen atom. This atom then combines with another oxygen

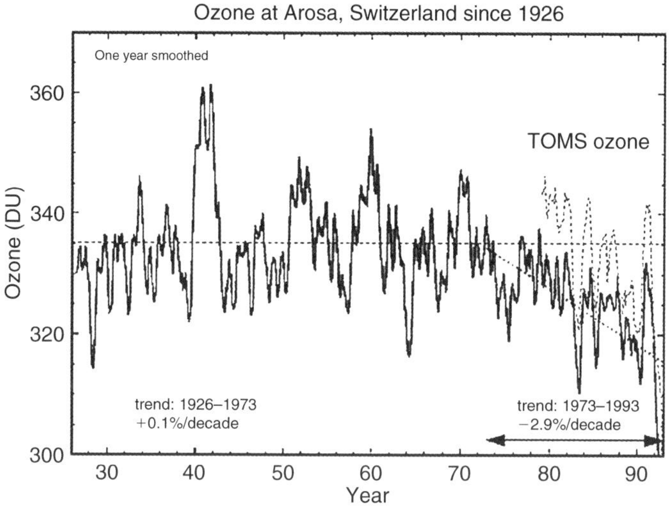

Figure 11.3 Ozone at Arosa, Switzerland from 1926 to 1993. TOMS ozone refers to stratospheric ozone levels measured with the satellite based total ozone monitoring spectrometer. Such measurements are only available since 1979. Courtesy of NASA, SEES: Studying the Earth's environment from space. June 2000. Available at Ref. 5

molecule to regenerate ozone. The result of these reactions is the conversion of solar UVB into heat energy [8].

Stratospheric ozone effectively forms a shield around the Earth, absorbing sufficient UVB radiation to maintain "safe" ambient levels, consistent with the evolutionary development of plant and animal life.

2.1 The supersonic transport issue

In the late 1960s and early 1970s the race to develop a fleet of supersonic transport airliners (SST) was on. Britain and France had developed the Concorde, the USSR the TU-144. In the United States a larger and higher flying SST fleet of 800 aircraft (by 1985–1990) was planned. These aircraft would fly through the stratosphere, and fears of damage to the stratospheric ozone layer were raised. These fears centered initially on water vapor and then oxides of nitrogen emitted from the SST [9]. An atmospheric physicist, James McDonald, calculated in 1970 that the projected US fleet of 800 SST would cause a 4% decrease in stratospheric ozone. Using a more conservative figure of a 1% decrease, McDonald calculated an increase of 5,000 to 10,000 new cases of skin cancer a year in the United States [10]. In addition, McDonald pointed out that increasing levels of UVB because of ozone depletion would affect not just humans but plants and animals that may have less adaptive capacity – over too short a time for evolution to fashion safe adaptation [10].

By the mid-1960s journal articles were appearing connecting the rising incidence of skin cancer with increasing exposure to UVR in sunlight [11]. Increasing incidence rates of skin cancers, documented for melanoma since 1955 [12], were probably related to increased sun-seeking behaviors [13] and changes in clothing conventions [12] from the middle of the first half of the 20th century. But by the mid-1970s scientists were already theorizing links between ozone depletion, increasing UVR, and skin neoplasms [14]. The potential skin cancer hazard from stratospheric ozone depletion was noted in the 1975 report of the Climate Impact Assessment Program (CIAP) examining the effects of SST.

While economic factors were probably most important in the eventual cancellation of the SST program, the issues of ozone depletion and the related dangers to public health have been likened to "the straw that broke the camel's back" [10].

2.2 The chlorine menace

The possibility of atmospheric chlorine as a catalyst for the destruction of stratospheric ozone first received attention in 1972 by scientists working with the US National Aeronautics and Space Administration (NASA), when it was recognized that the rocket boosters of the proposed space shuttle would inject chlorine, in the form of hydrochloric acid, directly into the stratosphere [15]. In 1973, Stolarski and Cicerone concluded that chlorine compounds in the stratosphere, from sources such as the space shuttle or naturally from volcanic eruptions, could be potent destroyers of stratospheric ozone [cited in 15]. One year later, Molina and Rowland identified chlorofluorocarbons (CFCs) as sources of plentiful atmospheric chlorine and potent ozone-depleting chemicals [16].

In the early 1920s, scientists had developed the chlorofluorocarbons as nontoxic, nonflammable replacements for the toxic and flammable refrigerants then in use, such as ammonia. The safety and excellent thermal properties of the CFCs meant that they rapidly replaced ammonia as refrigerants and were further developed for use in the automotive industry and subsequently as propellants for aerosol cans. In the United States, annual sales of aerosol cans had risen from 5 million in 1940 to more than 500 million a decade later, with a peak in 1973 of 2.9 billion – about half the world total [10]. Unlike refrigerants, CFCs in aerosols were directly and deliberately released into the atmosphere as propellants for a wide range of compounds: cosmetics, deodorants, hair sprays, waxes and polishes, insecticides, cooking sprays, and disinfectants. In 1973, 75% of all CFC emissions came from spray cans, 15% came from refrigeration and air conditioning systems, and the remainder were associated with a variety of other uses of CFCs, including foam blowing (to produce styrofoam cups and insulating material for buildings) [10]. CFCs released into the lower atmosphere rise into the stratosphere over a number of years.

Molina and Rowland theorized that at the very low temperatures in the stratosphere, ozone would be catalytically destroyed by the action of free chlorine atoms released from atmospheric CFCs after reaction with incoming UVR. Unlike the usual reactions of ozone with UVR, in which there is regeneration of ozone molecules, this catalytic destruction would destroy ozone molecules but regenerate the active free chlorine radical, which would then be available to destroy further ozone molecules [9,16].

In the four years following its publication, there was fierce debate between industry, scientists, and politicians (with active media input) on the Molina and Rowland theory, its relevance to health and environmental concerns, and the necessity for regulation of CFCs. For a full account, see *The Ozone War* [10].

In September 1976, the National Academy of Science (NAS) released its report on the ozone layer, concluding, "It is inevitable that [fluorocarbons] released to the

1892	First production of CFCs in Belgium.
1928	Discovery in US that CFCs are a stable, non-toxic and effective refrigerant (CFC-11, CFC, 12).
1960s	Papers appear in the public health literature pointing out probable links between sunlight and skin cancer
1971	Scientists hypothesize that nitrogen oxides from supersonic aircraft may reduce stratospheric ozone.
1974	Lovelock discovers presence of CFCs in the atmosphere. Molina and Rowland propose a theory of depletion of stratospheric ozone by atmospheric chlorine released from chlorofluorocarbons (CFCs).
1975	National Academy of Sciences (US) and United Nations Environment Program (UNEP) launch research into causes of stratospheric ozone depletion. A US federal task force concludes that CFC use may have to be restricted.
1977	Coordinating Committee on the Ozone Layer established by UNEP. US Government announces its intention to phase out most CFC use as aerosol propellants.
1978	CFCs in domestic aerosol products banned in United States.
1979	Canada, Sweden, Norway and Denmark ban use of CFCs in aerosols; West Germany agrees to a one-third reduction in CFC use in aerosols.
1980	EEC requires member nations not to increase CFC production and to reduce CFC use in aerosols from 1976 levels.
1981	UNEP working group begins to elaborate a framework convention for the protection of the ozone layer.
1985	March: 28 countries adopt Vienna Convention for the Protection of the Ozone Layer. May: Discovery of Antarctic loss of stratospheric ozone.
1987	46 countries adopt the Montreal Protocol on Substances that Deplete the Ozone Layer.
1988	Ozone Trends Panel concludes that CFCs are responsible for the Antarctic ozone hole.
1989	Montreal Protocol enters into force.
1990	London Amendments to the Montreal Protocol.
1991	World Meteorological Organization "On the whole, the 1991-92 winter can be classified among those with the most negative deviation of systematic ozone observations, which started in the mid-1950s." [19]
1992	Copenhagen Amendments to the Montreal Protocol.
1994	Total phase-out of halons in developed countries.
1995	Vienna Accord.
1996	Total phase-out of CFCs, carbon tetrachloride and methyl chloroform in developed countries, and of hydrobromofluorocarbons (HBFCs) in all countries.
1997	Montreal Adjustments.
1999	Beijing Adjustments.
2000	Largest Antarctic ozone hole on record.
2001	Adoption of a $573 million funding package to halve the consumption and production of CFCs in developing countries by 2005.
2002	Total phase-out of bromochloromethane in developed and developing countries. Ozone hole smaller than 2000 and 2001 (related to weather patterns and may not reflect recovery of the ozone layer). Declining levels of atmospheric chlorine.
2005	Total phase-out of methyl bromide in developed countries.
2010	Total phase-out of CFCs, halons and carbon tetrachloride in developing countries.
2015	Total phase-out of methyl chloroform and methyl bromide in developing countries.
2030	Total phase-out of HCFCs in developed countries.

Figure 11.4 Timeline of developments in the story of stratospheric ozone depletion. Source: Adapted from Ref. 6

atmosphere do destroy stratospheric ozone". In addition, "It would be imprudent to accept increasing [fluorocarbon] use, either in the United States or worldwide" [10]. In May 1977, the US Food and Drug Administration, Environmental Protection Agency, and Consumer Product Safety Commission (CPSC) jointly announced a schedule to phase out CFCs in aerosols, beginning with a ban on manufacture of such chemicals in 1978, followed by a ban on the use of existing supplies and in 1979 the banning of interstate shipments [10]. Notably, the American public had already taken action and voluntarily reduced their use of aerosol cans; sales dropped by 50% even before the legal bans were enforced [17].

At this stage there was no evidence for a sustained decrease in levels of stratospheric ozone – public and political action was taken on the basis of scientific theory. By limiting the use of CFCs in aerosols, legislation targeted the main immediate source of atmospheric CFCs. Refrigerants were seen as being more essential, with fewer alternatives available than the "wasteful luxuries" of aerosols [10].

Global policies on CFC use were slow to change, but in 1977, the United Nations Environment Programme (UNEP) established the Coordinating Committee on the Ozone Layer [18] and adopted the World Plan of Action on the Ozone Layer. In the early 1980s, the European Economic Community (EEC) proposed that there should be a voluntary cut in the usage of CFCs and some Scandinavian countries joined Canada and the United States in banning CFCs as propellants for nonessential aerosols. Between 1981 and 1983, UNEP worked with Scandinavian countries to draft a global convention to decrease production and use of CFCs [18]. Two years later, in March 1985, the Vienna Convention for the Protection of the Ozone Layer was signed by 20 nations and the EEC, to declare the intention of cutting the use of CFCs and providing a framework for international research and monitoring (see Fig. 11.4).

Surprisingly, it was not until May 1985, two months after the signing of the Vienna Convention, that the first reports of measurable ozone loss were made by British scientists working in the Antarctic [20]. Global action had been taken, and the Vienna

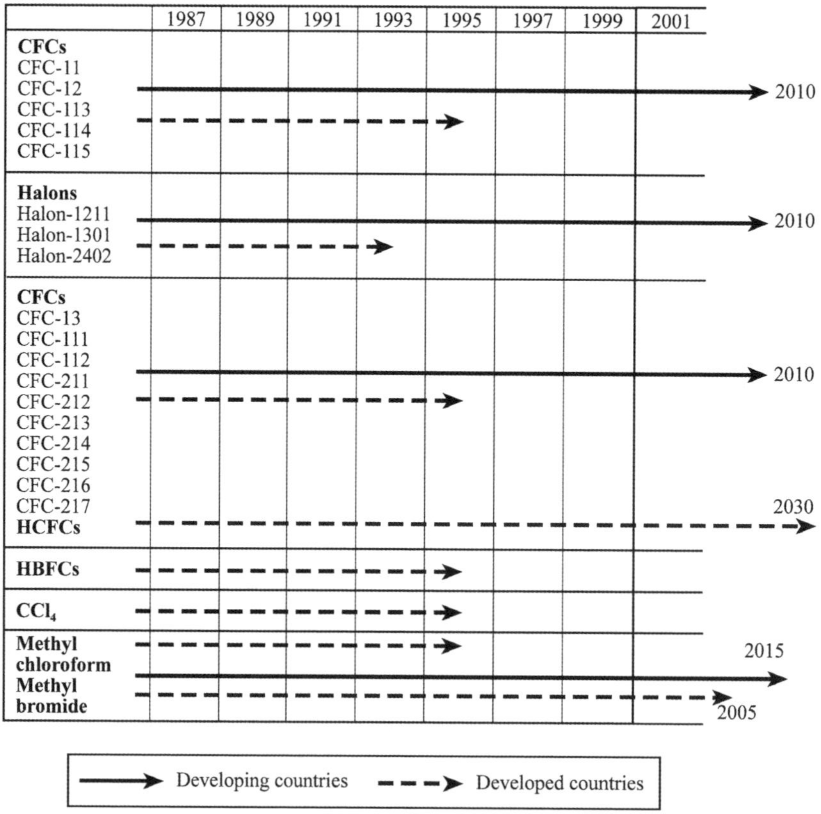

Figure 11.5 Summary of phase out schedule for halocarbons under the Montreal Protocol and Amendments

Convention signed, on the basis that it was theoretically likely that CFCs could cause depletion of stratospheric ozone. Indeed, this action had been taken two months before the first concrete evidence of changes in the ozone layer was released!

In 1987, 36 nations signed the Montreal Protocol on Substances that Deplete the Ozone Layer to limit the production of ozone-depleting chemicals, with an aim to halve

Table 11.1 Atmospheric lifetime (years) and ozone-depleting potential of selected halocarbons

Compound	*Relative ozone-depleting potential*[a]	*Atmospheric lifetime (years)*[b]
CFC-11	1.0	50
CFC-2	1.0	108
CFC-113	0.8	88
CFC-114	1.0	180
CFC-115	0.6	385
HCFC 22	0.055	13
HCFC-123	0.016	1.4
HFC-134a	0	18

Notes
a From the Montreal Protocol.
b From Ref. 21.

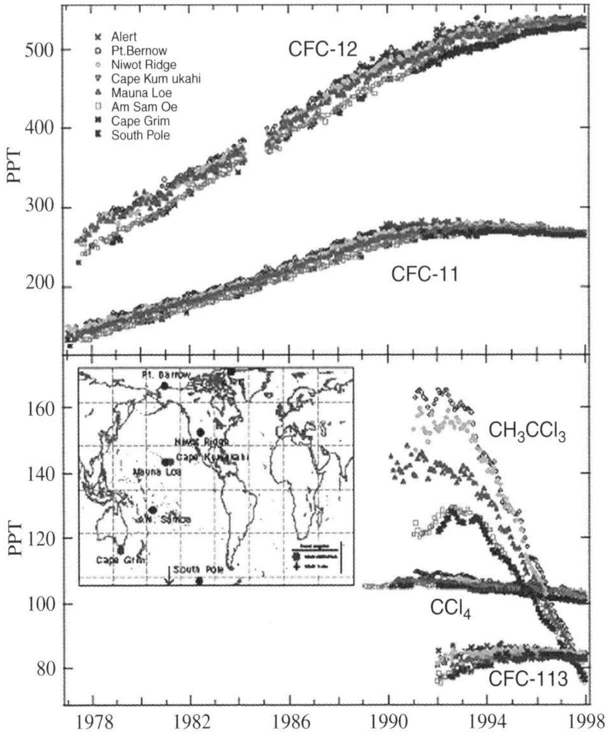

Figure 11.6a Accumulation of atmospheric CFCs and destruction of stratospheric ozone.
Source: Data from Ref. 22

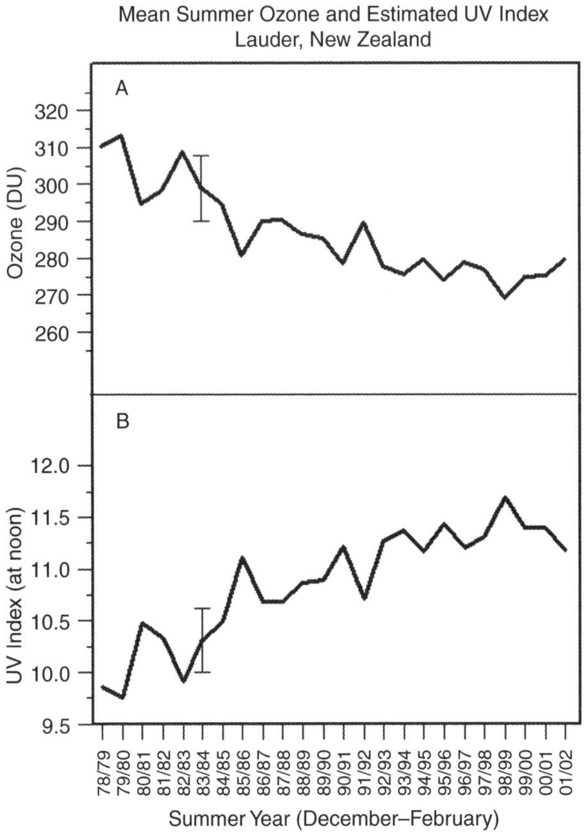

Figure 11.6b The upper panel shows average summertime ozone measured at Lauder (45S, 170E, 370 m alt), Central Otago, New Zealand. Here the summer period is defined as December through February. The lower panel shows the corresponding noontime UV index for clear skies. A clear decrease in ozone and increase in UV, which is larger than the year to year variability, can be seen over the period 1980 to 2000. There is a suggestion of a recovery since then, but the time is too short to detect it unambiguously over the natural variability. Data from Richard McKenzie, personal communication

CFC emissions by 2000 [9]. The Montreal Protocol was amended in 1990 (London) and in 1992 (Copenhagen) to accelerate the phaseout schedules for CFCs, halons, and hydrochlorofluorocarbons (HCFCs). Further adjustments to phase out the production of other ozone-depleting substances such as methyl bromide were recommended in the Vienna Accord of 1995 and the Montreal Adjustments of 1997. In 1999, the Beijing Adjustments recommended the phaseout of production of HCFCs by developed countries by 2020 and by developing countries by 2040 [21]. A summary of the phaseout schedule is shown in Fig. 11.5.

CFCs have extremely long atmospheric lifetimes (see Table 11.1). This means that even with the relatively prompt international response to limit CFC production and consumption, we would expect to see increasing accumulation of atmospheric CFCs and destruction of stratospheric ozone for a number of years (see Fig. 11.6). Levels of atmospheric CFCs (as measured at Cape Grim, Tasmania) increased from 1976 until 2000 (approximately 5% per year), but since 2000 a slow decline (1% per year) has been observed [23]. The ozone hole over Antarctica in 2000 was the largest (in area) ever recorded (30 million km^2), and was of similar size in 2001 (26.5 million km^2). The 2002 ozone hole was smaller (similar to 1988); it split into two parts and disappeared earlier than usual, by late October. However, this smaller ozone hole probably reflected planetary weather systems rather than lower atmospheric chlorine levels [24]. Larger ozone holes were seen again in 2003 and 2004. Ozone recovery from stratospheric CFCs will not occur until there have been substantial reductions in atmospheric CFC levels [23].

3. HEALTH RISKS AND BURDENS OF UVR EXPOSURE

3.1 Introduction

When considering the health effects of stratospheric ozone depletion, we are primarily concerned with effects of increased exposure to UVR in the UVB range. UVA passes through the atmosphere and is not absorbed by stratospheric ozone. It penetrates the skin more deeply than UVB, but it appears that it is radiation in the UVB range that is absorbed by DNA. Subsequent damage to DNA appears to be a key factor in the initiation of the carcinogenic process in skin [25,26]. UVA may play an important role in photoaging of the skin [27], and there is some evidence that it plays a role in the causation of malignant melanoma [28].

UVB penetrates only superficially into the body, so the direct health effects are on exposed surfaces – the eyes and skin. However, exposure to UVB also has beneficial systemic effects in the production of vitamin D by human skin and thus the prevention of rickets and osteomalacia [29]. There is some evidence that UVR exposure, either directly or through enhanced levels of vitamin D, may have beneficial effects: on the immune system both locally and systemically (promoting self-tolerance and thus decreasing the risk of a number of autoimmune disorders); on the cardiovascular system; decreasing the risk of mortality from various cancers, e.g. breast, prostate and colon [1]. The following discussion, however, concentrates on the deleterious effects of excess exposure to UVB as the side effect of ozone depletion caused by the human production of ozone-depleting substances.

It is important to note that the *possible* individual dose of UVR exposure is determined by the ambient level of UVR. Season, latitude, and the level of stratospheric ozone in turn determine this ambient level. In addition, human skin pigmentation differs, such that deeply pigmented skin has a natural sun protection factor (SPF) of 13.4 [30]. For any level of UVR exposure, the biologically damaging effect will be much greater on fair skin than on more deeply pigmented skin. Actual individual UVR exposure is then determined by individual differences in exposure habits.

Associations between UVR exposure and human health are difficult to study. Epidemiological studies examining links between UVR and diseases such as skin cancer and eye disease – typically diseases of mid- to late adulthood – rely on recalled patterns and amounts of sun exposure, including sunburns and hours in the sun over a lifetime. Most health effects of UVR exposure occur after chronic exposure over many years, while others may be set in train after some critical exposure in early life [31,32]. Ideally, prospective studies could be carried out on individuals wearing personal UVR monitors for several decades to clarify the complex dose-response relationships and patterns of exposure.

3.2 Effects on the eyes

The eye is the only part of the human body not shielded from harmful UVB radiation by the protective layer of the skin. The vulnerability of the eye to environmental hazards is the price we pay for being able to see. Each part of the eye may be affected by UVR.

3.2.1 *Cornea and conjunctiva*

The cornea and conjunctiva can be affected by acute, high dose UVB exposure, resulting in acute inflammation of these surfaces (photokeratitis and photoconjunctivitis), as well as by chronic exposure, which may be associated with a fleshy, wing-shaped growth on the surface of the eye (a pterygium). Acute exposure to intense direct or reflected solar radiation for as little as 30 seconds may cause photokeratitis, commonly seen as snow blindness or temporary blindness associated with exposure to welders' arcs [33]. Experimental work indicates that this damage is primarily caused by UVB [34]. UVR exposure has been implicated in the causation of pterygium for many years [35]. Recent studies have confirmed that sun exposure (with wavelengths in the UVB range probably being most important) accounts for almost half of the avoidable risk for pterygium development [36].

In addition to pterygium, there is mounting evidence that squamous cells of the cornea and conjunctiva can undergo malignant change in response to chronic UVR exposure.

3.2.2 *The lens*

There is compelling epidemiological and biological evidence of a causal relationship between certain types of "senile" cataract and UVR exposure, with exposure to UVB being of prime importance [37]. Cataracts occur mainly in older age groups, causing various levels of visual impairment up to complete blindness. They are extremely common and may be associated with an increased risk of mortality in developed as well as developing countries [38,39].

3.2.3 *Retina*

Under normal circumstances, the anterior parts of the eye and the vitreous humour filter out most UVB radiation. However, a retinal "sunburn" can occur if the sun is viewed directly. Also known as phototoxic retinopathy or eclipse retinopathy, solar retinopathy is well recognized as a cause of acute vision loss.

3.3 Effects on the skin

Because it is directly exposed, skin is the organ most susceptible to UVR. The main action of UVB appears to be through absorption by, and subsequent damage to, cellular DNA. Exposure of skin to UVR causes both acute and chronic skin disorders.

Sunburn is the immediate result of overexposure of the skin to sunlight, and UVB is three to four times as effective as UVA in causing erythema in humans [40]. Sunburn is extremely common, even in high latitude countries: according to a recent paper from Sweden, 55% of the study population, aged 13–50 years, reported sunburn in the previous 12 months [41].

Chronic exposure to UVR is implicated in the causation of cutaneous malignant melanoma (CMM), basal cell carcinoma (BCC), and squamous cell carcinoma (SCC) of the skin [25]. These diseases generally are most common in regions where pale-skinned populations are exposed to high levels of UVR, such as that observed at lower latitudes [32]. Lesions of BCC and SCC are found on those areas of the skin that are usually exposed to the sun, and epidemiological studies confirm an increased risk of each of these diseases with a history of sun exposure, although the importance of specific patterns of sun exposure varies for each disease [42].

3.4 Immune system effects

In addition to effects on DNA, other molecules, or chromophores, in the skin absorb UVB radiation. UVR-induced changes in some of these chromophores lead to changes in both the local and systemic immune systems [43]. UVB exposure appears to dampen the activity of T helper 1 lymphocytes (T_H1), which are important in the body's reaction to simple chemicals, to intracellular infections such as those caused by viruses, and to tumor growth [1].

Locally, an important effect of UVB exposure is that immune responses to abnormal cells are turned off, which in turn allows the development of skin cancers [43]. Systemically, UVR exposure may be involved in turning off the immune response to "self" – a lack of UVR exposure may thus be associated with the development of autoimmune disorders such as multiple sclerosis (MS), type 1 diabetes, and rheumatoid arthritis [44].

There is a well-known association between incidence of MS and latitude, with increasing incidence at higher latitudes (lower ambient UVR) [45]. Epidemiological evidence is accumulating of a link between low sun exposure and increased risk of MS [46,47].

There is a wide variation in the incidence of type 1 diabetes across Europe that is not explainable in terms of climate, temperature, or genetics [48]. UVB exposure is necessary for the synthesis of vitamin D precursors in the skin, and there is some evidence that lack of UVB exposure may explain seasonal and geographic variation in the incidence of type 1 diabetes, either through effects on vitamin D levels or through a direct dampening down effect of UVB on the immune system [49].

Hypothetically, UVR-induced immunosuppression may increase susceptibility to a range of infections [50] and decrease the immune response to vaccination. Such effects could have profound implications on patterns of infectious disease, particularly in countries with high UVR exposure. A recent review of the evidence indicates that while there are effects of UVR exposure on response to vaccination, there is no current evidence that these are of clinical significance in humans [51]. Immune suppression is a known risk factor for non-Hodgkin's lymphoma and it has been hypothesized that increasing UVR exposure may account for the rising incidence of this disease [52].

4. THE SHAPING OF PUBLIC HEALTH POLICY

By the mid-1960s there was already scientific and public awareness of an association between skin cancer and UVR [10,11]. In 1970, SSTs were linked with stratospheric ozone depletion [10]. Theoretical risks of increased harmful UVB exposure due to stratospheric ozone depletion were recognized well before such an increase in UVB could be demonstrated [53,54]. The first national and international responses to stratospheric ozone depletion, with the banning of CFC use in domestic aerosols in the United States, Canada, Sweden, Norway, and Denmark, and the subsequent signing of the Vienna Convention occurred in response to this theoretical risk [6].

It is remarkable and unprecedented that a global agreement could be made on the basis of scientific theory before physical proof of the problem. Though much of the early research was undertaken in the United States as a spinoff of the SST program, it was soon obvious that this was a global problem and would require an international response [9]. Although the SST program had been cancelled in the United States, other countries were continuing with their SST programs, with the likelihood that such programs would have effects on stratospheric ozone.

While there was increasing scientific support for the theory of ozone depletion caused by CFCs [6], the public health message that depletion of stratospheric ozone could cause an increased risk of skin cancers personalized the problem and mobilized the public [9]. Skin cancer was relatively common, and "cancer" carries strong emotional overtones. This was not some theoretical problem to be debated by politicians, but the concern of the ordinary person – the use of nonessential products such as aerosol sprays could pose a threat to health. And this was not only a threat to health, but also a threat from an exposure that was invisible, largely unavoidable, a hazard to future generations, and possibly globally catastrophic.

When the depletion of ozone over Antarctica was discovered in 1985, the theoretical risks became real. The ozone hole could be detected by satellite and pictures could be widely displayed. It was dramatic and measurable, and it confirmed the dire predictions made by scientists more than 10 years earlier.

Under the Montreal Protocol and its subsequent amendments, CFC production and use has not continued to increase. In 1996, Slaper and colleagues estimated the increase in skin cancer incidence under the various CFC restriction protocols [55]. Under a "no restrictions" scenario, there would be up to a quadrupling of skin cancer incidence by 2,100 – 6,530 excess cases per million per year in the United States. Under the restrictions of the Montreal Protocol, the increase would be up to a doubling – 1,958 excess cases per million per year in the United States. Under the tighter restrictions of the Copenhagen Amendments, with the production of 21 ozone-depleting chemicals reduced to zero by the end of 1995, there would be an ozone minimum around 2000 and a peak relative increase of around 10% in the incidence of skin cancer 60 years later [55].

In addition to engaging public support to promote policies aimed at reducing emissions of CFCs, public health messages addressed measures to adapt to and ameliorate the effects of increasing exposure to UVR due mainly to behavioral practices but likely to worsen with ozone depletion. Reports of the development of sunscreens first appeared in the literature in 1968 [56], and an active industry has increased their efficacy, duration of action, and acceptability over the last 30 years. It is difficult to distinguish to what degree public health initiatives were prompted by concerns generated by behavioral increases in UVR exposure rather than fears of the effects of ozone depletion.

Health promotion has focused on highlighting the dangers of sun exposure, particularly to the young. Efforts have been made to try to change the common perception that a tanned look is synonymous with a healthy look. Wearing hats and sunglasses and protecting the skin with clothing have been promoted as simple and unobtrusive ways of protecting oneself from harmful ultraviolet rays [57].

The following case studies outline the different responses to the threat of ozone depletion in the United States, Europe, and Australia.

4.1 United States (based on Ref. 10)

The increased risk of skin cancer subsequent to stratospheric ozone depletion first made headlines in 1971 during the debate on the environmental costs of the SST. Harold Johnston's calculations that 500 SSTs flying for 7 hours a day would deplete half the world's ozone layer in less than a year were reported in the New York Times on May 17, 1971.

The public furor in regard to the ozone layer did not really begin until late 1974, when the Associated Press asked, "Is the homely aerosol spray can and its charge of propellant gas sowing the seeds of doomsday, threatening to destroy earth's ozone shield and bake the planet barren with solar radiation?"

By the end of 1974 and early in 1975 there was considerable media coverage of the aerosol/ozone issue, with dire prophecies as an inevitable result of inaction. New Times magazine reported, "Aerosols have probably doomed more people than were killed by the atomic bomb dropped on Hiroshima"; "Farmers would have to plow what fields survived at night", and (confusing ultraviolet with infrared radiation) "Glaciers thousands of feet thick, leveled cities across Europe and North America". The Philadelphia Inquirer suggested, "The world will end – not with a bang, not with a whimper, but with a quiet pfft ... . The earth may already have committed partial suicide or at least self-mutilation".

Opinion polls in early 1975 suggested that a significant number of people (29%) not only were aware of the threat of ozone loss with consequent damage to health or the environment but also were using aerosols less.

By 1977, the EPA, FDA, and CPSC were sufficiently concerned about the spraycan threat to the ozone layer and human health to introduce bans on the manufacture and use of nonessential domestic aerosols.

In 1987, the American Public Health Association (APHA) developed a policy statement on "Depletion of the Stratospheric Ozone Layer" [58]. In this policy statement (formulated before the Montreal Protocol) it was noted that only the United States, Canada, and some Scandinavian countries had banned the use of CFCs for nonessential aerosols. Despite this and because of increasing use of CFC-11 and CFC-12 in other applications (air conditioning, refrigeration, foam rubber, insulating board and packaging materials), by 1987 demand for these CFCs had nearly reached pre-ban levels. The APHA supported a global ban on CFCs in aerosols and a phasing out of CFCs in other applications within 10 years both nationally and globally, and also supported research for environmentally safe alternatives. In 1990, a further policy statement also recommended public education on adverse health impacts related to ozone depletion and increased levels of UVR [59].

In 1998, after 2 years of development, the EPA launched the SunWise School Program, providing education on protection from UVR to schools. This is a partnership program with schools, aiming to raise community awareness of stratospheric ozone depletion, UV radiation, and sun protection [60]. In 2002 5,000 schools were registered for the program, from all 50 states. In addition, the UV Index is forecast daily in 58 cities across the United States to provide information allowing appropriate sun-exposure decisions for the informed public [60].

4.2 Europe

In 1973–1974, Western Europe and the United States were each responsible for about half of the worldwide production of CFC-11 and CFC-12, but Europeans seem to have been more skeptical about the threat CFCs posed to the ozone layer.

In April 1976 the British Department of the Environment Report stated, "Limited mathematical models suggest maximum (ozone) depletion in about 100 years. There are, however, many uncertainties concerning the mechanisms for destruction in the troposphere and in the theoretical predictions of stratospheric effects" [61]. Scientists in the United States also searched for tropospheric sinks for atmospheric ozone, but the National Academy of Science (NAS) report released in September 1976 concluded, "It is inevitable that [fluorocarbons] released to the atmosphere do destroy stratospheric ozone". Despite careful analysis scientists could find no evidence of tropospheric sinks to contain CFCs and prevent their migration to the stratosphere [10]. The NAS report specifically cited the increased risk of skin cancer from increased ambient UVR due to ozone depletion.

British scientists seemed more critical of the evidence for such a link. Specifically, there was concern that no changes in the ozone layer had been noted that could be blamed on atmospheric CFCs. There is considerable variability in the ozone column from day to day, seasonally, and annually – but although stratospheric ozone was being regularly measured (since 1926 in Switzerland), there had been no convincing evidence of a downward trend [61].

In addition, even if there really were a decrease in ozone and an increase in UVR, this needed to be discussed in the context of already rising rates of skin cancer due to changes in behavior. At a conference on CFCs in the stratosphere in 1978, it was noted [61] that:

> predictions of 15% steady-state ozone depletion would imply, at worst, about a 30% increase in skin cancers. This should be looked at against the background of much larger increases over the past 30 years in some places and in some communities. One looks for the possible reasons why this should have been the case, particularly among white communities in North America and to some extent in Europe, including the British Isles. In this last case the Clean Air Acts have probably contributed more than any likely ozone depletion would. It is known that there has been a great increase in sunbathing, and that the personal habits of people must be involved because the increase in skin cancers is far greater in New York than in Colorado, although the amount of ultra-violet in Colorado is much greater because of the higher altitude and the cleaner air.

Of reactions in Europe, it was stated that "the continental European industry ... does not share the opinions which have led the United States Authorities to what is considered, in Europe, as premature and hasty regulatory action" [61]. Industry focused on the many contradictions in the 1977 NAS report and highlighted the new Climate Impact Assessment Program report of 1977 that disproved ozone depletion by the SST program [9].

In 1978, global ozone was noted to be still increasing and the Royal Society (UK) reported that a further 4 years of research was necessary at virtually no risk. In 1979, CO_2 in the stratosphere was implicated in causing increases in stratospheric ozone for at least 17 years [61].

4.3 Australia/New Zealand

In 1966, studies in Queensland indicated a high rate of skin cancers, particularly of the nonmelanoma type. Skin cancers of all types were most common in those with Scots or Irish ancestry and least common in those with Asian, South American, or Australian Aboriginal ancestry. This high incidence of skin cancers was unrelated to ozone depletion; rather it was a result of a mismatch between skin type and ambient UVR as well as a preference for increased personal sun exposure, a tan, and a warm climate encouraging minimal clothing [61].

In Australia, the "slip, slop, slap" campaign was launched in 1981 – slip on a shirt, slop on some sunscreen, slap on a hat – aimed at reducing sun exposure [62]. Since the mid-1970s, public health workers in Australia have worked on education programs to discourage excessive sun exposure and encourage sun protective measures – not because of depletion of the ozone layer, but because of the high levels of ambient UVR observed in Australia, a country of clear skies and relatively low latitude.

In 1990, the Public Health Association of Australia (PHAA) released a statement on stratospheric ozone depletion, recognizing the hazards to Australians of the combination of naturally high incident UVR and proximity to Antarctic ozone depletion [63]. Schools have a "no hat, no play" policy – not only for summer but also for spring, when the ozone hole is generally largest [64]. The PHAA advocates increased research, intervention programs to reduce human UV exposure, and continued national and international efforts to phase out ozone-depleting substances.

Despite these programs, the latest National Secondary School Students Sun Protection Survey (1999) in Australia showed that 80% of students had been sunburnt at least once in the previous summer (up from 72% in 1996 and 68% in 1993) [65].

Perhaps what public health has failed to realize is that some sun exposure is essential to harness the beneficial effects of UVR exposure for vitamin D production and possibly directly for the immune system. Coupled with a lifestyle in which many young people spend little time outdoors, strict sun avoidance could cause a rise in rickets, osteomalacia, and osteoporosis [66]. Unlikely as this may seem, even in a sunny country like Australia there is a moderately high prevalence of vitamin D insufficiency within the normal population; 23.4% of an adult population sample in Queensland, Australia in 1997–1998 had vitamin D insufficiency [67].

Health concerns contributed to efforts to restrict the production and consumption of CFCs and related compounds to try to limit damage to the ozone layer. Meanwhile public health bodies have sought to educate the public to ways in which they can adapt to living in an environment where UVB may be increased and create a hazard to health.

5. HOW IS THE OZONE STORY DIFFERENT FROM THE CLIMATE CHANGE STORY?

Climate change and stratospheric ozone depletion, each representing a form of global environmental change caused by human activity, are essentially different processes. In simple terms, the greenhouse gases that cause climate change (primarily carbon dioxide, methane, nitrous oxide) are produced as the result of human activities – the burning of fossil fuels, agricultural practices, and the destruction of forests – on a scale not previously encountered in history. As these gases accumulate in the troposphere, beneath the stratosphere, they accentuate the natural greenhouse effect and are predicted to cause global climate change with consequent warming and changes in rainfall, natural disasters, and sea level.

Climate change is a result of traditional individual lifestyle practices modernized and applied to a vastly increased global population. Humans have always used fires for warmth and cooking. The modern version of this is the burning of fossil fuels, not just for individual warmth but also to provide electricity on a large scale to make life more comfortable. For example, electricity and heat production accounted for 32.7% of the total greenhouse gas emissions in Australia in 2000 [68].

Humans are accustomed to convenient mobility, in former times perhaps by horse, but with a freedom to go where and when travel is desired. In the 20th century the motorcar replaced the horse, but we still wanted to retain our individual mobility – road transportation accounted for 12.9% of the total greenhouse gas emissions in Australia in 2000 [68].

Humans have long farmed animals and grown crops – but with an increasing global population and the consequent food requirements, this is now on a greater scale than ever before. Emissions from agriculture, principally methane from enteric fermentation by livestock and nitrogen oxide from the cultivation of agricultural soils, accounted for 18.4% of greenhouse gas emissions for Australia in 2000 [68].

Despite the basic separateness of process, there are several links between climate change and stratospheric ozone depletion:

- CFCs are also greenhouse gases.
- Rising global temperatures may increase human behaviors typically associated with warm weather – reducing clothing and enjoying the outdoors, thus increasing UVR exposure [69].
- There may be an increased incidence of UVR-induced skin cancers as increased temperatures cause acceleration of tumor genesis [70].
- Pollution of the lower atmosphere may cause a decrease in ground level UVR due to filtering by tropospheric ozone [71].
- Warming of the troposphere due to greenhouse gas accumulation may cause stratospheric cooling and delay recovery of the stratospheric ozone layer by a decade or more [72,73].
- Climate change may affect cloud cover and cloud optical thickness, which would in turn affect ground-level ambient UVR [73].

Notwithstanding these interconnections, there are eight types of differences between the stratospheric ozone depletion and climate change stories.

5.1 History of emissions

The history of the practices that are now causing global climate change is different from the processes that have caused stratospheric ozone depletion. While both stratospheric ozone depletion and climate change are the result of human activities, stratospheric ozone depletion is due to a relatively more acute exposure (CFCs) and is being addressed by a rapid and progressive management strategy. In contrast, climate change is a more chronic environmental stressor with slower onset and less dramatic effects, and strategies to address the problem are controversial and difficult to implement. Ozone-depleting substances were introduced in the late 1920s, while climate change represents the gradual outstripping of the Earth's ability to cope with the increased greenhouse gases produced as a result of human activities over several centuries [74]. In the past there has been a balance between the quantity of greenhouse gases emitted and the amount that can be absorbed by natural sinks in terrestrial vegetation and the oceans. But in the last few hundred years the burning of fossil fuels for energy, increases in emissions from industry and agriculture for an increased population, coupled with deforestation, have meant that emissions now exceed the capacity of the natural system to maintain balance [74].

5.2 Individual behaviors and benefits

A significant portion of greenhouse gas emissions is caused by the behavior and perceived "needs" of individuals (road transport, energy industries); reduction of emissions therefore requires behavioral change by individuals (both in the modes of preferred transport and perceived needs from energy industries), perhaps without obvious individual benefits but with increased costs. For stratospheric ozone depletion, the deleterious effects affect individuals, but profit-making industries developed and promoted the causative chemicals. Health is a precious commodity to individuals, and the deleterious effects of UVR (and the likelihood that stratospheric ozone depletion would cause increased UVR) on individual lifestyles and health are a clear and powerful stimulus to demand action. By contrast, dealing with the problems of climate change will require behavioral changes by individuals, constraining individual lifestyle freedom to avoid nebulous outcomes, many of which could be adapted to by the use of technology, e.g., air conditioning. Repairing the damage to the ozone layer has not required significant lifestyle changes by individuals; rather the brunt of change has been borne by industry in the development of adequate substitutes. This is unlikely to be the case with climate change.

5.3 Measures of process and impact

In examining the problem of stratospheric ozone depletion, we developed tools to measure directly the effects. Satellite monitoring and ground level monitoring of total ozone can be combined with various other geographic and atmospheric parameters (surface albedo, altitude, date and time, surface pressure, atmospheric temperature) to forecast levels of UVR. This UV Index is reported daily to the public in several countries [57,75], providing information to guide personal sun exposure behavior. That is, we can directly, simply, measure and communicate the effect of stratospheric ozone depletion and the likely hazard potential. In addition, skin erythema and sunburn are immediate, clear-cut effects of too much UVB. For climate change we can measure the causes, atmospheric pollution, levels of greenhouse gases, etc., but the effects remain much more difficult to quantify. Consequences of global climate change are projected impacts over future years, and there is little agreement about the exact nature and time course of these effects. This is especially true when one asks how a local region will be affected. It may be more difficult to "see" the hazard potential of a theoretical 1–2°C rise in temperature than to appreciate the satellite pictures of an increasingly large ozone hole.

5.4 Specificity, certainty, and commitment

There is an impression that there is more "give", more procrastination, in the global environmental response to greenhouse gas emissions than in the response of stratospheric ozone to CFCs. The problem of stratospheric ozone depletion is vivid – one can imagine or portray chlorine atoms in the stratosphere vigorously destroying ozone. Although initial action on ozone depletion was taken before physical evidence, since 1985 satellite pictures have shown an enlarging ozone hole over both the Arctic and the Antarctic. Ozone depletion is a "here and now" problem that requires an urgent solution. There is – rightly or wrongly – a much lesser degree of urgency sensed in regard to global climate change. Some regions will get a little warmer, some a little colder, some a little wetter, some a little drier. The IPCC [74] concluded that the globally averaged surface temperatures increased by 0.6 ± 0.2°C over the 20th century and globally averaged sea level is projected to rise 0.09 to 0.88 m by 2100. Changes have already been happening for at least a century, and while the chemical composition of the global atmosphere has undergone marked changes in the last 100 years [74], there have been no dramatic or visible changes in global climate in that time. It is only on examination of past trends in climatic variables and the accelerating changes forecast for the future that climate change becomes an evident problem.

5.5 Categories of health impact

From a health perspective, the biological and health hazards posed to human populations by increased UVR exposure are easier to specify and understand than are those known or thought to be due to the more complex and environmentally diverse processes of climate change. A flow diagram of the effects on health of excess CFCs and thus stratospheric ozone depletion would be quite simple (see Fig. 11.7), while the flow diagram of the effects of greenhouse gas emissions on health is much less well-defined and more wide ranging (see Fig. 11.8).

When theories of stratospheric ozone depletion were being elaborated, there was already concern about rising rates of skin cancer and investigations were beginning into other possible effects on the immune system and the eye [76,77]. Effects on health from UVR exposure were already being seen. Such effects were very concrete, common, and perceived as being specific to UVR exposure; i.e. although other exposures

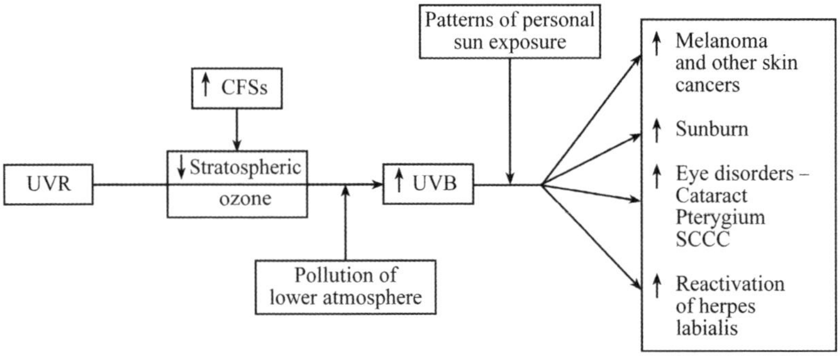

Figure 11.7 Effects on health of depletion of stratospheric ozone

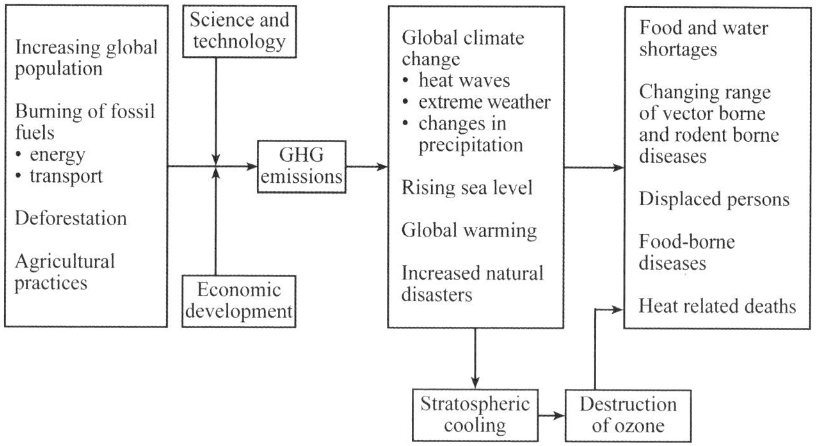

Figure 11.8 Health effects of global climate change

can cause skin cancers, these diseases are commonly viewed as being UVR-specific. The potential health effects of climate change, on the other hand are less clear-cut – there may, for example, be a greater likelihood of malaria in previously malaria-free regions, there may be more heat-related deaths (in some areas, and less in others), and there may be an increase in the toll of death, injury, and disease due to natural disasters such as droughts and flooding. But these effects are not specific to climate change – changes in their incidence will be relatively subtle and easily overlooked unless trends over long periods of time are examined.

5.6 Political considerations

From an economic perspective, the greatest risks of skin cancer were in fair-skinned populations living in temperate zones (mid- to high latitudes). These populations also happen to comprise most of the world's developed, rich, and powerful nations. They were a strong voice to force change. The developing nations were less likely to produce or use CFCs and thus were smaller contributors to the problem. The international agreements on ozone-depleting substances allowed developing nations to delay the phase out of the use of ozone-depleting substances beyond that of industrialized countries, but with an agreed schedule and recognition of the need for economic support to enable compliance. By contrast, the Kyoto Protocol, which seeks to limit the production of greenhouse gases by industrialized nations, remained stalled for several years, because it was unable to reach ratification targets [78]. The United States is the largest single contributor of greenhouse gas emissions (36.1% of 1990 CO_2 emissions by Annex 1 Parties to the Kyoto Protocol) and yet it has abandoned the Kyoto Protocol, citing huge losses to the US economy under that agreement [79]. The protocol does not impose emission limits on the developing countries, including large polluters such as China and India. Yet it is likely that it is the developing countries that will be most affected by climate change, with the least ability to invest in adaptive or preventive strategies.

5.7 Technical substitution

From a technical perspective the solution to stratospheric ozone depletion was relatively straightforward. Alternative, less ozone-destructive gases were quickly developed for industrial, refrigerant, and agricultural uses. The task therefore entailed a direct substitution, with relatively little economic penalty to national economies. Substitution of greenhouse-gas-producing technologies has been much more difficult, since much of our industrial, transport, energy-generating, and domestic infrastructure is predicated on fossil fuel combustion. Industrialized economies did not have a fundamental dependence on ozone-destroying gases in the way that they depend on fossil fuel combustion for the provision of energy. To constrain or eliminate the latter dependency will require much more radical transformations in social values, economic practices and technologies.

5.8 Adaptation versus mitigation

The IPCC [74] report notes "The extent and urgency of action required to mitigate emissions depend on our vulnerability". Our vulnerability in turn depends on our potential and the potential of our environment to adapt to the effects of climate change. But mitigation can only be effective if the entire global community participates.

5.8.1 *Potential for adaptation*

In contrast to climate change, the deleterious effects of stratospheric ozone depletion on human health are mediated through a single pathway – increased exposure to UVB. This means that adaptation to avoid harm to human health is theoretically quite simple: avoid exposure to UVB. Thus a whole industry has grown around products to provide sun screening, particularly from UVB – sunscreen, SPF-labeled clothing, and shade structures. There is a strong commercial incentive in the promotion of avoidance of UVR exposure. While some segments of the public have not taken the health messages related to sun avoidance seriously and others have over-reacted to the risks of exposure and incurred disease related to under-exposure [80], the potential for adaptation to stratospheric ozone depletion is great. Analysis of stratospheric ozone depletion with regard to the determinants of adaptive capacity and requirements for public health prevention outlined in Chapter 2 indicates a high level of potential for both adaptation and prevention (see Table 11.2).

On the other hand, the effects on health from climate change are as wide-ranging as the effects on the climate. The displacement of persons from low-lying atolls consequent on sea level rise is not an easy adaptation. Other effects such as heat waves are perhaps easier to adapt to by widespread use of air conditioning, primarily an adaptation of developed countries. Natural disasters by their nature are sudden and unpredictable and provide limited avenues for adaptation (early warning systems, evacuation protocols, etc). The spread of vector-borne diseases may be ameliorated by the development of insecticides or better treatment of those affected, but such adaptations are likely to be available to developed countries rather than the less developed countries that will be most affected.

Adaptation strategies suggested by the Australian Commonwealth Scientific and Industrial Research Organization (CSIRO) include better forecasting and changing the varieties, fertilizer regimes, and cropping mixes for the wheat industry [81].

Table 11.2 Determinants of adaptive capacity and requirements for public health prevention[a]

Determinants of adaptive capacity	*Requirements for public health prevention*
Range of available technological options (5)	Awareness that problem exists (5)
Availability and distribution of resources (4)	Sense that the problem matters (5)
Structure of critical institutions (5)	Understanding of the causes (5)
Human capital (5)	Capability to deal with problem (5)
Social capital (2)	Political will to influence (4)
Access to risk-spreading (1)	
Ability to manage information (5)	
Public's perceived attribution (5)	

Notes

a Scale = 1 (low) to 5 (high).

Range of available technological options: When the CFC controversy began in the 1970s, alternatives to CFC-containing aerosols were already available for a number of products, e.g., other propellants, pump sprays. For other CFC-containing aerosol products and refrigerants, industry relatively quickly sought safer alternatives. In addition, a new industry concerned with sun protection developed to aid (human) adaptation to the effects of stratospheric ozone depletion.

Availability of resources and their distribution across the population: While adaptation to stratospheric ozone depletion by human populations is primarily in the protection of individuals from excess exposure to UVB, other ecological systems are vulnerable. Less developed countries may have limited access to educational resources that elucidate the hazard.

The structure of critical institutions, the derivative allocation of decision-making authority, and the decision criteria that would be employed: Recognition of the hazard of stratospheric ozone depletion by human activities sparked the formation of critical international decision-making bodies.

The stock of human capital including education and personal security: The messages of the hazards of ozone depletion were relatively simple to understand and demonstrate. Personal security was clearly at risk and personal adaptive capacity high.

The stock of social capital including the definition of property rights: The global nature of stratospheric ozone depletion meant that each individual could contribute to amelioration of the hazard, but no individual would be immune from the consequences. It is not clear that social capital was important in considerations of adaptive capacity in the stratospheric ozone depletion setting.

The system's access to risk-spreading: Risk-spreading was unavoidable with stratospheric ozone depletion, but did not decrease any individual's risk.

The ability of decision-makers to manage information, the processes by which these decision-makers determine which information is credible, and the credibility of the decision-makers, themselves: There was relatively rapid scientific consensus on the science, the causes and the likely effects of ozone depletion. This consensus meant that critical decision-makers did not have to weigh different theories and decide what was credible. The discovery of the ozone hole and the rising levels of atmospheric CFCs were concrete demonstrations of the hazard.

The public's perceived attribution of the source of stress and the significance of exposure to its local manifestations: As outlined in the text, there was considerable media hype surrounding the release of CFCs to the atmosphere from aerosols, ozone destruction, and skin cancers. This chain of events was simple, believable, and frightening.

Awareness that a problem exists: As the text outlines, surveys in the United States indicated that a significant proportion of the population was aware of the ozone problem. There was considerable media coverage of the topic.

A sense that the problem matters: The link of ozone depletion to the common disease, skin cancer, as well as media portrayal of "loss of the Earth's protective shield" emphasized that this was a problem that mattered.

Understanding of what causes the problem: Scientists quickly moved to gain greater understanding of the theorized problem. The subsequent demonstration of the ozone hole was a graphic demonstration of the cause of the hazard (excess ultraviolet radiation). In addition, the scientific consensus on the cause of ozone depletion (CFCs) meant that the public was given a consistent message on the problem.

Capacity to deal with the problem: There was a clear path to both mitigation (through decreasing production and use of CFCs) and adaptation (through use of sunscreening technologies), until mitigation could take effect.

Political will to control the problem: The initial political will to control ozone depletion, with cancellation of the SST program, was probably more economically driven than environmentally driven. The subsequent bans on production and use of CFCs may have been driven by the public and the science community. But with the rapid development of replacement technology, there would be little to lose politically and much to gain in public opinion (saving the Earth's shield) by acquiescence to subsequent global agreements.

Rankings are based on a current global assessment of determinants present (but not necessarily perceived at the time) over the last 20 years.

In New Zealand, adaptation includes designing new roads, drainage, and water supply systems to take account of increased risks of flooding, landslides, and soil erosion [82]. Adaptation may be planned or can occur in a knee-jerk response to an event. Planned adaptation is more likely in situations where investment in the adaptation and the operating costs for the adaptation are relatively short term and time limited.

5.8.2 *Potential for mitigation*

Stratospheric ozone depletion ranks high in its potential for mitigation. It is a single effect with limited causes that have relatively simple, relatively inexpensive solutions. Mitigation is popular with the public both because it minimizes the lifestyle changes they must make and because a solution that allows a return to "safety" is foreseeable.

Climate change has probably been ongoing for many decades. There is a huge economic cost to mitigation and no easy alternatives that would cut greenhouse gases, and the necessary changes to individual lifestyle would be politically unpopular. Climate change is a slow phenomenon – cuts to greenhouse gases will be slow, as will any reversal of the effects of their excess. The potential for mitigation is low, and yet the potential for adaptation to most of the health effects is also low.

Perhaps the most important difference between ozone depletion and climate change that affects their potential for mitigation is the difference in time frame. It is likely to take several centuries to reverse the projected effects of climate change [74]. Compare this to the length of an elected term of office, the lifetime of a factory, or even a human lifetime. Stratospheric ozone depletion was recognized only 20 years ago, steps were taken to treat the problem with increasing rigor over that time, and atmospheric monitoring is providing the first signs that a cure may be occurring, with decreasing levels of some atmospheric CFCs.

6. CONCLUSION

In retrospect, the story of the global response to stratospheric ozone depletion appears to be a shining light of international cooperation to limit and repair the damage to a life-protecting atmospheric mantle around the Earth. Even so, it was a hard-fought battle, conducted over more than 20 years. The parallels to the need for protection of Earth's other natural life-sustaining systems are clear, and yet the international community has generally been reluctant to recognize and act on the causes. The reasons for this difference in response are complex and diverse. Compared to finding substitutes for ozone-destroying gases, the larger economic consequences and greater technical difficulties involved in finding adequate alternatives for the causal processes underlying global climate change, biodiversity losses, and freshwater depletion loom ominously large.

How might the response to greenhouse gas emissions be different if there were a common, life-threatening health problem clearly and specifically caused by those emissions in the same way that skin cancer is associated with increased UVR due to stratospheric ozone depletion? It is likely that the recognition of a potential major public health hazard in relation to first the possibility and then the reality of stratospheric ozone depletion was an extremely important stimulus to international cooperation, leading to the Vienna Convention, the Montreal Protocol, and the subsequent amendments. Elucidating the health effects of global climate change with more certainty could conceivably convince both individuals and governments of the corresponding importance of mitigation activities.

While adaptation strategies to lessen the effects of climate change may be possible in the short term and for wealthier individuals and wealthier nations, such strategies have limits. Eventually all will have to face the need for mitigation of climate change, i.e. for primary prevention of humanity's disruption of the global environment.

Stratospheric ozone depletion is dramatic, easily measurable, and clear-cut, and it is relatively simple to understand the general concepts of ozone loss and the consequences to life on Earth. In contrast, global climate change is a slow-moving, subtle phenomenon, and the likelihood and level of changes due to greenhouse gas emissions have been debated over many years. While there are still "doubters" of the stratospheric ozone depletion story, there has been much more skepticism in the climate change story. In contrast, the success story of the international cooperation to ban the production and use of ozone-depleting substances is a story of collaboration – between public health personnel; chemical, biological, physical, and social scientists; politicians; and industry. Such collaboration is still in its infancy for climate change.

What are the lessons to be learned from the ozone depletion story?

- There may be unintended consequences from technological advances. Little did we imagine that the development of safe and effective alternatives to ammonia in refrigerators could cause global environmental damage.
- There are difficulties in making changes in individual behavior compared to more structural changes, particularly when the alternatives for individuals are less acceptable in terms of cost or convenience.
- We cannot rely just on technological fixes, but must keep monitoring to check that the "cures" are proceeding on the desired trajectory.

REFERENCES

1. Lucas, R.M. and Ponsonby, A.L.: Ultraviolet Radiation and Health: Friend and Foe. *Med. J. Aust.* 177 (2002), pp.594–598.
2. McMichael, A.J.: The Thinning Ozone Layer. In: *Planetary Overload*. Cambridge University Press, Cambridge, UK, 1993.
3. Jablonski, N.G. and Chaplin, G.: The Evolution of Human Skin Coloration. *J. Hum. Evol.* 39 (2000), pp.57–106.
4. Space Telescope Science Institute's Office of Public Outreach, Formal Education Group. http://amazing-space.stsci.edu/light/ems-frames.html.
5. Morris, G.: Morphology of Ozone. Chapter 3 in: *Stratospheric Ozone: An Electronic Textbook*. NASA's Goddard Space Flight Center Atmospheric Chemistry and Dynamics Branch, 2000. http://see.gsfc.nasa.gov/edu/SEES/.
6. UNEP: Action on Ozone. Ozone Secretariat, United Nations Environment Programme, Nairobi, Kenya, 2000, pp.1–22. http://www.unep.org/.
7. Newman, D.P.: An Introduction to Stratospheric Ozone in Stratospheric Ozone: An Electronic Textbook. NASA's Goddard Space Flight Center Atmospheric Chemistry and Dynamics Branch, 2000. http://see.gsfc.nasa.gov/edu/SEES/.
8. Fraser, P.: Chemistry of Stratospheric Ozone and Ozone Depletion. The Health Consequences of Ozone Depletion Conference, Hobart, Tasmania, Australia, 1996.
9. Morrisette, P.: The Evolution of the Policy Responses to Stratospheric Ozone Depletion. *Nat. Res. J.* 29 (1989), pp.793–820.

10. Dotto, L. and Schiff, H.: *The Ozone War*. Doubleday & Company, New York, 1978.
11. Blum, H.F.: On Hazards of Cancer from Ultraviolet Light. *Am. Ind. Hyg. Assoc. J.* 27 (1966), pp.299–302.
12. Magnus, K.: Incidence of Malignant Melanoma of the Skin in Norway, 1955–1970. Variations in Time and Space and Solar Radiation. *Cancer* 32 (1973), pp.1275–1286.
13. Rehfuess, E.: Ozone Depletion and Public Health: Challenges for Policy-Makers and Health Professionals. The Combined Effects of Stratospheric Ozone Depletion and Climate Change on Human Health. WHO European Centre for Environment and Health, Rome, European Environment Agency, Ministero dell'ambiente e della tutela del territorio, Orvieto, Italy, 2001.
14. Schulze, R. and Kasten, F.: Effect of the Atmospheric Ozone Layer on the Biologically Active Ultraviolet Radiation on the Earth's Surface. *Strahlentherapie* 150 (1975), pp.219–226.
15. Kowalok, M.: Common Threads: Research Lessons from Acid Rain, Ozone Depletion, and Global Warming. *Environment* 35 (1993), pp.12–20, 35–38.
16. Molina, M. and Rowland, F.S.: Stratospheric Sink for Chlorofluoromethanes: Chlorine Atoms-Catalysed Destruction of Ozone. *Nature* 249 (1974), pp.810–812.
17. Ungar, S.: Knowledge, Ignorance and the Popular Culture: Climate Change versus the Ozone Hole. *Public Underst. Sci.* 9 (2000), pp.297–312.
18. The European Commission: History of the Montreal Protocol. http://europa.eu.int/comm/environment/ozone/history.htm.
19. Rowlands, I.: The Fourth Meeting of the Parties to the Montreal Protocol: Report and Reflection. *Environment* 35 (1993), pp.25–34.
20. Farman, J.C., Gardiner, B.G., and Shanklin, J.D.: Large Loss of Total Ozone in Antarctica Reveal Seasonal ClOx/NOx Interaction. *Nature* 315 (1985), pp.207–210.
21. Parker, L.: IB97003: Stratospheric Ozone Depletion: Implementation Issues. CRS Issue Brief for Congress, 2000.
22. Bell, G.D., Halpert, M.S., Ropelewski, C.F., Kouskey, V.E., Douglas, A.V., Schnell, R.C., and Gelman, R.C.: Climate Assessment for 1998. Figure 18. American Meteorological Society, Boston, MA. http://www.cpc.ncep.noaa.gov/products/assessments/assess_98/fig18.gif.
23. Dunse, B.L., Fraser, P.J., Krummel, P.B., Derek, N., Porter, L., and Oram, D.E.: CFC Measurements at Cape Grim, 1978–2002. *Cape Grim Baseline Air Pollution Station, Annual Scientific Meeting 2002: Abstracts*. CSIRO Atmospheric Research, Aspendale, Australia, 2002.
24. NASA: Waves in the Atmosphere Batter South Pole, Shrink 2002 Ozone Hole. Goddard Space Flight Center, Greenbelt, MD, 2002. http://www.gsfc.nasa.gov/topstory/20021206 ozonehole.html.
25. WHO: Environmental Health Criteria 160 – Ultraviolet Radiation. World Health Organization, Geneva, 1994.
26. Horneck, G.: Quantification of the Biological Effectiveness of Environmental UV Radiation. *J. Photoch. Photobio. B.* 31 (1995), pp.43–49.
27. Griffiths, C.E.: Dowling Oration Delivered at the Royal College of Physicians, London, Friday 5 June 1998. Retinoids: Renaissance and Reformation. *Clin. Exp. Dermatol.* 24 (1999), pp.329–335.
28. Woodhead, A.D., Setlow, R.B., and Tanaka, M.: Environmental Factors in Nonmelanoma and Melanoma Skin Cancer. *J. Epidemiol.* 9 (1999), pp.S102–114.
29. Holick, M.F.: McCollum Award Lecture, 1994: Vitamin D—New Horizons for the 21st Century. *Am. J. Clin. Nutr.* 60 (1994), pp.619–630.
30. Halder, R.M. and Bridgeman-Shah, S: Skin Cancer in African Americans. *Cancer* 75 (1995), pp.667–673.
31. Kricker, A., Armstrong, B.K., English, D.R., and Heenan, P.J.: Does Intermittent Sun Exposure Cause Basal Cell Carcinoma? A Case-Control Study in Western Australia. *Int. J. Cancer* 60 (1995), pp.489–494.

32. Armstrong, B.K. and Kricker, A.: The Epidemiology of UV Induced Skin Cancer. *J. Photoch. Photobio. B.* 63 (2001), pp.8–18.
33. Bergmanson, J.P. and Soderberg, P.G.: The Significance of Ultraviolet Radiation for Eye Diseases. A Review with Comments on the Efficacy of UV-Blocking Contact Lenses. *Ophthalmic. Physiol. Opt.* 15 (1995), pp.83–91.
34. Sliney, D.H.: Estimating the Solar Ultraviolet Radiation Exposure to an Intraocular Lens Implant. *J. Cataract Refract. Surg.* 13 (1987), pp.296–301.
35. Cameron, M.: *Pterygium Throughout the World.* Thomas, Springfield, IL, 1965.
36. McCarty, C.A., Fu, C.L., and Taylor, H.R.: Epidemiology of Pterygium in Victoria, Australia. *Brit. J. Ophthalmol.* 84 (2000), pp.289–292.
37. Taylor, H.R., West, S.K., Rosenthal, F.S., Munoz, B., Newland, H.S., Abbey, H., and Emmett, E.A.: Effect of Ultraviolet Radiation on Cataract Formation. *N. Engl. J. Med.* 319 (1988), pp.1429–1433.
38. Taylor, H.R., McCarty, C.A., and Nanjan, M.B.: Vision Impairment Predicts Five-Year Mortality. *T. Am. Ophthal. Soc.* 98 (2000), pp.91–96.
39. West, S.K., Munoz, B., Istre, J., Rubin, G.S., Friedman, S.M., Fried, L.P., Bandeen-Roche, K., and Schein, O.D.: Mixed Lens Opacities and Subsequent Mortality. *Arch. Ophthal.* 118 (2000), pp.393–397.
40. IARC: IARC Monographs on the Evaluation of Carcinogenic Risks to Humans – Solar and Ultraviolet Radiation. International Agency for Research on Cancer, Lyon, France, 1992.
41. Boldeman, C., Branstrom, R., Dal, H., Kristjansson, S., Rodvall, Y., Jansson, B., and Ullen, H.: Tanning Habits and Sunburn in a Swedish Population Age 13–50 Years. *Eur. J. Cancer* 37 (2001), pp.2441–2448.
42. Armstrong, B.K. and Kricker, A.: How Much Melanoma is Caused by Sun Exposure? *Melanoma Res.* 3 (1993), pp.395–401.
43. Clydesdale, G.J., Dandie, G.W., and Muller, H.K.: Ultraviolet Light Induced Injury: Immunological and Inflammatory Effects. *Immunol. Cell. Biol.* 79 (2001), pp.547–568.
44. Garssen, J., Vandebriel, R.J., De Gruijl, F.R., Wolvers, D.A., Van Dijk, M., Fluitman, A., and Van Loveren, H.: UVB Exposure-Induced Systemic Modulation of Th1- and Th2-Mediated Immune Responses. *Immunology* 97 (1999), pp.506–514.
45. McMichael, A.J. and Hall, A.J.: Does Immunosuppressive Ultraviolet Radiation Explain the Latitude Gradient for Multiple Sclerosis? *Epidemiology* 8 (1997), pp.642–645.
46. Freedman, D.M., Dosemeci, M., and Alavanja, M.C.R.: Mortality from Multiple Sclerosis and Exposure to Residential and Occupational Solar Radiation: A Case-Control Study Based on Death Certificates. *Occup. Environ. Med.* 57 (2000), pp.418–421.
47. van der Mei, I.A., Ponsonby, A.L., Blizzard, L., and Dwyer, T.: Regional Variation in Multiple Sclerosis Prevalence in Australia and its Association with Ambient Ultraviolet Radiation. *Neuroepidemiology* 20 (2001), pp.168–174.
48. Eurodiab: Variation and Trends in Incidence of Childhood Diabetes in Europe. EURODIAB ACE Study Group. *Lancet* 355 (2000), pp.873–876.
49. Ponsonby, A., McMichael, A.J., and van der Mei, I.: Ultraviolet Radiation and Autoimmune Disease: Insights from Epidemiological Research. *Toxicology* 181–182 (2002), pp.71–78.
50. Norval, M., Garssen, J., Van Loveren, H., and el-Ghorr, A.A.: UV-Induced Changes in the Immune Response to Microbial Infections in Human Subjects and Animal Models. *J. Epidemiol.`* 9 (1999), pp.S84–926.
51. Termorshuizen, F., Garssen, J., Norval, M., Koulu, L., Laihia, J., Leino, L., Jansen, C.T., De Gruijl, F., Gibbs, N.K., De Simone, C., and Van Loveren, H.: A Review of Studies on the Effects of Ultraviolet Irradiation on the Resistance to Infections: Evidence from Rodent Infection Models and Verification by Experimental and Observational Human Studies. *Int. Immunopharmacol.* 2 (2002), pp.263–275.

52. McMichael, A.J. and Giles, G.G.: Have Increases in Solar Ultraviolet Exposure Contributed to the Rise in Incidence of Non-Hodgkin's Lymphoma? *Brit. J. Cancer* 73 (1996), pp.945–950.
53. Latarjet, R.: Carcinogenic Risks Associated with Radiation Pollution. *IARC Sci. Publ.* 13 (1976), pp.179–190.
54. Kopecky, K.E.: Ozone Depletion: Implications for the Veterinarian. *J. Am. Vet. Med. Assoc.* 173 (1978), pp.729–733.
55. Slaper, H., Velders, G.J., Daniel, J.S., de Gruijl, F.R., and van der Leun, J.C.: Estimates of Ozone Depletion and Skin Cancer Incidence to Examine the Vienna Convention Achievements. *Nature* 384 (1996), pp.256–258.
56. Anderson, F.: "Sylvasun" and Sunburn Protection. *Med. J. Austral.* 1 (1968), pp.802–803.
57. U.S. EPA: The Sun, UV and You: A Guide to SunWise Behavior. U.S. Environmental Protection Agency, Washington, DC, 1999. http://www.epa.gov/sunwise.
58. APHA: Policy Statement 8709: Depletion of the Stratospheric Ozone Layer. American Public Health Association, Washington, DC, 1987. www.apha.org.
59. APHA: Policy Statement 9007: Increased Ultraviolet Radiation Exposure due to Stratospheric Depletion. American Public Health Association, Washington, DC, 1990. www.apha.org.
60. U.S. EPA: SunWise School Program. U.S. Environmental Protection Agency, Washington, DC, 2002. http://www.epa.gov/sunwise/index.html.
61. Sugden, T. and West, T. (eds): *Chlorofluorocarbons in the Environment: The Aerosol Controversy*. Ellis Horwood Ltd., Chichester, UK, 1980.
62. The Cancer Council of Victoria: Slip! Slop! Slap! Targets a New Generation. The Cancer Council of Victoria, Australia, 2002. http://www.accv.org.au/cancer1/whatsnew/mediareleases/2002/20020807a.htm
63. PHAA: Stratospheric Ozone Depletion. Public Health Association of Australia, Curtin, 1990. http://www.phaa.net.au/policy/stratos.htm.
64. CHDF: UVR and the Risk of Skin Cancer, Virtually Healthy. Children's Health Development Foundation, North Adelaide, South Australia, 2001. http://www.chdf.org.au/icms_wrapper?page = 433.
65. Lemon, R.: Taming the Trend: Are Teenage Girls Still Seeking that Flawless Tan "To Die for"? *Healthlink* Autumn (2002), p.9.
66. Vieth, R.: Vitamin D Nutrition and its Potential Health Benefits for Bone, Cancer and Other Conditions. *J. Nutr. Environ. Med.* 11 (2001): 275–291.
67. McGrath, J.J., Kimlin, M.G., Saha, S., Eyles, D., and Parisi, A.: Vitamin D Insufficiency in South-East Queensland. *Med. J. Australia* 174 (2001), pp.150–151.
68. AGO: National Greenhouse Inventory Report 2000. Australian Greenhouse Office, Commonwealth of Australia, Canberra, 2002.
69. McMichael, A.J., Lucas, R., McMichael, A., Lucas, R.M., Ponsonby, A-L, and Edwards, S.J.: Stratospheric Ozone Depletion, Ultraviolet Radiation and Health. Chapter 8 in: A.J. McMichael, D.H. Campbell-Lendrum, C.F. Corvalán, K.L. Ebi, A. Githeko, J.D. Scheraga and A. Woodward (eds): *Climate Change and Human Health. Risks and Responses.* World Health Organization, Geneva, 2003, pp.159–180.
70. van der Leun, J.C. and de Gruijl, F.R.: Climate Change and Skin Cancer. *Photochem. Photobio. Sci.* 1 (2002), pp.324–326.
71. Madronich, S., McKenzie, R.L., Bjorn, L.O., and Caldwell, M.M.: Changes in Biologically Active Ultraviolet Radiation Reaching the Earth's Surface. *J. Photochem. Photobio. B* 46 (1998), pp.5–19.
72. Shindell, D.T., Rind, D., and Lonergan, P.: Increased Polar Stratospheric Ozone Losses and Delayed Eventual Recovery Owing to Increasing Greenhouse-Gas Concentrations. *Nature* 392 (1998), pp.589–592.

73. Kelfkens, G., Bregman, A., de Gruijl, F.R., van der Leun, J.C., Piquet, A., van Oijen, T., Gieskes, W.W.C., van Loveren, H., Velders, G.J.M., Martens, P., and Slaper, H.: Ozone Layer-Climate Change Interactions. Influence on UV Levels and UV Related Effects. National Institute for Public Health and the Environment, 2002.
74. IPCC: Climate Change. Intergovernmental Panel on Climate Change, Geneva, 2001.
75. Marks, C. and McKenzie, R.: UV Index Forecasts and Data for New Zealand on the Internet. *Water Atmos.* 5 (1997), pp.13–14.
76. Fisher, M.S. and Kripke, M.L.: Systemic Alteration Induced in Mice by Ultraviolet Light Irradiation and its Relationship to Ultraviolet Carcinogenesis. *P. Natl. Sci. USA* 74 (1977), pp.1688–1692.
77. Pitts, D.G.: Glenn A. Fry Award Lecture—1977. The Ocular Effects of Ultraviolet Radiation. *Am. J. Optom. Physiol. Opt.* 55 (1978), pp.19–35.
78. UNFCCC: A Guide to the Climate Change Convention and its Kyoto Protocol. UNFCCC Climate Change Secretariat, 2002. http://unfccc.int/resource/convkp.html.
79. SMH: US Rejects Possibility of Signing Kyoto Protocol. *Sydney Morning Herald*, Australia, 2002. http://www.smh.com.au/articles/2002/10/25/1035416938187.html.
80. Zlotkin, S.: Vitamin D Concentrations in Asian Children Living in England. Limited Vitamin D Intake and Use of Sunscreens May Lead to Rickets. *BMJ* 318 (1999), p.1417.
81. McAlpine, G.: CSIRO Solutions for Greenhouse. Commonwealth Scientific and Industrial Research Organisation, 1999. http://www.csiro.au/csiro/ghsolutions/ index.html.
82. Wratt, D.: Climate Change New Zealand: Impacts, Vulnerability and Adaptation. *Pac. Ecologist* 1 (2002), pp.9–14. http://www.pirm.org.nz/pe1_ccnz.html.

12 The adoption of adaptation measures

Stephen H. Linder

ABSTRACT: Public health interventions range from military-style campaigns to sophisticated, community-based programming. A large portion, however, achieve only mixed or temporary success, especially when it comes to changing people's minds, manners, or morals. Still, there are lessons to be learned. As adaptation measures under consideration by the climate adaptation community come to depend on public action and support for their adoption, these lessons will become increasingly instructive. approaches to adaptation policy that rely on technocratic planning and stakeholder involvement may well meet the same fate as the formalized approaches to lifestyle change that foundered in the mid-1980s. This chapter explores these parallels and offers suggestions for avoiding obvious pitfalls. Since success in adoption efforts – whether for disease prevention or loss prevention measures – follows no sure recipe, the analysis here emphasizes ways to avoid more predictable failure. This chapter examines four principal approaches employed by the public health community for understanding and altering health-related behavior, paying special attention to their assumptions about people and how the problem of adoption is framed. It also provides some lessons on what appears to affect whether the public adopts or ignores adaptation-promoting interventions.

1. INTRODUCTION

While there is still some disagreement over how science improves and advances, the case of most professions is relatively clear – they learn. Of course, it is the people who belong to them that learn; but there is evidence that the problem-solving practices within professions are adjusted over time based on experience [1]. As a result, they happen to get better at solving their core problems, even though new ones keep coming along, testing their inferences from prior experiences. These notions challenge the conventional wisdom that superior technical rationality is the key. Schon [1] makes the argument that the best learners among the professions rely not only on experience but also on piecemeal trial and error. Making small adjustments under complex conditions permits them to isolate what works from what does not and to reduce the scale of possible failures. More important, the experiential learners get better at learning as they go along by attending to how learning itself occurs and how it can be improved. That is to say, there is critical self-reflection on the lessons of experience. If we examine the track record of the prevention community's population-based interventions, several cycles of learning appear.

In the early to mid-1980s, a series of large-scale interventions for preventing cardiovascular disease through behavioral changes were launched in the United States and UK. They were designed to be context-insensitive, standardized, and formulaic for ease of application. Moreover, they were deployed as you would an experimental intervention – all at once. By the late 1980s, the results were in. There were few detectable differences in behavior attributable to this style of intervention. By the early 1990s, interventions were community-based, and their formulaic character was made responsive to local conditions and local beliefs. There was some mixed success in this round. By the late 1990s, community-based interventions were involving their subjects in the design and implementation of mutually reinforcing change strategies. What changed over these cycles was not so much the technical rationality underpinning these interventions; rather it was the assumptions that informed or misinformed professional practice. The interventions were eventually done differently because they had not worked, even though the theory predicted that they would. Assumptions were changed and practices were altered. Aside from paying attention to what has been learned about how one does effective interventions to encourage adoption of behavioral adaptation measures, there are lessons here about learning itself.

The time scale for getting things right is always too short. If effectiveness comes from cycles of piecemeal testing and adjustment, then there must be room for making mistakes. If the best fixes are drawn from experience, they will rarely be quick. The design of interventions should include processes that support learning as we go along; otherwise, all of the mistakes will become obvious at once, once the intervention fails. Similarly, the assumption that one size can fit all – that there are standardized solutions for problems that appear familiar – denies the need for adjustments to local conditions and local knowledge. This, too, has led to failures in making designs workable and effective. The climate adaptation community is facing pressure to provide quick fixes that can be readily disseminated and emulated. The lessons from public health experience in these circumstances, however, suggest that effectiveness will prove illusive until several rounds of learning from experience are permitted to occur. In any event, the way public health has structured adaptation and responded to its demands offers lessons from its experience, perhaps sparing others some of the trouble.

The first round of knowledge transfers between the public health and climate adaptation communities rightly emphasizes the technical side of bolstering local capabilities for adapting to the possible hazards of climate change. Here, the emphasis is on interventions that require only passive cooperation from citizens. The spotlight focuses on the roles of experts in disease control and in disaster management, doing what they do best: surveillance and detection, planning and assessment, and the design and deployment of protective measures. Although citizens may be consulted indirectly through their elected office-holders or local officials, for the most part, the implementation of adaptation measures involves professionals acting in some authoritative way to protect lives, property, and well-being. The interventions themselves cover a wide range of activities, but most depend on public expenditures rather than private purchases for their support. When these interventions work, they assume the classical features of a public good, whether in greater infrastructure resilience or in reduced incidence of disease. Everyone benefits without individuals, alone, having to bear the costs or the responsibility.

At some point, however, the technical side of adaptation reaches the limits of its capability. On the one hand, tacit public support begins to wane, due to rising opportunity costs, apparent success, or both; this is the well-known paradox of preventative action – when the intervention works, the dangers that justified it appear to fade away, eroding support for its continuation. The case for intervention, then, must be remade in political, not technical, terms, so that the risks are not just understood but actually move higher on the public's agenda. On the other hand, achieving greater effectiveness can call for the public's involvement in the intervention itself. A portion of the responsibility for adaptation, in this circumstance, falls to individuals and communities. Perhaps, new beliefs must be accepted, everyday practices altered, and new behaviors adopted. While technical knowledge remains important, it is no longer adequate to the task of inducing change that must rely on individual cooperation in personal acts of adaptation. Hence, the behavioral concerns of enlisting and managing public involvement become a prelude to successful adaptation.

A second round of knowledge transfers should attend to this behavioral side, beyond the more familiar technical focus on altering the landscape and interrupting agents of harm. Unlike the technical side, what the lay public thinks and does, and how it responds to messages, signals, and exhortations, helps define the behavioral side. More important, it is a *particular* public, whose stock of beliefs and habits is bound up in the local context. For the designers of interventions, this particularity entails far more than proper sizing and fit. Beyond the adjustment of means, the goals of adaptation efforts and the values that underlie them may well be subject to challenge and mutual adjustment. Attention shifts from the design of adaptation measures to concern over processes to ensure the adoption of such measures.

There has been some place-based experience in the climate adaptation community with strategies to enhance adoption. Much of this, with a few notable exceptions, has entailed interactions with stakeholders, ready to negotiate priorities for governmental action without much citizen involvement. Modeling work on adoption likewise attends principally to stakeholders, but has been more accommodating toward individuals – so long as they respond to price signals and incentives in rationally predictable ways. While the public health community has some experience with both of these strategies, it also has substantial experience with the behavioral complexities of adoption, drawn from direct engagement with the lay public in community settings. Not all of this engagement has been salutary or successful, however. Rather than advancing any one example of successful, community-based work as a reliable recipe for promoting adoption, my purpose is to sound a note of caution to those who seek global solutions in local experience.

To understand how the adoption of adaptation measures works within the public health community, I begin by categorizing the rationales behind these measures into distinct models and approaches. Although public health has a special affinity for the biomedical model, its broad definition of health has forged alliances with a wide range of academic disciplines. Rather than attempting an exhaustive summary, the next section offers a schematic depiction of four selected models, drawn from a spectrum of approaches. These models, while not necessarily representative, reveal contrasting premises about who adopts adaptation measures and why. To be sure, as these premises vary, so do the designs of adaptation programs.

2. FOUR PERSPECTIVES ON ADOPTION

For our purposes, we can array the relevant approaches along a spectrum ranging from reductionist to holist. Reductionist, at one extreme, represents human subjects – our adopters – in terms of their physical and biological states. Holist, at the opposite extreme, portrays subjects in relation to aggregate states of populations. The former isolates subjects from their social relations; the latter identifies them with those relations. None of the four, considered here, falls at either extreme. Moving from more reductionist to more holist, we have preventive medicine, psychosocial behavior, community health, and social determinants approaches. Each is considered briefly, in turn. It should be noted at the outset that most efforts at this kind of synoptic characterization are doomed by exceptions. These approaches overlap, so the most one can hope for is to capture a few themes as central tendencies. Accordingly, our schematic treatments can be neither exhaustive nor definitive in the space permitted.

Comparisons focus principally on how each approach constructs the person as a subject at risk and frames the adoption of any adaptation measures in therapeutic terms. These promise to be the most relevant and enduring points of contact with the adaptation concerns of the climate adaptation community, once attention turns to engagement with the lay public. For the moment, I overlook the substantial methodological and procedural differences that distinguish these four approaches and put aside the many hybrid forms that crop up in public health work.

2.1 Preventive medicine

A preventive approach to medical care alters the orientation of the conventional clinical encounter from retrospective diagnosis and treatment to avoidance of prospective illness. The subject, in this case, is a patient whose risk of disease can be assessed from a complex profile of biological markers, physical traits, and likely exposures. A regimen for avoidance is prescribed and may entail pharmaceutical therapy and immunization, along with education in proven ways of reducing vulnerability. Adaptation, in effect, is a matter of coming to grips with one's physical limitations, inherited susceptibilities, and autobiography of exposures. Successful adaptation usually depends on three kinds of measures: adhering to a schedule of normalizing medications, if necessary; submitting to subsequent screening; and exercising self-discipline and control over one's exposures. To the extent that prudential judgment appears to play a role, adaptation fits into a moral economy of responsible self-care. You adapt to compensate for your revealed deficits in resilience; otherwise, you shoulder a measure of the blame for your own untoward health outcomes. Adoption, then, follows from patient education and the presumed merits of early detection and preventive treatment, but is socially constructed as a personal responsibility.

Much of this remains nominally voluntary, as responsibility follows from choice. In many institutional settings, however, coercion plays some role. Given the standing of physician directives, and the cultural premium placed on health maintenance, failure to adopt these measures invites sanctions from anyone with a stake in one's continued health and productivity. For this approach, adoption constitutes compliance and adherence to authoritative imperatives relative to dire, seemingly self-destructive, consequences. Even though population assessments provide the basis for calibrating health risk, the burdens of risk are personalized and linked to individual action and inaction, choices and consequences. Popular disease narratives account for incidence by reference to habits not changed, screening not done, and orders not followed. In short, you should know your portfolio of risk factors and do something to reduce them. The exceptions to such intimations of fault occur when responsibility can be shifted to divine providence, fate, or the negligence of others.

To be sure, few screening procedures are foolproof or immune to criticism. New findings periodically challenge prevailing practices, as with mammography and aerobic stress tests, eroding public confidence. And the expense of some procedures, such as full body scans, may not be worth the information that they provide. The most valuable tests detect conditions before symptoms appear, as with screens for some neoplastic diseases, for example, cancer of the colon or cervix, and for blood pressure, where timely intervention can prolong life. Still, getting patients to adhere to a schedule of periodic screening, especially when they are asymptomatic or uninsured, remains difficult. A similar picture of reasonable success attends the use of pharmaceuticals for primary and secondary prevention. Here, the principal issues are side effects and cost, and maintaining compliance in the face of these two. Finally, physician efforts at counseling patients on lowering their risk factors have found moderate success in some areas, such as smoking cessation in adults; in other areas, such as diet and exercise, the results have been less notable [2]. Outside the context of the research study, physicians seldom routinely counsel, and patients bear the expense of any subsequent programming that appears necessary for success.

In each instance, attention focuses on people's bodies, as opposed to other aspects of their selves; in effect, they are judged in social isolation and autonomy. The key risks they face involve the body breaking down prematurely, rather than dangers lurking in their social context. For the most part, risks are not only personal, affecting each body individually, but also private and shrouded in confidentiality. In this approach, as mentioned earlier, risks are constructed either as bodily deficits and susceptibilities or as threats to proper functioning that are largely self-imposed. Those that are self-imposed, the logic goes, can ultimately be self-remediated. To get started, however, people need to be willing to follow physician directives. Belief in the physician's authority, at least when it comes to bodily processes and risks, becomes critical. Furthermore, people need to be able to sustain the motivation for the necessary degree of self-control. Social norms that personalize risk, valorize expertise, and reinforce self-discipline help in this regard. Appeals to conscience may also play a role. Nevertheless, there are cross currents in social values that support their opposites: beliefs in the predominance of involuntary exposures, skepticism about allopathic medicine, and shifts in blame from the self to the purveyors of excess consumption. Motivation to adopt recommended regimes, even assuming that such regimes prevent disease, remains a complex problem. By most accounts, it prompted the development of the psychosocial approach, built on motivational concerns.

2.2 The psychosocial approach

The psychosocial approach begins with the basic parameters set by preventive medicine. First, people are at risk by virtue of their biological endowment and the lifestyle choices they make. Second, they can reduce these risks by making healthier choices and by following the regimes of screening and medication prescribed by their physician. In other words, there are rules of conduct that need to be followed to prevent disease. Unfortunately, not everyone who should comply does so. The question behind the psychosocial approach is, "why"? The answer comes in two versions. The health belief version focuses on getting the cognitive part right, while the health promotion version adds in the omitted noncognitive portions.

As patients, people are expected to respond positively to their physician's exhortations; the patient role has receptivity to expert advice and compliance with its dictates as its defining characteristics. As consumers, however, people's responses are thought to be mediated by a broader range of psychological factors that operate whenever choices about what to do or to buy come into play. When depicted as consumers, they become more autonomous than they do as patients, since the physician-patient relationship no longer structures motives and authority. They emerge as individuals facing a field of forces, armed only with instrumental reason and self-concern. Consumers can be induced to make more healthful choices, the argument goes, if they hold the right beliefs about the consequences of making and not making those choices. As with buying decisions, there are pluses and minuses, or costs and benefits, to be weighed before choosing. In psychological terms, these costs and benefits are perceived differently by people, hence the difference in choices, even if their preference for health over disease is the same. And these perceptions are derived from beliefs about consequences – beliefs that can be altered. Change the beliefs, this version claims, and you can change the choice of behaviors [3].

The relevant beliefs form a logical sequence in an idealized cognitive model. First comes receptivity to the message about proper choices. To pay attention, people must be convinced that the message applies to them. Since the core of the message is about the risks of dire consequences and what to do about them, people must accept the premise that they themselves are vulnerable to these consequences. In other words, they face the risk. They have the problem, and it is serious enough to warrant close attention. Next, they must be convinced that the remedies proposed, typically in the form of behavioral choices, actually reduce their risks. Having been convinced of their vulnerable condition and the efficacy of proposed remedies, the perception of insurmountable barriers may still sap motivation to change. Accordingly, ways to overcome these barriers must be highlighted and rehearsed. In the belief model, barriers are typically framed as cognitive rather than physical or financial. And so, thinking differently about them can potentially make them go away. To the extent that other psychic and social factors intrude, changing beliefs alone will not work.

Although the logic of the health belief approach endures – people appear to act as consumers, and their beliefs shape this action – there has been an effort over the past decade to expand this model to include both affective and relational elements [4]. The health promotion version has recrafted the consumer into the self-conscious, identity-seeking actor more familiar to modern advertising. In place of radical autonomy and rational calculation, this consumer operates in a web of complex relationships and responds to social approval, status insecurity, and feelings of efficacy more often than to tallies of prospective costs and benefits. The difficulty, of course, is that these psychic factors offer more intricate pathways to behavior than knowledge-based beliefs alone and appear more sensitive to social context, affiliations, and life course. Consequently, not only do health messages and other behavioral interventions have to be very carefully targeted to their intended audience, they must also find ways to reinforce their intended effect beyond a narrow focus on the individual's choice of behaviors. School-based programs, for example, must be reinforced in the home as well as in the community to be successful. The perils of not extending interventions beyond the individual are discussed in a later section.

On the whole, a number of these multilevel interventions have found some success in reducing risk factors, as recent reviews claim [5]. The more successful among them, however, exploit numerous pathways of influence as well as multiple levels of coordinated reinforcement. Unfortunately, what makes these interventions successful – that they are highly customized, carefully targeted, diversified, and fully coordinated – also makes them frightfully expensive, time intensive, and almost impossible to replicate. More complex designs also pose serious problems in measurement that can confound efforts to decide whether any verifiable success has actually been attained. Finally, unless the social and cultural influences on overconsumption and risk-taking behaviors have been effectively countered, once and for all, any observed changes simply will not be sustainable.

The alternative to these behavioral designs is to move upstream toward more inclusive campaigns that attend to whole communities and social groups rather than to individuals. Part of this effort employs mass marketing techniques to disseminate messages over print and electronic media in hopes of encouraging screening or reducing behavioral risk factors [6]. An example of this type of campaign appears in Chapter 11 of this volume. In this instance, ultraviolet exposure given the depletion of atmospheric ozone was documented to cause elevated levels of eye damage and skin cancer. Media campaigns were developed to convince the public to limit their sun exposure. The success of these efforts was confounded by the introduction of a technical substitute for ozone-depleting substances.

Reliance on the mass media for conveying this information means confronting a very different consumer than one finds in the physician's examining room. This consumer is inundated with messages of all kinds, most aimed at persuading them to choose and to buy. After years of these messages, consumers may develop a cynical, distrustful attitude that challenges marketers to be creative and timely in how they present their content. New formats quickly become clichéd and open to parody. Messages about health compete for attention in a crowded field of commercial messages. Because they offer counsel against the tide of consumption messages, they are especially susceptible to parody. Since popular culture changes so quickly, taking advantage of particular connotations or trendy appropriations means built-in obsolescence. For ads that persist, despite outdated references, there can be a "boomerang effect" [7]. Consumers respond by doing the opposite of what the ad counsels.

The most expensive public health communications campaign in history (almost $2 billion over 5 years, with $1 billion in federal money, matched in-kind by the media industry) is under way under the auspices of the US Office of National Drug Control Policy. Ads have been crafted, targeted, and pretested by top advertising firms. Extensive outreach efforts have been made to enlist the support of the news media and entertainment industries. Preliminary results appear to hold promise but, at this point, are based only on survey responses about perceived effectiveness [8]. More direct measures will come later. It remains to be seen whether the massive scale alone will be sufficient to overcome the contextual factors that promote drug use in adolescents. Its focus on individual awareness and resistance to social norms has more in common with earlier health belief models than upstream designs that cross social levels. Its success, if proven, may change thinking on these matters, but it is unlikely to offer a transportable design that could be scaled down and adapted to other purposes.

The expense and unsure merits of social marketing – and to some, its manipulative character – have helped rekindle interest in face-to-face, unmediated communications, relying on trust and community-based initiative to support the adoption of measures to improve health. Interest in "the community" has also come from a trend toward democratic participation in planning processes and a recognition of the centrality of mediating institutions to political reform in the post-communist world. Finally, the collective properties of communities, their assets, and social capital have drawn the attention of those interested in social and economic factors as upstream sources of health and disease.

2.3 Community health and participation

By most accounts, whether as partner, as reservoir of "proximal social relationships", or as socioeconomic context, community offers an alternative to the central role of the individual in models of health behavior. In effect, subjects emerge as members of a collective, perhaps as citizens or as neighbors, in place of their more reductionist treatment as patients or consumers. Similarly, the focus shifts from individual behaviors to involvement in social relationships and cultural practices. Every individual act takes place in a larger social and temporal context that defines its meaning and significance. In these circumstances, choices are made relative to a complex web of social expectations, responsibilities, and purposes that defy simple characterization as rational or irrational. Further, what appeared to be a free choice among health behaviors instead is taken to represent adaptation to structural and other influences that remain in the background of daily life. In other words, from a community health perspective, there is a social ecology of influences on health that transcends the cognitive and replaces notions of strictly individual agency with more collective determinations.

This logic of the individual as a social actor, bound within a particular context, finds expression in several approaches to community. And while they may share conceptions of the person as a community member, each approach employs a distinctive normative frame that shapes its efforts to enhance adoption. Still, their knowledge interests converge on building community resilience and reducing the vulnerability of its members through some notion of collective change. Consider, first, approaches that engage the community as partner. What is meant by "partner", in this instance, differs in degree. At one extreme, partnering is nominal and intended to supply the minimal conditions necessary to avoid resistance and induce cooperation. At the other extreme, the community is urged to assume creative control: interventions emerge from a collaborative process of assessment and design that relies for its content and implementation on the engagement of community members. Part of this commitment to partnering follows an old script from the annals of community development, where self-sufficiency and other civic virtues are to be built or recovered. In the course of defining its own problems and devising its own solutions, so the argument goes, the community strengthens its existing capabilities and develops new ones. Recent scholarship has retrieved many of these notions under the rubric of communitarianism, offering a moral basis for efforts to enhance civic life and health within communities.

Similarly, an enduring Jeffersonian tradition in the United States links grass-roots activism and local control to the moral claims of participatory democracy. Under these claims, the practice of participatory democracy not only is worthwhile, intrinsically, but also becomes the key to enhancing civic life. Active community involvement and deliberation, then, moves to the center of any proposal for social change. The term used currently in public health, however, is "empowerment" rather than democracy. The presumption is that people can assume responsibility for their own communities once they acquire certain rudimentary tools for planning and learn to work together to solve their problems. Once confined to the left, ideas about the need for mobilizing grass-roots efforts now find special resonance on the political right, which places stock in reducing the role of the public sector in community affairs. For reasons beyond the scope of this discussion, community organizing seldom poses the radical threat it once did to economic and political power-holders, as long as these efforts can be directed away from oppositional stances against those in power and toward planning for self-reliance. The nominally "empowered" community, then, seeks to recognize and remedy its own shortcomings rather than turning outward to confront any external causes of its prior state of powerlessness. Empowerment as self-help can be an effective tactic for the channeling and redirection of public dissatisfaction [9]. Hence, the term empowerment has taken on a pejorative meaning for some because of its association with continued "power over" rather than a "power to" as it originally promised.

Community participation also finds support on pragmatic grounds. Grass-roots involvement lends legitimacy to projects that promise local benefit and respect local control. In the parlance of health planners, without obvious signs of local involvement, there will be no "buy-in". From this perspective, there is a process requirement tied to any community project that involves external direction or assistance. Failing to meet this requirement can have a substantial, adverse impact on feasibility as well as on levels of cooperation and support. "Buy-in" extends beyond the well-known role of trust in fostering the public's uptake of risk communications. Individuals may well believe what authorities are saying, but if the message requires a voluntary and coordinated response by community members, some measure of local organization and commitment will be needed. The way to build commitment – in the absence of legal or moral obligation – is through voluntary, collective consent. The alternative to achieving "buy-in" is, of course, to apply coercion. Finally, without community involvement, up to some elusive threshold, programming cannot be tailored to local needs and circumstances, rendering its prescriptions irrelevant at best and unappealing or objectionable at worst. This is the enduring lesson of market segmentation that underlies modern advertising – messages need careful targeting in order to be heeded and given a chance to work. Even when uptake occurs, without the proper social reinforcement, programmatic effects are unlikely to be sustainable.

Community participation is now a familiar requirement for program funding from both government (regardless of the party in power) and philanthropy. The justification involved, whether pragmatic, political, or moral, appears to matter less among various advocates than the question of "how much is enough?" To the extent that maintaining control over program direction is a priority for funders, participation is likely to be nominal and largely symbolic – perhaps a town-hall meeting or a public hearing, depending on the funder's accustomed style of public "outreach" (or their public-relations sophistication). As the need for contextual information becomes critical to program implementation, more detailed input from the community will be sought and will be done so more systematically. Finally, as funders are willing to devolve control to the community itself, perhaps out of concern for sustainability, respect for local knowledge, or commitment to independence, participation will extend from priority-setting to mobilizing local resources for collective action.

Beyond the question of how much involvement is enough, there is the problem of means – knowing how best to involve the community. Some models for guiding these efforts are flexible and can accommodate different degrees of involvement and control, while other are keyed to a certain level. The selection of a particular model, then, cannot be made independently of assumptions about control. There are basically two models in use: the focus group (and a deliberative variation) and the stakeholder conference (and a collaborative variation). The next section considers these more closely.

2.3.1 *Focus groups and stakeholders*

The focus group, drawn from marketing research, engages small groups in conversations about their needs, values, and preferences. One variation on this model moves away from casting participants as potential consumers of services and interventions toward engaging them as citizens. When cast as citizens, participants are viewed as capable of deliberating, as jurors do, and of forming positions on issues and sorting through candidate proposals, as active voters do. The focus group is a mainstay of qualitative needs assessment, while the deliberative assembly (science jury, citizen panel, public forum, etc.) has been applied to land-use planning, interest articulation, and environmental decision-making [10]. The validity of the former rests principally on what the participants bring, that is, how widely held their views and experiences are and how open they are to sharing them. The validity of the latter rests more on the process itself, and whether collective judgment can be formed from the give-and-take of open dialogue. The focus group assumes that people already know what their problems and priorities are; the task is to reveal them to outsiders and perhaps to each other. The deliberative assembly, in contrast, assumes that these features will be formed and changed as the process unfolds. The logic behind adoption for the focus group participant rests largely on relevance and appropriateness; in effect, adoption favors the solution that each regards as appropriate to a problem that each perceives as important. For the deliberative assembly participant, adoption comes out of collective commitment; it is based on a mutually agreed-upon course of action to a consensual definition of the problem. Hence, the framing of adoption takes advantage of the role expectations assigned to each kind of participant. In other words, although each ultimately chooses whether to adopt any recommended changes, the grounds for this choice shift between the two roles. The consumer participant decides based on personal risk and benefit, while the citizen participant consents to be bound by a collective outcome.

A second model, the stakeholder conference, comes out of the dispute resolution literature and views participants as neither consumers nor citizens, but as disputants with clearly defined, material interests at stake. Participants are selected on the basis of their stakes, or their claims for relief, not for their community membership. The relevant stakes to be included are typically delimited by funders or program sponsors, who organize stakeholder involvement from the outside. The development of this model in the United States was prompted by the steady growth in litigation involving regulatory agencies. By the late 1970s, there had been a vast expansion of plaintiffs' standing to sue polluters and rulemakers in federal court. Efforts to reform regulation in the early 1980s, both to lower regulatory costs and to keep the government out of court, took advantage of a parallel effort by the Bar to reduce the civil caseload through mediation and other forms of alternative dispute resolution. Stakeholder forums are now widely used to structure local involvement in waste-disposal cleanup efforts under the Superfund Act, as well as to resolve controversies over proposed governmental regulations [11]. Negotiated rulemaking has been institutionalized in the United States by amendments to the Administrative Procedures Act, making the multilateral settlement among (qualified) stakeholders a legitimate alternative to conventional notice-and-comment rulemaking processes.

The basic intent behind stakeholder involvement is to resolve disputes, often among potential litigants, before they end up in court or otherwise disrupt the organizer's plans. Juridical notions of adversely affected interests, then, influence the selection of stakeholders, even though in the case of rulemaking the alleged effects are prospective. Not surprisingly, the language and roles resemble those of a contractual negotiation among adversarial interests, willing to give up only some minimal stake to avoid the costs of escalating the conflict. Rather than employing neutral third parties to resolve disputes through some kind of mediated settlement, the government or program funders typically organize and preside over stakeholder involvement. Further, the intended product is not just conciliation, but a public commitment or formal agreement among the participants to abide by the details of their settlement. Depending on the dispute, a wide array of interests may be included and efforts directed at facilitating compromises. Nevertheless, as with pluralist notions of group politics, only the organized and visible get to the table, unless substantial efforts are made beforehand to mobilize other voices. Not only is there a power differential among the participants, some being better prepared and more experienced than others, but also the stakeholders selected to participate are likely to reflect the unequal distribution of power in the community at large. As a consequence, the resolutions that emerge will seldom challenge either the power or the stakes of those best able to dominate the negotiations. At their worst, stakeholder forums build in a bias against any change perceived as a threat to well-established interests.

At their best, stakeholder forums provide a vehicle for consultation and collaboration – but only when goals are shared among the parties. Adoption (of participant roles) follows from the recognition of collective advantage. Paradoxically, these cooperative vehicles appear to perform best when they least resemble the classical stakeholder forum. In this instance, the motives for stakeholder involvement shift from conflict management to building local support and cooperation. Depending on the organizer's investment, they can resemble the symbolic outreach efforts mentioned earlier or support the formation of coalitions to assist the organizer's efforts. In public health, the most familiar of these are the health collaboratives involving service providers rather than community members and focusing on coordination and funding issues [12]. Here, the stakes differ as well. In place of material interests, each participant brings a service specialty and a target area or clientele to the table. Since their services tend to be complementary rather than competitive, and they often share service ideals, there is a natural cohesiveness at work that separates them from adversarial stakeholders, preoccupied with negotiating for advantage. The successful collaboratives have been those well funded, well led, and supported by stable membership [13].

2.3.2 *Community as research setting*

Before leaving community health, it is important to recognize a second approach that constructs community, not as a partner but as either a source of health-mediating factors or a collective subject. The more extensive set of studies posits community as a source of social and cultural influences on people's health status. Most involve population-based interventions built around preventive health services for a particular community. What distinguishes them from the psychosocial studies mentioned earlier is that they typically deploy more channels of influence, accommodate a wider range of social factors within the community, and rely on multiple outcomes for establishing effectiveness. The smaller set of studies considers the community itself as a unit of analysis and looks for positive treatment results, along the lines of a clinical trial. Fortunately, for our purposes, both kinds have been the subject of extensive recent reviews. The results of the latter set of studies appear mixed, largely because of the difficulties of maintaining the scientific integrity of these multiple-community designs, once they are in the field. Moreover, any intervention effects are often eclipsed by secular trends, affecting both treatment and control communities [14]. In short, no template for intervention design has emerged from these studies that can assure success in improving health status [15]. In contrast, the former set of within-community studies registers some notable successes.

The Task Force on Community Preventive Services, an advisory panel to the US Centers for Disease Control and Prevention, has undertaken systematic reviews of studies that test the effectiveness of various preventive services in community settings [16]. Their findings are posted and updated regularly on a dedicated web site (www.thecommunityguide.org) and published as a series of supplements to the *American Journal of Preventive Medicine*. Their recommendations, so far, cover several areas of risk behavior (tobacco use and motor vehicle safety); the utilization of certain preventive services (immunizations, dental screening, and sealants); one target of health promotion (physical activity); and several social factors implicated in population health (early childhood development and family housing). For our purposes, adoption of the desired habits and behaviors by community members most reliably comes about by employing multiple channels and modes of influence: some informational, some social, and some environmental. The more successful, say tobacco cessation interventions, engage the full range of factors in a given social ecology, from media and social support to regulatory measures. This entails not only coordinated policy changes but also a legion of skilled providers with sufficient funding to sustain the anticipated effects. The paradox here is that while most of the recommended interventions are targeted at community leaders and local service providers, they were designed, tested, and funded by authorities from outside the target communities – even when community involvement was instrumental. It remains unclear whether, when left to their own devices, communities can afford to replicate these successes, even if the know-how is transferable. The cost-effectiveness of most of these interventions has yet to be established.

2.4 Upstream determinants of health

Our final approach to public health, a focus on social determinants, is the most holistic of the four. Although social determinants increasingly find their way into complex models of population health that span levels of analysis, much of the research that supports their importance relies on claims about social structures [17]. The central structural feature of interest is stratification. That is to say, societies differentiate among people through an unequal distribution of power, wealth, and status, effectively positioning them in a hierarchical order, where privilege increases as one moves up the ordering. In Western industrialized countries, these hierarchies typically take on a pyramid-like shape, with strata separated by levels of income and education, type of occupation, race, and other aspects of social background and experience. And while poverty and the other disadvantages characterizing people at the bottom of these hierarchies have long been associated with more disease and shorter lives, recent evidence has established a robust gradient of steadily improving (deteriorating) health and longevity as one moves up (down) the hierarchy [18]. This has two major implications.

First, inequalities in social rewards and other privileges translate directly into inequalities in health. The fact that some people live healthier, longer lives than others do can be accounted for, in large part, by impersonal, aggregate features of their societies. Moreover, this relationship applies across countries as well as across strata; societies that have steeper pyramids of social inequality will manifest more dramatic inequalities in health and well-being. In other words, health status varies not only by social stratum but also by relative differences across strata: not only will people at the bottom of a steep hierarchy have poorer health status than those above them, they will also have poorer status than those at the bottom of a more graduated hierarchy [19]. Further, the effects are felt, not just at the bottom, but also throughout the hierarchy. Although wealth was traditionally assumed to confer health benefits on a population, it apparently does so only to a certain threshold; after that, economic inequality begins to have a pernicious effect. And the greater the relative inequality, the more substantial is the effect [20].

Second, these population-level inequalities in health appear to have little to do with the individual person's risk behaviors or access to medical care. The conventional targets of public health intervention have focused either on exposure reduction, typically involving chemical, biological, and physical controls, or on actions that individuals need to take, such as screening, immunization, and behavior changes. If structural factors such as stratification are the principal determinants of health in populations, then the focus shifts to measures that either affect the overall pattern of stratification and inequality in a society, such as labor market reforms and redistributive social policies, or compensate for its pernicious effects, such as early childhood education and family support [17]. Individuals can make choices to reduce their risk, but social relations, context, and certain structural features of the society in which they live and work mediate these choices and their effects. From this perspective, interventions focusing on individual-level changes deal only in temporary, symptomatic relief, since the roots of poor health are principally structural and systemic.

This is not to suggest that any determinism, economic or otherwise, is at work, ignoring the intentional component of social action, but rather that, if ignored, the structural component of inequality can operate at cross purposes to any intervention addressed to changing habits or behaviors. Moreover, the impact of structural determinants on health is not limited to vulnerable segments of underserved groups, but is experienced by all, relative to those higher in the social and economic pyramid. Framed in this way, reducing health risks must begin with the distribution of valued goods in the society and devise ways to alter it toward greater relative equality. One difficulty is that these are upstream effects that extend to factors beyond the normal purview of the health care community. Another difficulty is that social stratification mirrors and reinforces the prevailing distribution of power in society. Efforts to alter patterns of inequality, then, are likely to face substantial resistance. Still, there has been some programmatic headway in advocating and implementing redistributive reforms in Great Britain and Canada, motivated by social determinants research [18, 21]. Here, the targets for adoption shift from the patient and consumer to the politician and policymaker.

3. LESSONS ON ADOPTING ADAPTATION ADVICE

While lessons for those attempting to guide the public adoption of various regimens of improvement and protection have been mentioned throughout this chapter, the central ones bear repeating. First, the successful campaign for the adoption of behavior change is likely to be complex, difficult, costly, and, some would add, lucky. Once we move beyond simple motives and stimulus-response assumptions, interventions require information on relevant psychosocial mechanisms, an array of values and beliefs, social relations, context, and community involvement before a feasible design can be developed. Then, multiple channels should be engaged, if possible, at multiple levels with, as the Institute of Medicine notes, patience and persistence. To quote from a recent report [5, p.4]:

> To prevent disease, we increasingly ask people to do things that they have not done previously, to stop doing things they have been doing for years, and to do more of some things and less of other things. Although there certainly are examples of successful programs to change behavior, it is clear that behavior change is a difficult and complex challenge. It is unreasonable to expect that people will change their behavior easily when so many forces in the social, cultural and physical environment conspire against such change.

Even well-designed and amply funded demonstrations can falter, despite the latest techniques and the best of intentions. National trials in the 1980s, such as the Multiple Risk Factor Intervention Trial in the United States [5] and the Beating Heart Disease project of the Health Education Committee in the UK [3], were deemed ineffective because they focused on individual-level changes in behavior and standardized compliance regimens that, at the time, were the state of the art. In other words, they were built on the preventive medicine approach that narrows in on people as patients and promotes their adherence to prescribed changes, based on clinical authority. Contemporary demonstrations are more likely to be community-based, with substantial public involvement in design and implementation, and to accommodate multiple social and contextual factors that mediate both the meaning and significance of change for individuals. If we take tobacco cessation and initiation prevention as a prototype for this kind of design, the addition of regulatory and policy changes to this multilevel mix has been critical to its record of relative success [22].

Second, when it comes to dealing with humans in their social milieu, there are normative assumptions behind even the most rigorously scientific models. Every program for social change begins with some notion of their human subjects, their circumstances, how they behave, and why. In public health efforts centered on the adoption of expert advice, people have been constructed as patients, consumers, research subjects, citizens, community members, and social classes. Each construction conveys certain expectations about individuals, not only about the social roles they play but also about what it takes for them to make a change in their behaviors (habits or practices). With these expectations come social norms of appropriateness and implicit moral judgments concerning what normal people ought to do when faced with expert claims about the need for change. That is to say, they convey a sense of responsibility, if not for assuming the health risk itself then for its remediation. When constructed as consumers, for example, people tend to be held responsible for choices that presume both autonomy and a certain moral symmetry: people choose behaviors that put themselves at risk and can likewise choose behaviors to reduce that risk (so long as they have enough information and cues to do the right thing). In contrast, when constructed as community members, norms are assumed to be refracted by local institutions and social conditions. Responsibility, then, is more diffuse. Changing behaviors now entails changing social practices and the norms that reinforce them rather than simply individuals controlling their consumption. Finally, as members of a social class, expectations will vary by social location, as will responsibility for affecting change. Like the structure of social privilege, responsibility will increase as one moves up the pyramid. Interestingly, in this portrayal, experts can no longer be outsiders. Hence, the roles we assign to people are our roles as well.

Part of the difficulty in adoption arises because people may not see themselves in these designated roles or bound by these expectations. Instead of remaining hidden as common sense in the background of daily life, the norms (and the responsibilities they convey) then enter as foreign and coercive. Being socialized to these norms over the life course through familiar institutions is certainly quite distinct from having them offered de novo by experts. Further, the tendency to classify people in nominal groupings, based on demographic aggregations or physical and ascriptive characteristics, may itself be objectionable. These groupings and the benchmarks of normality derived from them bear no necessary resemblance to people's sense of themselves and their social identity. Hence, community is important, not just because of the need for input and cooperation, but also because it is the locus of social relations, norms, and identities. Without it, there can be no lasting connection to the people one hopes to influence.

REFERENCES

1. Schon, D.: *Reflective Practitioner: How Professionals Think in Action.* Basic Books, New York, 1990.
2. Institute of Medicine: *Promoting Health: Intervention Strategies from Social and Behavioral Research:* B.D. Smedley and S.L. Syme (eds). National Academy Press, Washington, DC, 2000.
3. Farrant, W. and Russell, J.: *The Politics of Health Information: Beating Heart Disease as a Case Study.* Health Education Council, London, UK, 1986.
4. Green, L.W.: Health Education's Contributions to Public Health in the Twentieth Century: A Glimpse Through Health Promotion's Rear-View Mirror. *Annu. Rev. Publ. Health* 20 (1999), pp.67–88.
5. Institute of Medicine: *Health and Behavior: The Interplay of Biological, Behavioral, and Societal Influences.* National Academy Press, Washington, DC, 2001.
6. Goldberg, M.E., Fishbein, M., and Middlestadt, S.E. (eds): *Social Marketing: Theoretical and Practical Perspectives.* Lawrence Erlbaum Associates, Mahwah, NJ, USA, 1997.
7. Fishbein, M., Hall-Jamieson, K., Zimmer, E., von Haeften, I., and Nabi, R.: Avoiding the Boomerang: Testing the Relative Effectiveness of Antidrug Public Service Announcements Before a National Campaign. *Am. J. Public Health* 92 (2002), pp.238–245.
8. ONDCP: Evaluation of the National Youth Anti-Drug Media Campaign: Fifth Semi-Annual Report of Findings, 2002. U.S. Office of National Drug Control Policy. www.mediacampaign.org/publications/westat5/.
9. Petersen, A. and Lupton, D.: *The New Public Health: Health and Self in the Age of Risk.* Sage Publications, London, UK, 1996.
10. Abelson, J., Forest, P-G., Eyles, J., Smith, P., Martin E., and Gauvin, F-P.: Deliberations about Deliberation: Issues in the Design and Evaluation of Public Consultation Processes. McMaster University Centre for Health Economics and Policy Analysis Research Working Paper 01-04, June 2001.

11. EPA: Stakeholder Involvement and Public Participation at the US EPA. EPA-100-R-00-040. U.S. Environmental Protection Agency, Washington, DC, 2001, January. www.epa.gov/stakeholders.
12. Israel, B.A., Schulz, A.J., Parker, E.A., and Becker, A.B.: Review of Community-Based Research: Assessing Partnership Approaches to Improve Public Health. *Annu. Rev. Public Health* 19 (1998), pp.173–202.
13. Sabol, B.: Innovations in Collaboration for the Public's Health through the Turning Point Initiative: The W.K. Kellogg Foundation Perspective. *J. Public Health Manag. Practice* 8 (2002), pp.6–12.
14. Hancock, L., Sanson-Fisher, R.W., Redman, S. et al.: Community Action for Health Promotion: A Review of Methods and Outcomes 1990-1995. *Am. J. Prev. Med.* 13 (1997), pp.229–243.
15. Sorensen, G., Emmons, K., Hunt, M.K., and Johnston, D.: Implications of the Results of Community Intervention Trials. *Annu. Rev. Public Health* 19 (1998), pp.379–416.
16. Truman, B., and 15 coauthors: Developing the Guide to Community Preventive Services – Overview and Rationale. *Am. J. Prev. Med.* 18 (2000), pp.18–26.
17. Tarlov, A.R. and St. Peter, R.F. (eds): *The Society and Population Health Reader: A State and Community Perspective*. The New Press, NY, 2000.
18. Acheson, D. (Chair): Independent Inquiry into Inequalities in Health Report. The Stationary Office, London, UK, 1998.
19. Wilkinson, R.G.: *Unhealthy Societies: The Afflictions of Inequality*. Routledge, London, UK, 1996.
20. Kawachi, I. and Kennedy, B.P.: *The Health of Nations: Why Inequality is Harmful to Your Health*. The New Press, New York, 2002.
21. CIAR: Accomplishments of CIAR's Population Health Program. Canadian Institute for Advanced Research, March 2003. www.ciar.ca/.
22. Warner, K.: The Need for, and Value of, a Multi-Level Approach to Disease Prevention: the Case of Tobacco Control. In: *Promoting Health: Intervention Strategies from Social and Behavioral Research*. National Academy Press, Washington, DC, 2000, pp.417–431.

13 International public health policy case study

Michael Sharpe

ABSTRACT: While the subject of climate change has been around in the climate science community for many years, the health sector is only beginning to organize itself to understand and manage the potential implications of climate variability and change for public health. This case study presents the strategic framework that the health sector is following to engage the national and international public health structures around the issue of climate variability and change. In addition, the framework leads the health sector to reach out to include other social and economic sectors in the research, assessment, and policy development mechanisms. To understand and encourage the relationships of the health sector to nonhealth sectors, a model of health-development-environment cause-effect relationships is integrated in a structured fashion with the determinants of health from the population health approach to visually demonstrate their relationships.

1. THE PROBLEM

Many of our social and economic decisions made at the individual, local, and national levels have either a direct or an indirect effect on our health or on the health of others. Climate variability and change are accelerating physical environmental conditions that also have the potential to have varying and widespread impacts on human health, depending on geographic location. Therefore, when we consider how the policies and practices of nonhealth sectors can adversely affect or benefit health outcomes, such as changes in agricultural output or transportation that affect human health, we realize that human health becomes a driving factor. When we consider how the quality of human health can adversely affect or benefit other social and economic sectors, such as lost productivity and community building, we realize once again that human health becomes a driving factor. The health sector must follow its convention of assessing and managing the risks posed to human health by various exposures, while reaching out to include other sectors in its risk management processes. It must also get involved with the decision-making processes of other sectors for a larger understanding and resolution of the potential problems from climate variability and change.

The question addressed by this case study is how the health sector is organizing itself to increase its understanding of such a complex issue in order to modify its policies and practices as well as those of other sectors and health partners. This case study presents the strategic framework that the health sector is following in some countries to engage national and international public health structures around the issue of climate variability and change. In addition, the framework allows the health sector to include other social and economic sectors in the research, assessment, and policy development mechanisms, an essential step to better appreciate and influence the actual evaluation of trade-offs that will be made.

1.1 International management of climate change and health

The World Health Organization (WHO) is involved with other agencies on the climate change and health issue [1]. WHO, together with the World Meteorological Organization and the United Nations Environment Programme (UNEP), have formed the Inter-Agency Network on Climate and Human Health to address the fourth of the four main areas of work of the larger Inter-Agency Committee on the Climate Agenda (IACCA), namely, (a) dedicated observations of the climate system; (b) climate services for sustainable development; (c) new frontiers in climate science and prediction; and (d) studies of climate impact assessment and response strategies to reduce vulnerability. The network is looking at studies of climate impact assessment and response strategies to reduce vulnerability through work on capacity building, information exchange, and research promotion. The network has undertaken a number of initiatives under each area of work, including national and regional assessments, burden of disease assessments, adaptation strategies, translation of Intergovernmental Panel on Climate Change (IPCC) findings, guidelines and training, human resources development, work with vulnerable regions, implementation of early warning systems, establishment of collaborating centers, expansion of the network, and research promotion.

Signatory nations to the United Nations Framework Convention on Climate Change (UNFCCC) are obliged to conduct regular national assessments of the impacts of climate change, including those on human health, every seven years and to report the results to the UNFCCC. Such assessments are intended to be of help to national governments in identifying, assessing, and managing the risks to human health and well-being that may result from climate variability and change at the community, national, and regional levels. The London Conference of Health and Environment Ministers in 1999 noted the clear absence of adequate assessment of the impacts on health and health systems in national assessments submitted to date at that time [2]. In large part, only the United States, Portugal, and Great Britain have, since then, significantly assessed part or all of their public health concerns in their national assessments. Small island state governments have also seen their health ministries become engaged in multisectoral policy responses to varying degrees on the issue.

It can be said in general, then, that health ministries and a large part of the health sector are not yet engaged at the policy level on climate variability and change but more at the technical level, if at all. In comparison, the climate science and environmental policy communities have been actively working collaboratively for 20 years or more on climate variability and change, which led to the creation of the UNFCCC in 1993 and the Kyoto Protocol in 1997.

Ministries of health and their health partners need to be engaged nationally and internationally on the climate change and health issue within two contexts: that of the requirements placed on signatory nations to the UNFCCC [3] and that of their role, responsibility, and capacity to manage a national public health system to protect and promote the health and well-being of their citizens. These two contexts influence how ministries of health and their health partners will assume their role of managing adaptation of public health infrastructure to the potential adverse effects on health and well-being from climate variability and change. Involvement of other social and economic sectors whose policies and practices have implications for population health must also be engaged by the health sector. Even though the determinants of health model recognizes the role of other sectors on influencing health outcomes, it is not the tradition of the health sector to engage nonhealth sectors in its health risk management processes, at least not in all areas or in all countries. For a complex physical environmental issue such as climate variability and change, there is little choice to doing things differently if the health sector wishes to be more effectively prepared.

A cross-sector collaborative effort is needed to support the identification of potential local, national, and regional health impacts resulting from climate change through the use of sound, applied research in national impact and adaptation assessments and to integrate the findings into adaptive public health policy that will reduce vulnerabilities and strengthen national capacity. Ministries of health need to give policy level oversight and facilitation to the national population health risk assessment process and research, provide a mechanism to include organizations involved with climate change and health, and use the national assessments as a lever to access Global Environment Facility (GEF) funds to aid in strengthening and increasing national and regional capacity. Improved health outcomes can be achieved through adaptation measures that modify or prevent future and current risks to health. They can also be achieved through the participation of other social and economic sectors whose policies and practices directly and indirectly affect health outcomes. Thus adaptation to climate variability and change by other sectors is equally important to the health sector.

1.2 Summary of impacts on public health

The international health community has identified potential health effects from climate change and other global and regional environmental changes that are anticipated to increase in the future [4,5]. These include, but are not limited to, health exposure issues related to increased smog episodes, heat waves, water- and food-borne contamination, vectors and the pathogens they carry, stratospheric ozone depletion, and extreme weather events, affecting vulnerable populations and associated socioeconomic impacts. Even as efforts are taken to reduce greenhouse gas emissions, climate variability and change have the potential to accelerate in this century, and adaptive policies will be needed to minimize the potential impacts of climate change on human health. Table 13.1 illustrates the range of health exposures to climate change, and examples of possible public health vulnerabilities [4,6].

Vulnerability to climate change impacts will vary depending on geographical location, national and local health service infrastructure, the economic and social systems and pressures of a country, and the underlying burden of disease. The vulnerabilities already exist thus Table 13.1 should come as no surprise to the health sector. It is a matter of incorporating climate variability and change as an additional variable that will increase the stress on current and future priorities and programs of ministries of health and their health partners. The health sector and other sectors that can affect health outcomes will need to ask themselves the following questions based on the strength of evidence and uncertainty: (1) What do we need to do more of (or less of) in the face of climate change? (2) What do we need to do differently?

Table 13.1 Health exposures and vulnerabilities

Health exposures	*Examples of health vulnerabilities*
Extreme temperature events	– Cold and heat related illnesses – Respiratory and cardiovascular illnesses – Increased occupational health risks
Extreme weather events	– Damaged public health infrastructure – Injuries and illnesses – Social and mental health stress due to disasters – Occupational health hazards – Preparedness and population displacement – Nutrition losses from food production failure
Air pollution	– Changed exposure to outdoor and indoor air pollutants and aeroallergens – Asthma and other respiratory diseases – Heart attacks, strokes and other cardiovascular diseases – Cancer
Water- and food-borne contamination	– Enteric diseases and contaminants – Sea water infiltration
Vectors and the pathogens they carry	– Changed patterns of diseases caused by bacteria, viruses, and other pathogens carried by mosquitoes, ticks, and other vectors
UV radiation	– Skin damage and skin cancer – Cataracts – Disturbed immune function
Rural and urban population vulnerabilities	– Seniors – Children – Poor health – Low income and homeless – Traditional populations – Disabled – Immigrant and destabilized populations
Health and socioeconomic change	– Changed determinants of health and well-being – Vulnerability of community economies – Health and social cobenefits and risks of greenhouse gas reduction technologies

2. A MODEL AND SOLUTION

2.1 Integrated population health risk management

It is generally understood that the state of the environment and human health are linked. This is one of the relationships defined by the principle of sustainable development. A sustainable development approach would define the state of the human environment in terms of the physical environment, the economic environment, and the social/community environment. The premise is that sustainable policies that are good for the state of the human environment would be good for human health. Conversely, unsustainable policies would be bad for human health, if not in the short term then certainly in the long term. In short, this is the base of conventional sustainable development policies and programs.

The WHO has expanded these human environment relationships [7]. It has measured the state of the human environment against the effects of common national driving forces of economic, social, and health systems that create societal pressures and behaviors, with their resulting outcomes on human health. The resulting model (in the left column of Fig. 13.1) integrates driving factors, societal pressures, state of the human environment, human health exposures, and health effects into a health, development, and environment cause-effect analytical framework. Actions that affect health outcomes occur at each stage of the causal chain according to the span of influence a policy has over a nation and the health of its population. The model is commonly known by its acronym, DPSEE.

Another element added to the framework of integrated health risk management comes from the population health approach, that of the determinants of health. The benefit of focusing on causes rather than effects has been less clear in public health, in large part because the links between some determinants and health outcomes are not always direct [8]. Determinants range from national or community level health services and social support networks to individual genetic endowment factors. There is general agreement that all the determinants play a role to varying degrees in the healthy lives of people wherever they may live in the world. By associating the determinants of health in a structured fashion to the integrated framework (the middle column of Fig. 13.1), it becomes clear not only how the determinants relate to each other but also where they relate to counterparts in the DPSEE model, making a rational linkage to the benefits and costs to health exposures and outcomes of policy decisions made in non-health sectors. Therefore, from a policy research and planning perspective, it becomes

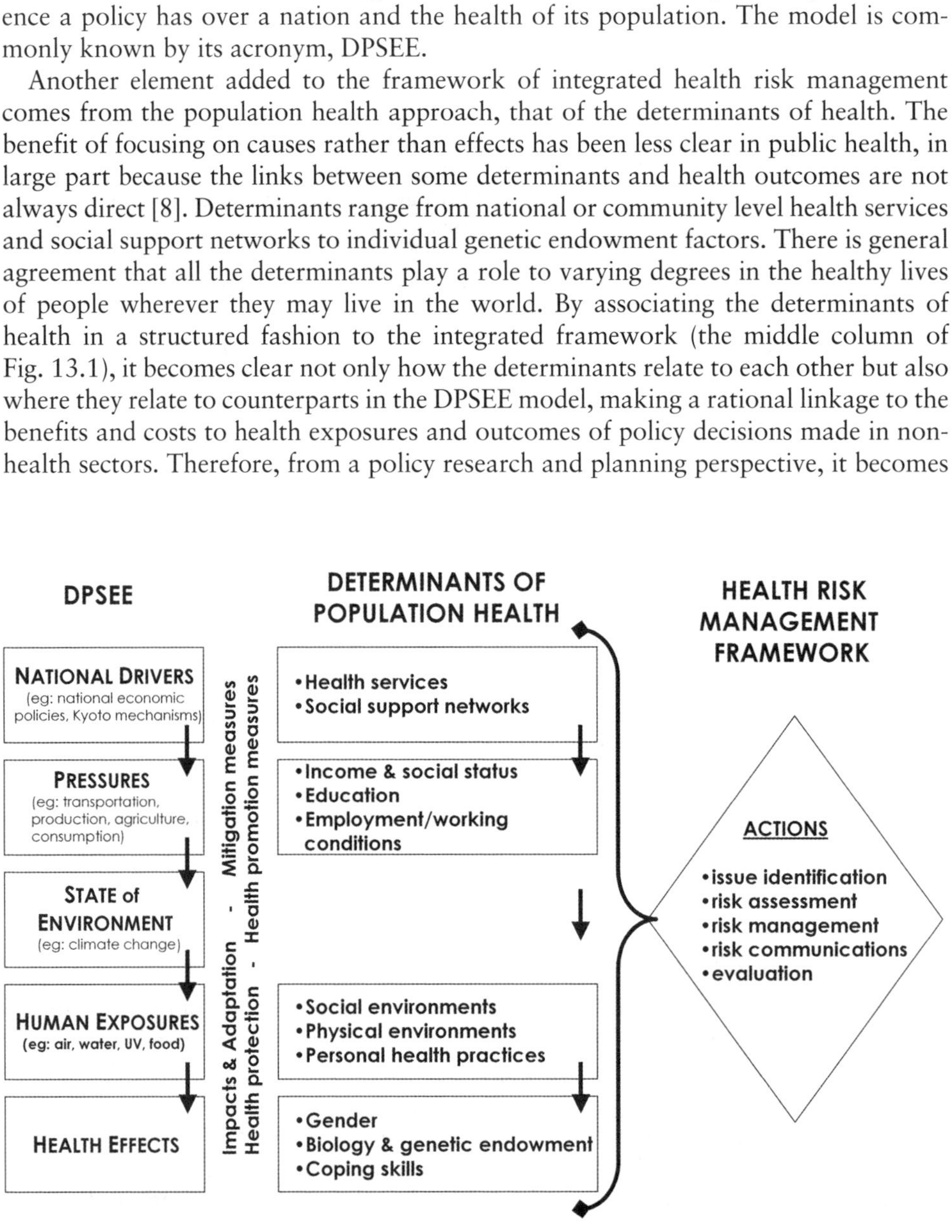

Figure 13.1 Integrated population health risk management model

easier to analyze the roles and responsibilities that other sectors have on health outcomes.

For the public health sector, the next step is to integrate the DPSEE model and the determinants of health into the policy actions of its decision-making framework for health risk management (in the right column of Fig. 13.1). Usually, issue identification, risk assessment, and risk management processes of health sector decision makers examine only cause and effect within the factors of either human exposures or within health effects, and usually interpret the problem narrowly. By associating both risk assessment and risk management processes with the factors in each of the DPSEE stages and with all the determinants of health, a greater range of options arises because the link to the cause from policies made by other economic and social sectors comes earlier in the analysis of cause-effect relationships. Policy solutions applied earlier in the chain of the integrated framework, usually by other sectors, always have the greatest benefits for health outcomes at less cost. And, of course, the converse is also true whereby unsustainable practices by various economic and social sectors can have highly adverse impacts on health outcomes.

The strength of integrated population health risk management is that it shows how policies ranging from agriculture to transportation can demonstrate either positive or negative effects on human health outcomes as well as on the physical environment. Many public health policies, therefore, are made by sectors other than health. The public health sector must recognize that it needs to actively participate in influencing other sectors' policy development processes to ensure sustainable policies and practices, minimize harm to human health, and enhance positive benefits to healthy living. This, of course, is a two-way street.

This model situates the health sector in the complex decision-making process along with partners from the other social and economic sectors rather than in isolation. Evaluating risks and benefits to health outcomes becomes a joint responsibility. Through a broader collaboration of sectors it becomes possible not only to reduce human exposure to hazards from climate variability and change as would be conventional in public health but also to create or induce significant change in a society or its economy to eliminate or mitigate the hazard itself, as well as create innovative and cost-effective adaptation measures that benefit health outcomes.

The strategic framework described later incorporates the use of an integrated approach to population health risk management.

2.2 Role of health and climate change

The public health community, outside of a few countries, has not yet been engaged at the policy research and decision-making level on the issue of potential impacts on human health from climate variability and change. Why is this important? Next to the potential impacts on physical and economic infrastructure, the potential impacts on health and social systems are likely to be equally large and much longer lasting. Table 13.1 tells the health story in climate variability and change. Numerous existing health problems and programs will potentially see additional stress placed on them over the next 50 to 100 years.

If the potential impacts are so great, why is the health sector not getting more engaged in the issue? As in all systems, there are higher current priorities to address with the limited resources at hand. Incorporating the issue into current health programs makes it less threatening organizationally and a pragmatic use of significant resources already allocated across programs. Plus, the public health sector is a very large entity to mobilize in every country and internationally. In short, the health research and policy processes will engage the climate variability and change issue only within their current health mechanisms before it will involve itself in policy processes of nonhealth sectors such as the environmental policy community. Both levels of engagement are essential to achieve the two-way street of policy development influence described above, but the health sector must organize the climate variability and change issue on its own terms before it can effectively contribute to and participate in external policy processes. Clearly all of the aforementioned discussion refers to the realities of government health policy research and decision-making. However, the health sector must get itself out of the old paradigm of analyzing and resolving issues in isolation.

Thus the education, influence, and commitment of health policy makers are needed to successfully mobilize all aspects of health research and policy analysis on the issue, to get beyond the piecemeal and unorganized involvement at the technical level. This explains the need for a strategic framework that structures the major research and policy activities needed to build and integrate the climate variability and change issue into current health research and programs, as well as participate in nonhealth sector policy processes.

2.3 Strategic framework

The goal of this strategic framework for health and climate variability and change is to create a cycle of four major types of activities that continuously provide input to the rest of the cycle activities depicted in Fig. 13.2. The activities are premised on the need for research-informed policy and for policy-guided research, leading to a balanced integration of science and policy.

The goal will be achieved through communication with and understanding of the many partners who have a role in the health story in climate variability and change. Achieving the goal requires a cycle of four basic mechanisms that provide opportunities for knowledge exchange and impetus for collaborative actions to generate knowledge; assess national, regional, and global risks to health; and manage and communicate the risks to health from climate variability and change. The cycle fundamentally follows a risk assessment and risk management process: scoping of the problem, assessment of the problem, and management of the problem [9]. In the

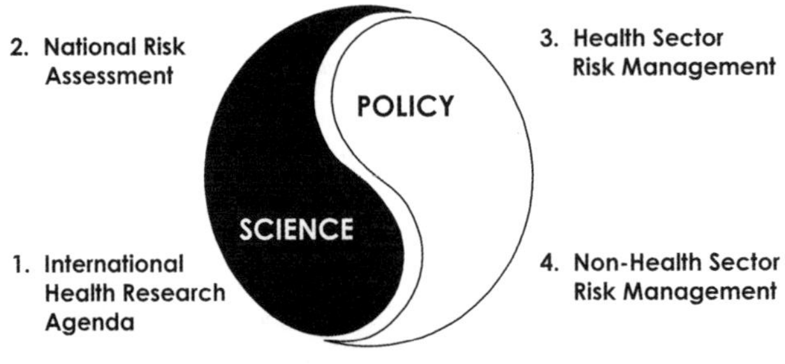

Figure 13.2 Strategic framework for research and policy processes

strategic framework two types of science mechanisms provide results to the two levels of policy mechanisms, which in turn provide feedback and policy questions to be solved to the two science mechanisms. The two science mechanisms of research and innovation plus national risk assessment represent the scoping and assessment of a policy problem. The two policy mechanisms of risk management by health partners and by other sectors represent management of the problem.

2.3.1 *Research and innovation*

The first mechanism is an international health research agenda for climate variability and change. This mechanism emulates national processes and is necessary to assess the strengths and limitations of current knowledge, methods, and science capacity on the topic. This is a crucial input to policy development. Assessing the gaps in the current knowledge and outlining interdisciplinary research questions and their prerequisites will mobilize the international research community to address the gaps and weaknesses. The result is an international health research agenda that can guide all involved, worldwide.

The research agenda becomes a tool around which to galvanize national and international research centers and individuals to work toward its achievement in multidisciplinary, collaborative research. The path to create and continuously revise the research agenda is via occasional events that pull together the science and policy research community around all eight health exposure issues (see Table 13.1) at the same time. The path also includes events that continually explore the eight health exposure issues individually and in greater detail.

The second aspect for research and innovation is the deliberate involvement of health science and policy researchers in international and national research agenda setting events held by other sectors on the subject of climate variability and change. Cross fertilization of knowledge, methods, and data is essential for new research collaborations to address complex problems in an integrated way.

2.3.2 *National risk assessment*

As mentioned earlier, some countries have conducted national assessments of the potential impacts on human health from climate variability and change. Many other countries have conducted a comprehensive review of the scientific literature to satisfy the 7 year national communication requirement of the UNFCCC. A national assessment can provide essential information for decision-making [11] to: better understand current vulnerabilities; evaluate a country and its communities' capacity to adapt to climate change by modifying the health infrastructure or by adopting specific adaptive strategies, policies, and measures; and determine essential knowledge gaps that must be filled to fully understand possible climate change impacts.

Common national assessment tools are essential for comparability of results, research promotion, and capacity building. The UNFCCC guidelines for the preparation of national communications have few references to human health [4,11]. A collaboration between the WHO, UNEP, and Health Canada has facilitated the development of methods for assessing vulnerability of climate change and human health [10, 11]. Both the assessment tools and the guidelines learn from international health research and through feedback from health ministries who use the national assessment tools to evaluate their own population and system sensitivities and vulnerabilities. These common tools will aid in producing comparable results for use at the local, national, regional, and international levels and provide the ability to communicate urgent needs for adaptation strategies, policies, and measures.

While the national health ministries are encouraged to take the lead role in facilitating and coordinating the population health risk assessment, the ministry must democratize the process, whereby the scientific community joins with the policy community to conduct the assessment. Furthermore, "all stakeholders — those parties with specific interests in the issue or measures to address it — should be given the opportunity to participate" from the beginning [12]. The stakeholders represent the affected communities at the local and national levels, and often sectors other than health as well as the health sector itself. However, no one group of stakeholders can be allowed to drive the assessment process in support or protection of its policies or practices [12].

For an assessment to be successful, the affected community, not just the scientific and policy communities and certainly not just the health sector, must agree that there is an identifiable problem. Thus for public health to influence other sectors it will have to expand its definition of stakeholders beyond its conventional health partners to include new partners from other sectors in the research, assessment, and policy making processes. The choice of other sectors to be involved in the assessment of each potential health exposure issue can begin by examining what is considered to be a potential cause of adverse health outcomes and who is capable of executing policy that has an impact on a determinant of health at the national driver level, at the social and economic pressure level, and at the state of the environment level. This is essential to achieve an integrated approach in the scoping of a problem, its assessment, and its management.

The participative approach is intended to motivate other sectors to include climate change and health as part of their thinking and policy decisions [12]. In return, other sectors can offer data, perspective, and research collaboration. This results in a sharing of the work and the cost of an assessment, and in an integrated assessment that has a greater chance of acceptance at the policy level because of its broader support in a society.

A population health risk assessment is intended to inform decision makers so that they can make appropriate decisions; thus, it must be relevant to their information needs. The difficult part of the facilitating and coordinating role of health ministries is that they must maintain the credibility, usefulness, and effectiveness of the assessment process for the results to be accepted in the risk management phase [12].

2.3.3 *Public health risk management*

A regular meeting of senior public health policy officials is an essential mechanism to organize and develop the health sector to realize its role and responsibility for implementing adaptation and core capacity building globally and within and between their countries, to assess and manage the potential risks to the health and well-being of their people from climate variability and change, and to manage climate change effects on health infrastructure. In short, the health sector must get itself organized on the issue of climate variability and change to collaborate effectively with other sectors and with its health partners. The participation of nonhealth sectors is of pivotal importance to expand health policy beyond single sector planning.

Such a regular policy mechanism is necessary to assess the strengths and limitations of current international health policies as well as identify and assess the risk posed by policy gaps. This is crucial feedback to an international health research agenda to give guidance on scientific and policy research questions. The challenge for a policy conference will always be to set priorities that will (1) produce research that policy-makers and others will want to use, and (2) engage the interests and commitments of the research community [13]. A regular policy conference must also share comparable results from national health impact assessments to better understand and collectively assess the degree of risk to global and regional health from climate variability and change, and its consequential effects on world trade and movement of people. Collectively, senior public health officials must also be prepared to discuss with the international community of health partners and other sectors national and international priority issues for successful adaptation measures for health impact assessment, training, capacity building, and risk management and communication. Policy recommendations to modify or adapt current public health practices could then be raised by the health ministries to the forum of the WHO's World Health Assembly for international health policy decision making.

2.3.4 *Other sectors' risk management*

The international community of the United Nations and other international organizations and sectors implicated in health or climate change also need to consider modifying or adapting their current policies and practices to improve health outcomes. While active participation in the international health and climate change policy conference mechanism outlined above is essential to inform and influence the health sector of the interests and concerns of other sectors, the inclusion of the health sector in the policy events and processes of the nonhealth sector is equally important to achieve improved health outcomes and socioeconomic and environmental benefits as well.

3. CONCLUSIONS

There are many risks posed during the development and execution of a strategic framework to solve a problem that will be with us for a long time. The first is always the role of leadership. The more complex a problem is, and environmental change from a changing and variable climate is surely a complex problem, the more complex the partnerships and the possible solutions. Population health risk management is never simple at the best of times, and an integrated approach to health risk management implicates all economic and social sectors.

The major parties in the nonhealth sectors must be involved early and often in the risk management and communication processes at the local, national, regional, and international levels to ensure success. Differing perspectives, values, and capacities, especially when they are involved early on in the integrated population health risk management model, will lead to two things: (1) more innovative solutions earlier in the cause-effect chain of relationships, usually with greater health benefits and lower costs; and (2) broader support in the society at large [14]. The health sector is in a unique position to increase the understanding of the general public of the implications of climate change in their lives through potential impacts on their health and well-being, and aid in understanding the importance of their contribution in prevention and mitigation of climate change. Broader support by the general public is needed to convince other sectors to take corrective action.

A key ingredient in the collaboration of multiple organizations, multiple disciplines, and multiple mechanisms is a cycle of activities that generate and exchange knowledge between the research and policy communities. Although this strategic framework concentrates at the international and national levels of activity, the purpose is to support the local, national, and regional needs for understanding and developing knowledge for policy making. The expected constraint of developing quality information at all levels is addressed in the development of a common and comparable set of assessment tools, as well as in the mechanisms to exchange new knowledge, interests, and concerns between the multisectoral, multidisciplinary research and policy communities on a regular basis to stimulate a common understanding of the range of problems faced.

In international organization terms, the strategic framework also supports three core functions of the WHO [5]:

- Assisting national health ministries, their community and research health partners, and other social and economic sectors in collaboratively assessing and managing the risks to health and well-being from the adverse effects of climate variability and change; in the process, establishing the evidence for cross-sectoral policy actions to improve health outcomes and to reduce future health care costs.
- Establishing and expanding collaborative and strategic research and information networks to create a milieu that can take advantage of innovative science and policy ideas for complex policy problems.
- Investigating local, national, regional, and international collaboration within the health sector and with other sectors to provide research-informed policy as well as technical development support for sustainable capacity building.

As with all environmental health problems, it is not possible to resolve it with the conventional approach in which each socioeconomic sector analyzes and plans in isolation. The public health sector can play a leadership role because of its size and importance, socially and economically. Combining the integrated population health risk management model and the strategic framework for research and policy processes is a long-term collaborative approach to include and influence policy considerations and processes of all socioeconomic sectors.

REFERENCES

1. WHO and Health Canada: *Meeting Report for Ministries of Health to Address Climate Change and Health*. World Health Organization, Geneva, 2002.
2. WHO: *Third Ministerial Conference on Environment and Health, London, UK*. World Health Organization, Geneva, 1999. http://www.who.dk/london99/welcome.htm.
3. UNFCCC: Convention Document. United Nations Framework Convention on Climate Change, 1993. United Nations Framework Convention on Climate Change, Bonn, Germany. www.unfccc.de.
4. McMichael, A.J., Haines, A., Sloof, R., and Kovats, S.: *Climate Change and Human Health*. World Health Organization, Geneva, 1996.
5. Health Canada: *Climate Change and Health & Well-Being: A Policy Primer*. Health Canada, Ottawa, 2001.
6. Health Canada: *Integrating Climate Considerations into Health & Well-Being Business Planning: A Collaborative Strategic Framework*. Health Canada, Ottawa, 2002.

7. WHO: *Health and Environment in Sustainable Development: Five Years After the Earth Summit*. World Health Organization, Geneva, 1997.
8. Corvalan, C., Barten, F., and Zielhuis, G.: Requirements for Successful Environmental Health Decision-Making. In: C. Corvalan, D. Briggs, and G. Zielhuis (eds): *Decision-Making in Environmental Health: From Evidence to Action*. World Health Organization, Geneva, 2000.
9. Health Canada: *National Health Impact and Adaptation Assessment Framework and Tools*. Health Canada, Ottawa, 2002.
10. WHO: *Methods of Assessing Human Health Vulnerability and Public Health Adaptation to Climate Change*. World Health Organization Regional Office for Europe, Copenhagen, 2003.
11. WHO Regional Office for Europe: *First and Second Meeting Reports on Development of Guidelines to Assess the Health Impacts of Climate Change*. Rome, Italy, 2001, 2002. World Health Organization Regional Office for Europe, Copenhagen. http://www.euro.who.int/.
12. Scheraga, J. and Furlow, J.: From Assessment to Policy: Lessons Learned from the U.S. National Assessment. *Hum. Ecol. Risk Assess.* 7 (2001), pp.1227–1246.
13. Hanney, S., Gonzalez-Block, M., Buston, M., and Kogan, M.: The Utilization of Health Research in Policy-Making: Concepts, Examples and Methods of Assessment. In: *Health Research Policy and Systems*. BioMed Central, London, 2003. www.healthpolicy_systems. com/content/1/1/2.
14. De Koning, H.W.: *Setting Environmental Standards, Guidelines for Decision-Making*. World Health Organization, Geneva, 1987.

14 Grounds for convergence

Stephen H. Linder

ABSTRACT This chapter extends the case for sharing useful knowledge (and painful lessons) put forward in earlier chapters, but goes beyond the assumption of parallel interests between the public health and climate change communities to uncover their family resemblances. Three kinds of resemblances appear in certain "neighborhoods" of the two communities: their value commitments, their conceptual premises, and the way they construct their problems. In other words, apart from surface differences, common beliefs and understandings at the level of background assumptions reveal shared roots and the possibility for more explicit convergence. To the extent that these resemblances permit different neighborhoods to view themselves as parts of a larger whole, attention can shift from divisions of labor and informational exchanges to the development of overarching frameworks and common terminology. For our purposes, it offers the beginnings of a rudimentary approach for comparing, and ultimately consolidating, the adaptation interests of both communities. To be sure, the treatment here remain schematic and suggestive, a series of conjectures that deserve closer scrutiny.

1. INTRODUCTION

The chapters in this volume share a commitment to expanding the dialogue between the public health community and the climate change community over adaptation issues, anticipating that more extensive exchanges will prove beneficial to the interests and missions of both. Dialogue between professional research communities seldom comes easily, however, even under United Nations (UN) auspices and with research questions of global import. These two particular communities have different institutional histories and modes of discourse and spring from very different research traditions. And while they can be thought of metaphorically as communities, both are subject to internal divisions and ambiguous or shifting boundaries. To be more precise, it is not public health per se that is involved in adaptation work, but only a small segment of those doing population-based prevention research. Likewise, adaptation occupies a relatively small group of researchers among those addressing climate impacts, and not all of them identify equally with the climate change community. In effect, what we have are not communities so much as neighborhoods. An expanded dialogue in this instance will not only take work, it will depend on the initiative of informal networks of "neighborhood" specialists rather than formal arrangements between professional communities. Still, there are reasons to be optimistic.

Up to now, the dialogue between the prevention and adaptation groups has effectively muted talk of similarities through a division of labor. The prevention researchers assumed responsibility for inquiry into health impacts and for other specialized analyses within the health sector. Meanwhile, adaptation researchers dealt with more generic aspects of adaptation and devised adaptation strategies that spanned sectors. This division has become more mutable, however, as areas of interest find overlap. On the prevention side, greater interest in public health infrastructure [1] generally has coincided with the concerns of adaptation researchers for disaster planning and management. The recent shift in focus among some adaptation researchers to vulnerability and policy studies [2] parallels the prevention group's interest in needs assessment and capacity building. Finally, both neighborhoods pay attention to the critical relationship between vulnerability and development, albeit from different directions. In short, there are signs that each is edging tentatively toward the interests and perspectives of the other. An expanded dialogue, then, seems likely, but may no longer follow the conventional division of labor.

My purpose, here, is to reinforce the case for an expanded dialogue, not by inventorying examples or offering suggestions, but by examining common foundations. In the sections to follow, three kinds of family resemblances between the prevention and adaptation neighborhoods are examined: their value commitments, their conceptual premises, and the difficulties they face. To the extent that the two neighborhoods can be viewed as parts of a larger whole, attention shifts from divisions of labor and informational exchanges to the development of overarching frameworks and common terminology. This would most likely change the face of adaptation work as it is currently known, opening up alternative possibilities for inquiry and program design. Before considering these conjectures on family resemblance, we begin with a refined notion of what an expanded dialogue might entail.

2. A CONVERGENCE DIALOGUE

For some, dialogue connotes tentative efforts at mutual understanding and implies a need for diplomacy to dispel myths and misconceptions, arising from either unfamiliarity or elemental differences in language, professional culture, and points of view. The path here involves initial overtures, patient listening, efforts at translation and symbolic exchange, and reassurance that certain ends may be shared. Persistent differences are to be expected and in some ways, to be celebrated — each community finding value in the knowledge and standing of the other while reinforcing its own. Once successful, jurisdictions might then be renegotiated, accommodation expected, and disputes avoided.

For others, dialogue suggests the pursuit of mutual advantage, each community acknowledging something it needs from the other. At issue here is not understanding or accommodation so much as the instrumental exchange of intelligence. Dialogue, in this instance, relates to the terms of this trade. Selected knowledge can then be transferred and put to each community's own uses. One internal benefit of this more focused dialogue is that it preserves the existing patterns of specialization within each community. In effect, the big picture is divided into multiple parts and assigned to different specialties; these, in turn, can readily be augmented by still other specialties without contesting anyone's standing. Benefits can also accrue from simply expanding the scope of coverage, enlarging jurisdictions, and, at the same time, disarming potential detractors who might otherwise claim neglect of some prized features. Certainly, advantages from these kinds of exchanges would most likely extend to the larger society, enhance research efficiencies, and so on. Still, there may be greater gains from dialogue in yet a third sense, one that begins with an awareness of the points of convergence between the communities rather than exchange or understanding.

A convergence dialogue builds on shared commitments and purposes in pursuit of collaborative enterprise, possibly yielding new arrangements that bridge existing differences. For the skeptic, the question, however, is whether such convergence is even plausible given the manifest differences between the prevention and adaptation neighborhoods. If plausible, then two other questions arise. First, there is the normative question of worth. Should convergence be more deliberately pursued? This would entail an examination of whether a convergence dialogue and its ends are, in some sense, superior to dialogues intent on understanding and mutual advantage. Second, there is the empirical question of identifying what convergence, if any, has already taken place and might be expected to continue. Although different angles are being pursued on adaptation issues, there may well be an increasing homogeneity in points of view and research practices. In effect, the neighborhoods may come to resemble one another in certain respects as a product of their parallel work on adaptation. In the analysis to follow, we pursue the shared perspectives that support plausibility: What are the conceptual grounds that underpin convergence? The normative and empirical questions that arise in making a case for convergence must be put aside for now.

3. OBVIOUS DIFFERENCES AND PARALLELS

From a surface inspection, public health and climate change professionals have remarkably little in common, aside from an interest in certain features of adaptation to variable climatic effects. From this perspective, they represent two distinct communities, separated by differences in priorities, norms, and logics, as well as by more superficial differences in disciplinary affiliations and institutional settings. Their histories, politics, social organization, and institutional affiliations are quite distinct. The diversity of activities and orientations within each defies any simple summarization. Many of the differences appear to be conceptual and methodological, and others, byproducts of the way scholars and practitioners organize themselves into communities of inquiry and practice. When it comes to adaptation, the climate change community has largely framed the relevant problems and ranged widely in its assessments across human and natural systems. The question of adverse effects on health and disease, however, was taken up by the public health community and altered to fit its epidemiologic orientation and commitment to prevention. The result has been specialized analyses that serve health planning and disease management, but do not necessarily connect to the adaptation work of many climate change professionals. Part of the intent behind this edited volume is to communicate those findings back to members of the climate change community in terms that will be both meaningful and useful to them.

Aside from this technical assistance function and support for a dialogue of understanding, there is another rationale behind this collection. Some of the authors find striking parallels between the problems posed by climate adaptation research and the prevention problems addressed by public health. These parallels suggest a basis for knowledge transfers moving in both directions. We might expect a volume, analogous to this one, to emerge from adaptation research that might offer insights relevant to the reduction of health risks or to policies that protect the health of vulnerable populations. If we take this insight about parallels a step further, a wider range of similarities emerge, pointing to areas of potential integration. The two-community idea might then give way to a more inclusive conception that builds on beliefs and understandings that certain neighborhoods hold in common. This notion treats the differences between the larger communities as more apparent than real; it also recasts knowledge sharing more as intramural learning and diffusion rather than as technical assistance or bilateral exchange. Instead of proceeding on a topic-by-topic basis through parallels, however, I take this path a step further and examine whether common beliefs and understandings exist at the level of background assumptions. To be sure, the treatment here remains schematic and suggestive, a series of conjectures that deserve closer scrutiny.

4. CONVERGENCE IN VALUE COMMITMENTS

In this context, value commitments refer to enduring beliefs about priorities and the appropriate ways to attain them. Here, fostering adaptation is the priority of interest, even though both communities have many other priorities as well. Clearly, there are a number of ways to foster adaptation, as reflected in the chapters in this volume, but several more general strategies are apparent and seem to be shared by both groups. The first is to trust in planning processes that combine technical expertise with public participation. The image of technocratic planners managing risks through top-down regulation [3] has given way to a new generation of practices that are intended to engage the public as active participants in ways that build consensus and ensure success in implementation. The second is to frame problems in ways that support systematic interventions as solutions. After considering the priority assigned to adaptation, I discuss each of these strategies in turn.

4.1 The importance of fostering adaptation

Foremost among values in this context is the belief that social adaptation matters. Adaptation generally connotes adjustments of some kind, intended to reduce the severity of consequences. Accordingly, it signals the desirability of awareness and of action over inaction (or adjustment over endurance). Moreover, it suggests priorities for assistance and support based on differential need and capacity to act; some systems are more sensitive to adversity than others, and some are less able to adjust and hence more vulnerable than others. In any event, informed responses to the effects of rapid climate change are assumed to become increasingly important, not only to scholars but also to policymakers and the public, as evidence of adverse consequences accumulates. Whether to act and where to act first are fairly settled within the prevention and adaptation neighborhoods; adaptation is needed now and should be targeted to those areas that are most vulnerable. In the larger communities, there is still debate over timing, targeting, and competing priorities.

4.2 A new role for technical expertise

Technical expertise connotes a scientifically based capability to address complex problems in systematic ways, usually involving well-defined methods and professionalized norms. The faith in expertise that helped characterize the modern era came from an attachment to the premise that the world works in certain ways that can be known, and that one's method made these knowable. And more important, faith was bolstered from a record of success based on this knowledge. When compared to other ways of knowing – common sense, intuition, practical experience, and collective wisdom – expertise attained a position of privilege; its claims were trusted over those from other sources. Since a part of the objectivity that the expert claims comes from detachment, for the most part, experts tended to be outsiders to the problem context. As Scott [4] has carefully shown, what we have then is the modern recipe for top-down prescriptions by experts, perhaps backed by public authorities, who offer solutions to an acquiescent population. The standing of local knowledge and local insight was thereby diminished. Public input on risk was channeled through formal hearings. Solutions, once successful, were assumed to have universal applicability.

For the adaptation neighborhood, dealing with a global issue, none of this seemed to apply: local participants were national governments and planning relied on consensus for its legitimacy. Within the prevention neighborhood, adjustments were made in planning processes to accommodate a changed context of citizen activism but also a record of unsustainable performance in programs that failed to accommodate local knowledge and concerns. Both draw heavily on stakeholder consultation and involvement as a principal planning strategy. Originally designed to keep civil disputes out of court, the involvement of selected stakeholders can generate local input and cooperation and, at the same time, defuse local opposition. When it comes to sorting through the burdens of adaptation, however, stakeholders may have the opportunity to shift these onto the general population. To the extent that elected officials are involved, some accountability may be present to constrain these shifts. For the most part, stakeholders are engaged to voice their stakes and not for divining the public interest. The alternatives, for example, grass-roots community involvement, by comparison are more time consuming and less predictable. Still, working from the bottom up has the advantage of creating diffuse support for interventions, while getting a more nuanced reading of the local context and practices that stakeholders miss. Bottom-up approaches depend on faith in nontechnical forms of knowledge and effectively reduce the privilege that technical expertise traditionally has claimed. The adaptation and prevention neighborhoods are among the more active segments of their respective communities in deploying these approaches.

4.3 The logic of designed solutions

In tension with the consensus-based approach to intervention planning is a design logic that favors certain ways of framing problems and structuring solutions. This logic, formalized in areas ranging from economic decision theory to health services research, attends to expected levels of performance and frames lapses or gaps in performance as problems to be solved. This requires some agreed-upon ends, that is, what the performance is supposed to accomplish is assumed, and focuses attention on the proper choice of means as a solution strategy. Problems naturally entail solutions, at least within ordinary language, and are invariably framed in ways that support a certain repertoire of solutions, or at least narrow down the plausible set. Furthermore, certain features of the problem will be deemed amenable to manipulation and control; these serve as the design parameters for intervening. This logic, then, is both means-oriented and interventionist. Where the construction of a problem remains contestable or the ends are in dispute, there is no solution to be designed; it remains a political conflict to be resolved. While the adaptation and prevention neighborhoods tend to face these kinds of situations, and engage stakeholders directly, elements of design logic still seem to enter into their discourse, especially around the issue of assessments.

Assessments in either case are typically about relative risks and the prospects for their reduction through exposure management. Norms are established through enumerations of similar cases, and performance expectations provide a criterion for judging vulnerability or resilience. The value of what is risked can be expressed in a common metric, say, lost productivity or replacement value, so that the end becomes prevention or minimization. In turn, this justifies and helps direct interventions, so that the right problem is solved. Certainly, this perspective does not rule out a determination that doing nothing is the best course of action; it does, however, make this a less agreeable prospect. There may also be some measure of indeterminacy in the nexus between problem and solution that introduces ambiguity and hence stalls intervention. Both are equally prone to this difficulty. The next section turns to the identification of a set of conceptual commitments that the prevention and adaptation neighborhoods also appear to hold in common.

5. CONVERGENCE IN CONCEPTUAL PREMISES

Conceptual premises, for our purposes, include background assumptions that frame understanding, support expectations of order, and guide the construction of facts and explanations. While there is substantial diversity within each community on these matters (Shockley [5] calls them epistemic styles) we find three premises in common to be of special interest for adaptation work. The first two relate to modes of reasoning and the third to ways of constructing facts. Each is discussed briefly.

5.1 Systems thinking

Although of ancient origin, systems thinking serves as the model of well-ordered, systematic reasoning in modern, technological societies. In many professions, it assumes the status of common sense, background knowledge that remains tacit and lies just beyond reflexive challenge. Furthermore, it operates at a level of abstraction sufficient to permit its extension across most material phenomena, from watersheds and humans to machines and weapons. As a representational strategy, it can vastly simplify a complex reality, highlighting certain key elements and relationships thought to account for observed patterns and dispositions. Besides providing a versatile form of representation, systems thinking also has desirable, heuristic qualities that guide exposition and specification. Boundaries and levels of interaction must be clearly delimited. Key elements must be identified explicitly and their functions and causal connections specified.

Although its advocates claim that all limitations are extrinsic rather than intrinsic, systems thinking entails certain habits of mind that construct reality in a particular way. And while, in the hands of an inspired designer, any representation can be converted into a complex, dynamic, multilevel system, there are features of everyday experience and concern that are routinely absent from systems depictions, especially when humans are involved. In fact, there are equally prominent ways of thinking, modeling the world, and doing systematic inquiry that neither construct phenomena as systems nor frame problems as deficits in their performance. Consider, for example, the traditions of critical realism, pragmatism, and hermeneutics. Certainly, not all of the world fits easily into a systems mold, and not all systems deserve to have their performance improved. Similarly, materialist accounts are not always the most compelling ones, nor do instrumental calculations always capture the most meaningful aspects of value. That aside, systems accounts play a special role in both the public health and the climate change communities. Their language is the language of systems, and their concepts are grounded in systems concerns. Before considering these parallels in more detail, it is useful to consider, if only in schematic form, what systems thinking entails.

To make sense as a system, phenomena need a telos, some orderly pattern to their existence, typically consistent with specifiable functions and, ultimately, purpose. Systems not only do things, they appear to do them for some reason. The admissible reasons, however, tend to be material ones, such as physical necessity, self-maintenance, or self-seeking, and thus are rarely normative, unless of course, the norms favor self-maintenance. Still, a system's actions or inactions are judged and not just observed; but they are judged principally relative to these admissible reasons. In other words, a system's actions are framed principally in instrumental terms. Accordingly, these actions may be deemed normal or abnormal, satisfactory or unsatisfactory, appropriate or inappropriate, and so on. When judgments like these are important, systems do not simply act or react. From the systems thinker's viewpoint, they perform. And when performance falls below expectations, or outside the range of normal functioning, this provides a sufficient rationale for intervention or redesign. Systems thinking, as was argued earlier, complements an engineering perspective that sanctions instrumental interventions as the way problems get solved.

5.2 The logic of performance deficits

For the most part, systems thinking assigns abnormality, disorder, and most forms of disruption to the system's external environment. If, in the absence of a disruption from its environment, a system's actions were to remain abnormal or unpatterned over a period of time, doubt would be cast on whether the system's structure had been properly articulated. Whatever had acted, in those circumstances, could not properly be called a system. When some environmental perturbation does occur, however, the system is expected to react in ways that ultimately protect its functioning. It does so largely through internal changes that reduce the most disruptive effects of these perturbations. From a systems perspective, this is what adaptation entails. Some systems do well at adapting, that is, they are able to maintain or even improve their performance, while others do poorly. This gap in performance, assuming equivalent perturbations, is typically attributed to structural disparities. In effect, the poor performers have internal deficits relative to the capabilities of the better performers. Improving system performance, then, translates into identifying and reducing these deficits.

Deficit logic, with its imperative to bring flagging systems up to standard, finds application in both of our communities, in pursuits ranging from health education to hazard management. A conception of the normal or the well-functioning system serves not only as the yardstick for measuring deficits but also as a template for remediation of the poorly performing. This presumes a basic universality or transferability in the yardstick and template alike; that is, if they work in one system, ceterus paribus, they should work in another. Further, it depends on the presence of similarity or the strength of analogy across like-functioning systems. Idiosyncratic features, for example, in human systems, the unique social conditions that support a given practice, at best play subordinate roles. Hence, once places, sectors, or populations are framed as systems, similarities in their internal features move to the foreground and idiosyncrasies move to the back. Other features, however, are typically assumed away. Among the latter, for example, are conditions sure to undermine self-sustaining functionality from the inside: contested meanings and subjectivities, agonistic relations, power struggles, ideologies, and value conflicts. Instead, attention focuses on the repertoire of responses, or the lack of them, available to counter the disruptions threatening from the outside.

5.3 Parallelism in exposure and risk

For the public health community, disruptions can stem from exposure to agents, say, pathogens or mutagens, capable of inflicting grave harm on human populations. Adaptation in this case involves measures to reduce exposure to these harmful agents, either by fending them off or by protecting the system against the harm that these agents cause. Here, epidemiology deploys systems thinking under the guise of an ecological model that highlights the complex interactions between human systems and their environments in the prevention and spread of disease. Once surveillance and assessment systems determine the presence of harmful agents, protective measures, proven in other systems, can then be deployed. The vulnerable and susceptible are thought to present a range of deficits (some physiological, some social) that increase their risk of harm. Although these groups may be singled out for special treatment, the major emphasis in public health's version of adaptation is on structural responses that cover the entire population.

Since the agents and mechanisms of particular diseases are not always known, there is a close working relationship between the biomedical sciences and public health. Most basic knowledge about the causes and mechanisms of disease comes from the clinic and the laboratory. Both employ elaborate systems metaphors but at increasingly more micro levels. Consequently, systems concepts at the level of populations find reinforcement in systems claims at several other levels. Systems thinking also finds its way into models of health beliefs and lifestyle change, health care delivery, and societal determinants of health. Each differentiates system from environment, establishes expectations based on normed patterns of interaction, and addresses the problem of adapting to external disruptions in functioning. This is not to say that public health is exclusively systems oriented. There are a range of activities, in neighborhoods and communities, for example, where systems thinking gives way to less instrumental forms of communication and interpretation. In these instances, the particular and the idiosyncratic replace the general and the normal as frames of reference.

For the climate change community, disruptions are induced by exposure to climatic changes and events, ranging from gradual warming to episodic heat waves and hurricanes. As with public health, adaptation involves measures to reduce both exposure and the damage that such exposure can induce. Unlike public health, however, the adaptation concerns here extend across a broad range of natural, human, and engineered systems. Although effects on human health and well-being are ultimately of concern, attention focuses more immediately on the stability of ecosystems and on the sustainability of human production systems, especially land and water use, in the face of disruptions from severe weather and climate change. The logic of stable, functioning systems coping with possible disruption is reinforced from the large-scale systems models of the atmospheric and ecological sciences. In place of reductionist sciences undergirding the micro-oriented, biomedical models of public health, we have holistic and integrative models that are macro-oriented, framing how the climate change community understands adaptation.

In place of vulnerable individuals who face higher risks from exposure to a particular disease, we have vulnerable place-based systems whose capacity to respond to risks has somehow been compromised relative to the norm. Healthy systems, like healthy individuals, manifest resilience that helps maintain or restore functioning following exposure. Both communities find the definition and measurement of relative vulnerability a key design criterion for adaptation measures and coping strategies. Similarly, understanding the links between exposure and the risks of damage is central to both. When it comes to human populations and their settlements, the fields of natural hazards and disaster management parallel the role of epidemiology in assessing vulnerability and coping capacity. Again, there is a confidence, characteristic of systems thinking, that predisposing factors for vulnerability and resilience can be discovered and will generalize across systems and populations. At that point, norms for assessing vulnerability will emerge along with a ready diagnostic for singling out deficits in need of remediation. That vulnerabilities and the deficits associated with them constitute problems to be solved relates to the next conceptual feature common to both communities, an engineering perspective.

Although these commitments are interrelated, the set identified above remains far from complete.

6. MUTUAL DIFFICULTIES IN ADAPTATION WORK

6.1 State sponsorship and the paradox of official positions

Both the public health and the climate change communities depend on the support and sponsorship of the State for their continued existence. Although their histories are different, both had their origins and responsibilities defined in law. Both rely on governmental agencies for their standing and much of their financial support. Consequently, both are accountable in complex ways to governmental authorities. There are several implications from this. Although, when it comes to adaptation, the interests of the State typically coincide with the interests of scholars and practitioners, when prospective investments or preventive measures are involved, the State may assume a more conservative stance than the evidence supports. After all, adaptation involves social change, some of which may challenge the support that a given regime enjoys from those with a stake in current social conditions. Likewise, when the State is the client, initiatives may be altered to meet political objectives that fall outside the scope of a typical adaptation assessment. As dependence on the State grows, the State's influence over priorities and the choice of interventions does as well, possibly jeopardizing the detachment necessary for scientific judgment.

The question then arises as to whether there is room for dissent, or whether dependence on the State has a chilling effect on criticism of any sort. The larger danger to inquiry is that a certain form of orthodoxy will come to prevail and will remain protected from the inside and largely immune to criticism from the outside. Certain questions will not get asked. Certain positions will be systematically ignored. The system of rewards and recognition will favor conformity to the received viewpoint. While orthodoxy is a hidden danger in any social organization of inquiry, it takes on special significance when the stakes become social and political and no longer just a matter of internal contention over status and privilege. Paradoxically, the first sign of orthodoxy taking hold is that it begins to appear part of the natural order, a set of background assumptions guiding appropriate inquiry. No one notices. The issue of orthodox thinking disappears. The difficulty is to balance State sponsorship against the need for critical reflection on research practices, priorities, and habits.

6.2 Lessons and place-based casework

Throughout much of this volume, we assume that lessons are worth having; that is to say, that experience in a particular instance has meaning across other instances. Further, there is an implicit asymmetry to that experience, such that more compelling lessons appear to come from failure than from success. In other words, trial and error generates lessons for change and adjustment; trial and success reinforces current practices and perhaps tempts their over-extension. Success is more liable to become the one size that fits all. These claims about experience place one squarely in the pragmatic tradition, at home in a professional culture of cost and regret minimization, piecemeal incremental change, and fallibilism. Failures are transferable, but successes call for corroboration. Both offer lessons, but not equally instructive ones. When it comes to adaptation, both communities place great stock in case studies, typically grounded in a particular context and locale. The unique features serve as cautions, while the experience itself is transferred as instructive guidance.

The asymmetry between cases of success and cases of failure also agrees with a political presumption against collective forms of social intervention: the burdens of proof and persuasion rest with those who claim to have defied the odds of failure, not with those who fail. After the fact, failures are typically easier to apprehend than successes, since they reinforce attachments to the status quo rather than disrupt them. (This echoes the logic of inferential statistics, employing significance levels to raise the bar against chance departures from null hypotheses.) In addition, sorting through evidence of what worked is much more complex in real-world settings, especially where people are the registers of an anticipated change. Asking people whether an intervention worked for them as a promotional device may be one's only purchase on an outcome. Nevertheless, success and failure in social programming may be transient states relative to a constant background of ambiguity. Discernible results, in this context, often seem mixed. At best, some signs of success and of failure may be found, but not all of them intended or attributable to one's program. The difficulty, then, resides in determining not only what case experience can be transferred and what remains unique but also what the results mean. Were the interventions successful? It is often hard to tell.

6.3 Structural conditions and stakeholders

As noted in the earlier section on value commitments, one of the key premises behind adaptation work is that, with proper guidance, adjustments likely to make a difference can be accomplished. This suggests that the correlates of vulnerability, whether to disease or extreme weather events, not only are subject to manipulation but also can be altered in a controlled way to reduce vulnerability. Perhaps the most familiar correlate in adaptation work is infrastructure, where directed investment can be ameliorative and, hence, protective. Additionally, behavioral correlates can be addressed to induce change in people's activities and decision making. There are, however, far less tractable correlates that create difficulties for both communities; these are the upstream factors that reproduce the conditions of vulnerability on an ongoing basis and militate against change. In effect, changes in infrastructure or behavior may represent only sufficient conditions for adaptability, leaving the necessary conditions untouched. The most prominent of these is social inequality.

Much has been written in the climate change community about the connection between vulnerability and the relative lack of social and economic assets. The interesting point here is that public health arrives at the same insight from a different direction. In this instance, vulnerability to premature death is attributed not to the absence of health services (infrastructure) or to lifestyle choices (behavior) but rather to systems of social stratification in a given society. That is to say, social structure matters and can countervail any effort to alter more proximate conditions. At the same time, it represents a feature least amenable to modification, especially by small-scale, expertise-driven interventions. The difficulty for both communities is finding how to intervene in ways that support changes in these inequalities, even if they cannot be addressed directly. A related difficulty arises in attempts to involve local participants in the design and implementation of interventions. The desirability of local participation for reasons ranging from sustainability to cooperation is well rehearsed in both communities. Trouble begins with the choice of participants. If structures of inequality lie at the core of vulnerability, then involving those with the greatest stake in preserving those inequalities, since they reside in the upper strata, represents a categorical error. It is not so much an issue of initiating interventions from the top down or the bottom up; when it comes to the upstream determinants of vulnerability, these designations become less relevant. The question is, How do you build in local participation without reinforcing the social and political order that stands behind vulnerability? Both communities are struggling with this.

7. REPRISE: CONSOLIDATION

Given the range of shared assumptions and difficulties, social adaptation may be better served by consolidating treatments drawn from neighborhoods within the public health and climate change communities. Consolidation, in this context, challenges functional divisions of labor in two ways. First, it draws attention to areas of overlap between these neighborhoods rather than to areas of difference and uniqueness. Here, the emphasis shifts from knowledge transfer and informational exchange to uncovering common assets. Basic insights, for example, may be the same in both, despite terminological and conceptual distinctions. One implication is that these insights may represent more durable or core truth claims; another is that there can be practical advantages from marshalling the institutionalized authority of expertise that each community commands.

Second, for scholars and practitioners within each community, consolidation can legitimate vantage points external to both. To the extent that their central premises are shared, the communities effectively join an overarching body of approaches that offers an alternative point of view for judging strengths and weaknesses as well as similarities and differences. This capacity to "step outside" and look with fresh eyes upon one's mode of inquiry parallels the mental maneuver common to most forms of critique. For our purposes, it offers the beginnings of a rudimentary approach for comparing, and ultimately consolidating, the adaptation interests of the public health and climate change communities.

REFERENCES

1. Institute of Medicine: *Who Will Keep the Public Healthy?* National Academy Press, Washington, DC, 2003.
2. Burton, I., Huq, S., Lim, B., Pilifosova, O., and Shipper, E.L.: From Impacts Assessment to Adaptation Priorities: The Shaping of Adaptation Policy. *Climate Change* 2 (2002), pp.145–159.
3. Fischer, F.: *Technocracy and the Politics of Expertise*. Sage, Newbury Park, CA, USA, 1990.
4. Scott, J.: *Seeing Like a State*. Yale University Press, New Haven, CT, USA, 1998.
5. Shockley, S.: Epistemic Lifestyles in Climate Change Modeling. In: C. Miller and P. Edwards (eds): *Changing the Atmosphere*. MIT Press, Cambridge, MA, USA, 2001, pp.107–133.

15 Lessons learned and insights for adaptation policy

Ian Burton, Joel B. Smith, Kristie L. Ebi, Joel Scheraga[1]

1. INTRODUCTION

The preface of this book laid out two lines of questioning to be addressed in the case studies:

- What modifications to public health systems might be necessary to enhance adaptive capacity to climate variability and change?
- What lessons can be drawn from the history of managing environmental and other threats that can be applied to adaptation to climate variability and change?

We return to these themes in this concluding chapter, summarizing lessons learned from the case studies that may be applicable to all sectors – including public health – likely to be affected by climate change, then suggesting new directions to take climate variability and change more fully into account when formulating strategies, policies, and measures.

2. NEW CHALLENGES POSED BY CLIMATE CHANGE

Climate change is one of an emerging class of environmental problems that are markedly complex for both scientists and policymakers [1]. It may have serious environmental impacts, yet it is a difficult problem for policymakers because it is surrounded by significant scientific uncertainties and could be expensive to address. It will have to be solved globally, even as impacts will need to be dealt with regionally and locally. It has the potential to affect multiple sectors with different intensity across regions and time. And it involves significant interrelationships among physical, social, economic, political, and other factors.

Climate change has more to do with the success of the human enterprise than its failures. Industrial civilization, with its heavy reliance on fossil fuels, inadvertently created the problem of human-induced climate change. The emission of greenhouse gases due to human activities enhances the natural greenhouse effect, resulting in a human influence on climate that augments natural climate variability and change. The same process of scientific and technological development also created interconnected human communities of unprecedented size, with areas of high population density and

1 The views expressed are the author's own and do not represent the official policy of the US Environmental Protection Agency.

large-scale and rapid population movements. The potential for emerging and re-emerging diseases to spread quickly has greatly increased with the advent of modern air travel and the creation of large human settlements. The potential for such diseases to become endemic may increase.

Human societies have experience with adaptation, having responded with varying degrees of success to climate variability and extreme events. In addition, large-scale human migration has exposed people to very different climates, sometimes within a short time frame, as in the case of transoceanic or transcontinental migrations. It is a testament to human ingenuity and adaptability that viable societies have been established and flourished in a very wide range of climatic environments, given sufficient resources. However, anthropogenic climate change presents a substantial new challenge for the field of public health and for other sectors.

Coping with the potential effects of climate change will be a complex and ongoing process requiring action by individuals, communities, governments, and international agencies. One way to respond is through adaptation intended to increase the resilience of societies and ecosystems to change. Rapid social and technological adaptation is becoming an increasingly important challenge because climate is now projected to change at a faster rate. The effectiveness of adaptation will determine the degree to which individuals and communities will be affected by climate-sensitive health outcomes with a changing climate. Failure to invest in adaptation may leave a nation poorly prepared to cope with adverse changes and increase the probability of severe consequences [2]. Therefore, adaptation strategies, policies, and measures need to be considered as part of any larger policy portfolio.

The extent to which society is willing to expend resources to avoid the effects of climate change will depend in part on its perceptions of the risks posed by climate change, the perceived costs of the effort, ability to pay, and how much it is willing to risk possible negative consequences [3,4]. Because resources need to be shared among a variety of public health problems, along with other problems of concern to society, the ideal situation is to direct resources to their highest valued use to do the greatest public good [5]. This, of course, is a social choice, not a scientific decision. Stakeholders may have conflicting desires, and conflict resolution is likely to be required. Policymakers will have to consider issues of equity (e.g., a decision that leads to differential health impacts among different demographic groups), efficiency (e.g., targeting those programs that will yield the greatest improvements to public health), and political feasibility.

Policy makers outside the public health community are faced with similar challenges. Policy makers dealing with multiple social objectives (e.g., elimination of poverty, support for agriculture, promotion of economic growth, protection of cultural resources) and competing stakeholder desires must make difficult choices as they allocate scarce human and financial resources [5]. For this reason, the Intergovernmental Panel on Climate Change (IPCC) suggested that it is useful to view climate change as part of the larger challenge of sustainable development [6]. Climate policies – including those intended to protect public health – can be more effective when consistently embedded within broader strategies designed to make national and regional development paths more sustainable. The impact of climate variability and change, climate policy responses, and associated socioeconomic development will affect the ability of communities to achieve development goals. Conversely, the pursuit of sustainable development goals will affect the opportunities for, and success of, climate policies.

The development of adaptation strategies is complicated by the fact that significant uncertainties exist about the underlying science, the geographic and temporal scales of consequences, and the effectiveness of available response strategies. A major issue facing scientists and policymakers is how to communicate the scientific information that is available, the degree of certainty associated with this information, and the implications of uncertainties for the issues of concern to decision makers, resource managers, and other stakeholders. There are two dangers. One is that in the absence of sufficient and accurate information, or the absence of a compelling case for action, the public may take little interest and not respond effectively. This is one reason why governments and intergovernmental agencies spend time and money to raise public awareness and to inform the public as to potential risks and benefits. The other danger is over-reaction. Once an issue (a risk or a threat) becomes the focus of public attention, there can be a high demand for information, reassurance, and action. Given the complexities and uncertainties associated with climate change, and the aversion of the public to involuntary risks, facilitating an informed and thoughtful public response could become as difficult as managing the risk itself.

We next explore lessons from the history of climate adaptation and public health management in an effort to better understand how scientific knowledge and management practices can be combined to more effectively address the potential impacts of climate change.

3. LESSONS LEARNED FROM THE CASE STUDIES

Evaluating, designing, and implementing an effective adaptation strategy are complex undertakings. Policy makers should not be cavalier about the ease with which adaptation can be achieved, or the expected effectiveness of any policies implemented. Not only must the potential health impacts of climate change and options for responding to these impacts be identified, but barriers to successful adaptation and the means of overcoming such barriers also need to be evaluated [5].

The wide and varied experiences in coping with climate and other risks to human health provide a number of useful lessons on the process of adaptation that may have value to other sectors as they confront the challenges and opportunities of climate change. This section briefly summarizes some of the lessons from the case studies presented in this book that can provide guidance for the design and implementation of effective and efficient adaptation strategies, policies, and measures. The main lessons can be grouped into the following five themes:

- The design of adaptation strategies, policies, and measures must be based on an understanding of the multiple and interacting determinants of disease. Climate change may exacerbate or ameliorate disease determinants, with the possibility that thresholds or nonlinearities may be encountered. Increased understanding of the health impacts of current climate variability is likely to facilitate adaptation to future climatic conditions.
- Multiple political, social, economic, technological, and human factors determine whether adaptation strategies, policies, and measures are effective. Effective interventions are embedded in an understanding of human factors and are tailored to local situations. Also, maladaptation and unintended consequences of interventions can occur in many different ways.

- Surveillance and early warning systems, coupled with effective response capabilities, can reduce current and future vulnerability.
- Adaptation is a process that requires sustained commitment. Because health risks and their drivers change over time, monitoring and evaluating interventions are important. It is prudent to intervene on a small scale to test the effectiveness of solutions before large-scale deployment.
- Collaboration and coordination are required across sectors.

One of the underlying themes throughout the book is the need to establish an institutional structure with the responsibility to maintain vigilance in responding to climate change, and to commit sufficient resources on an ongoing basis to identify and respond to problems. The aphorism of acting in haste and repenting at leisure applies here, with the caveat that the consequences of a less than effective intervention can be severe in terms of human disease and death. There should be skepticism about adopting quick-fix solutions that promise to permanently solve problems. It is not that such solutions cannot happen, but that we must be wary of them because of incomplete understanding of problems and proposed solutions, inadequate research into appropriate approaches to implementation, and the fact that the situation can change and new or old problems can emerge. However, decisions must be made. The question is how best to make decisions despite the existence of uncertainties.

Below, we briefly discuss each main lesson with specific examples drawn from the chapters in this volume.

The design of adaptation strategies, policies, and measures must be based on an understanding of the multiple and interacting determinants of disease. Climate change may exacerbate or ameliorate disease determinants, with the possibility that thresholds or nonlinearities may be encountered. Increased understanding of the health impacts of current climate variability is likely to facilitate adaptation to future climatic conditions.

Climate change is only one of many factors influencing human health and social wellbeing. The ecological, socioeconomic, and climate systems are closely linked, which poses a challenge for policymakers. Before designing and initiating an intervention, it is critical to gather information to understand the multiple and interacting determinants of disease, including both the immediate causes of the disease (e.g., a heat wave can increase morbidity and mortality) and other drivers associated with the disease (e.g., increased urbanization can create an urban heat island that exacerbates the effect of a heat wave). Also, it is important to understand "contextual" determinants of the disease, including the larger societal drivers such as population growth, poverty, and ecosystem changes.

Poverty is an important driver of many diseases

In a number of the case studies, the widespread poverty of the population was a critical factor leading to high vulnerability to disease. One of the outcomes of poverty (although perhaps also a cause) is inadequate public health infrastructure and access to health care. The consequence is a reduced ability to prevent and treat diseases, resulting in situations where controlling the spread of disease is difficult. As is discussed in the case studies on campylobacteriosis (Chapter 4) and arsenic (Chapter 5), poverty is generally associated with inadequate sanitation, which can lead to exposure to water-borne diseases.

The role poverty plays in creating vulnerability to climate change cuts across almost all climate sensitive sectors. Wealth is one of the determinants of adaptive capacity [7]. Substantially reducing or eliminating poverty would greatly enhance the ability of societies to cope with many aspects of climate variability and change, and perhaps dramatically reduce vulnerability. However, reducing poverty alone will not eliminate vulnerability, as demonstrated by the many excess deaths during the European heat wave in 2003. Socioeconomic, political, and cultural considerations are examples of factors other than wealth that also influence vulnerability.

Ecosystem changes can facilitate the emergence and re-emergence of disease

Climate change can change or disrupt natural systems, making it possible for diseases to spread or emerge in areas where they had been limited or had not existed, or for diseases to disappear from areas that are no longer hospitable to the vector or the pathogen. Climate is one of multiple factors that determine the range of a disease; land use change is another major determinant. For example, campylobacteriosis is an "emerging" human gastrointestinal disease, with a dramatic increase in cases in the last few decades. In New Zealand, natural vegetation was replaced with pastoral farming. This increased sources for disease (animal waste) and reduced the ability of natural vegetation to remove the wastes from runoff. This has resulted in contamination of half of New Zealand's rivers and streams (see Chapter 4). In another example, Githeko and Schiff (Chapter 7) report that land use changes in Africa, such as deforestation in Tanzania and cultivation of swamps in Uganda, have been associated with the spread of malaria. Not all changes in land use have adverse consequences for human health; draining wetlands near human settlements can reduce habitats for disease-carrying mosquitoes.

This lesson also applies to sectors other than public health. For example, ecosystem changes may result in the spread of pests that can affect agricultural output. Using pesticides to control these pests can, in turn, increase the costs of agricultural production. And if climate change increases the frequency and intensity of rainfall events, there could be more runoff of pesticides from farms into rivers and streams, affecting water quality and wildlife. It is therefore critical that the effects of ecosystem changes be identified and that our understanding of the mechanisms through which these changes occur be improved.

Increased understanding of the health impacts of current climate variability is likely to facilitate adaptation to future climatic conditions

Climate change can be expected to change the location, frequency, and intensity of known health risks. Part of the preparation to address the potential impacts of climate change is to improve our understanding of risks from the current climatic conditions, such as those posed by heat waves (Chapter 8) and other extreme weather events (Chapter 9). In addition, as a number of chapters in this book point out, better understanding of potential risks from climate variability and change, and of the populations vulnerable to those risks, can help health authorities target surveillance, response, and other programs.

The importance of understanding risks from current climate and how the risks could change is borne out in other climate sensitive sectors. For example, the risks posed by climate variability to water resources are not new to societies, but changes in the frequency and intensity of precipitation might require modification of current water management practices.

Multiple political, social, economic, technological, and human factors determine whether adaptation strategies, policies, and measures are effective. Effective interventions are embedded in an understanding of human factors and are tailored to local situations. Also, maladaptation and unintended consequences of interventions can occur in many different ways.

Many of the chapters in this book demonstrate the need to design appropriate interventions tailored to local circumstances. Differences in culture, education, knowledge, availability and affordability of technology, and other factors mean that a "one size fits all" approach is likely to fail. The specific lessons learned include the following:

Changing the behavior of individuals may be difficult

Adaptations that require individuals to change their behavior may be difficult to successfully implement. Githeko and Shiff (Chapter 7) note that despite the demonstration that use of treated bed nets reduces the risk of malaria, their acceptability and affordability have been a barrier to adoption. Lucas and McMichael (Chapter 11) note that convincing individuals to change their behavior (reducing exposure to harmful UVB rays from the sun) can be challenging if the change involves inconvenience or increased cost, although information campaigns to reduce UVB exposure have had some success. (They note, however, that maladaptation could occur if the campaigns reduce sun exposure too much, limiting UVR exposure necessary to manufacture vitamin D.) In Chapter 12, Linder discusses the difficulties that can be encountered when encouraging or even coercing individuals to change behavior.

The fact that individuals do not readily change their behavior is a challenge for all sectors. Successful campaigns to change individual behavior such as anti-smoking or anti-drunk driving efforts often started with strong stakeholder involvement (such as Mothers Against Drunk Driving), took years to succeed, and required not just public appeals but also incentives such as higher cigarette prices or stricter drunk driving laws. This suggests that efforts to change individual behavior need to be undertaken, sometimes with incentives, in a sustained manner involving messages targeted to specific population subgroups. Better understanding of what has worked and why (and what has not worked and why not) can facilitate the development of more effective intervention programs.

Be wary of "simple solutions", particularly technological responses that have not adequately accounted for potential behavioral responses and other socioeconomic, cultural, and human factors

It can be tempting to adopt simple solutions that sound as if they would readily solve a problem, particularly when the proposed solution is a technological one. The tubewells installed to reduce morbidity and mortality from diarrheal diseases, particularly among children (Chapter 5), is a good example of such a solution. It was probably quite tempting to latch on to this technological solution without investigating such complicated matters as how people actually used water, their beliefs about water and disease, whether or not people would accept tubewells, and whether the solution would create other problems.

Effective interventions must be tailored to local situations

The discussion in Chapter 5 on arsenic in Bangladesh has a number of valuable lessons; among them is the importance of designing an intervention so that it is appropriate for the local situation. This is particularly the case for technological solutions. If the remedy is incompatible with local culture, it is unlikely to be adopted. Other factors also need to be considered, such as the local availability of resources; preferences that are a function of culture, traditions, religion, and other factors; the relative size of a vulnerable population (e.g., more elderly in one region than in another); equity considerations; and political feasibility. Also, as noted earlier, different stakeholders have different, and perhaps competing, objectives, which can make the choice of specific adaptation measures difficult.

Githeko and Shiff (Chapter 7) write about malaria:

> There has been a very high expectation of a magic bullet coming out of hi-tech laboratories (e.g., vaccines, genetically modified mosquitoes) and as a result, African malariologists have failed to optimize the use of existing tools such as selective indoor spraying of houses at the foci of transmission in the highlands. Sufficient knowledge exists to modify houses to make them less accessible to mosquito vectors, yet this knowledge has been completely unutilized .... It is our view that Africa can develop some simple homegrown and scientifically sound solutions for malaria transmission.

Tubewells were used in Bangladesh as a solution to polluted surface water supplies. However, many people refused to obtain water from them. This highlights the importance, not just of considering the culture one is operating within, but also of working with regional and local stakeholders in the design of interventions (see Chapter 9). This may mean more than simply consulting stakeholders, but empowering them to make decisions about appropriate responses.

This is an important message for adaptation in all climate-sensitive sectors. Whether they are responding to sea level rise, change in crop yields, loss of biodiversity, or many of the other impacts of climate change, adaptation strategies that do not consider local circumstances can lead to lack of use, misapplication, and unintended consequences.

Maladaptation and unintended consequences can occur in many different ways

Maladaptation occurs when current practices or behavior increase rather than decrease climate risks or risks from other stressors [8]. Maladaptation also occurs when interventions are implemented and scarce resources are used when adaptation was not necessary.

Addressing maladaptation where it already has occurred can produce a "win-win" outcome: the interventions can address a situation that is a problem under current climate and can improve the ability to adapt to the potential impacts of climate change. For example, Weinstein and Woodward note in Chapter 4 that maladaptation arose as a consequence of past management of water catchments, and that improving management, including revegetation with an emphasis on planting native species, would have multiple benefits, independent of climate trajectories, on reducing the prevalence of campylobacteriosis in New Zealand. Some of these adaptations would produce immediate benefits, such as reduced runoff and improved water quality.

The potential for maladaptation in other sectors has been studied [7,8]. One approach suggests that existing practices that yield unintended consequences (i.e., existing maladaptation) should be addressed now because they exacerbate or introduce risks, and there would be immediate benefits if corrected. Furthermore, such corrections will improve society's ability to adapt to future climate change (although additional adaptations may well be needed). Even if maladaptation is not occurring under current climate, it could arise over time with changing conditions. Systems that are well adapted now may become more vulnerable in the future as the climate changes. For this reason, ongoing monitoring and evaluation of implemented adaptation measures is needed.

It is important to understand the political, social, economic, and other forces that may affect implementation

Scientific information is only one input into policy decisions. Selection and implementation of adaptation strategies, policies, and measures are done within a political, social, and economic context. Understanding this context, and incorporating modifications to address this context, can facilitate the acceptability of desirable adaptations.

Among the major ethical issues facing public health is the matter of individual rights versus the good of society as a whole. One situation is where an individual can pose a risk to society as a whole (see Chapter 1). The typical public health response is to place the welfare of society above that of the individual. For example, the freedom of movement of an individual carrying an infectious disease can be limited to reduce the risk of spreading the contagion. Weinstein and Woodward (Chapter 4) raise another ethical dilemma: the problem of investing relatively large sums of money to protect a few individuals. They point out that drinking water quality problems may be greatest in poor isolated communities, where solutions on a per capita basis are expensive. Considerations of equity may demand investment in such cost-ineffective solutions.

Such ethical dilemmas arise in other sectors addressing climate variability and climate change. Certain individuals or subpopulations may have greater exposure to climate risks than society as a whole. A dilemma is whether it is better to invest large amounts of resources to protect the lifestyles of those individuals or better to take other measures such as to relocate them (or do nothing), which also has monetary costs.

Surveillance and early warning systems, coupled with effective response capabilities, can reduce current and future vulnerability.

Surveillance is an important resource for identifying risks

In Chapter 10, Wilson and Anker state, "Disease and weather surveillance are critical components of efforts to measure, recognize, evaluate, anticipate, and respond to climate effects on health". Surveillance systems are designed to provide early intelligence on the re-emergence of health risks at specific locations in time for effective responses to be mounted. Climate change can be expected to facilitate changes in the geographic range, seasonality, and duration of a variety of health outcomes such as malaria, dengue fever, and other vector-borne diseases (see Chapter 3). It may also introduce new risks.

Appropriately designed surveillance systems can identify risks to health, whether they are the result of climate change or other driving forces. However, to capitalize on this possibility, conventional surveillance systems need to account for and anticipate the potential effects of climate change. Surveillance systems will need to be implemented in locations where changes in weather and climate may foster the spread of climate-sensitive diseases and vectors into new regions. (Similarly, information about the timing, location, and potential severity of possible changes in climate-sensitive sectors other than public health can be used to prioritize the placement of surveillance and early warning systems to alert populations to changing risk conditions.) Increased understanding is needed of how to design these systems where there is limited understanding of the interactions of climate, ecosystems, and infectious diseases.

Early warning systems can reduce current and future vulnerability

Whereas surveillance systems are designed to detect and investigate disease outbreaks as they occur, early warning systems are designed to alert the population and relevant authorities that a disease outbreak is expected. Early warning systems can be very effective in preventing deaths, diseases, and injuries. One component of an early warning system, pointed out by Woodruff in Chapter 6, is disease prediction. The effectiveness of the prediction component depends on:

- an understanding of the mechanisms of disease transmission;
- reliable and up-to-date information on exposures and health outcomes; and
- a model that is accurate, specific, and timely.

As Woodruff discusses, knowledge of disease transmission requires an understanding of the causes of the disease. Being able to predict disease outbreaks requires good historical data on which to base the prediction, as well as reliable and up-to-date monitoring data. The model needs to be sufficiently accurate and valid for the use to which it is put. Models for early warning systems typically work best for episodic diseases rather than endemic ones.

Early warning is also important for extreme climate events. Street et al. in Chapter 9 point out how early warning can be critical for avoiding natural disasters such as hurricanes. Kovats and Koppe in Chapter 8 discuss the use of heat health warning systems to predict extreme heat waves that will pose a risk to health.

Risk prediction must be coupled with adequate response capabilities

An early warning of a problem will be inadequate if it is not accompanied by an effective response capability, including a specific intervention plan. In Chapter 6 Woodruff states, "Excellent predictive ability is of little value if the response capacity is not equally as good". Once a warning is given, the public health system must have the capability to take effective measures to reduce the predicted risks. Kovats and Koppe point out in Chapter 8 that the combination of a heat watch warning system and a response plan to reduce the exposure of vulnerable population groups to extreme heat led to a substantial reduction in deaths from extreme heat in Chicago in 1999. Street et al. in Chapter 9 discuss the importance of including communities in planning responses to extreme weather events.

Woodruff identifies the following necessary components of a response plan (applicable to all climate-sensitive sectors):

- details on thresholds for action;
- available and effective interventions;
- an economic assessment of the benefit and affordability of the system;
- a communication strategy; and
- involvement of all relevant stakeholders in the process.

Adaptation is a process that requires sustained commitment. Because health risks and their drivers change over time, monitoring and evaluation of interventions are important. It is prudent to intervene on a small scale to test the effectiveness of solutions before large-scale deployment.

This is a lesson of unwavering vigilance. Societies must not become complacent and assume that problems are solved forever, but must maintain efforts to monitor and evaluate the problems and their solutions.

Health risks and their drivers change over time

Effective public health management must address changes in drivers or the disease itself in a prompt and efficient manner. The burden of diseases such as malaria can increase with changes such as the development of drug-resistant parasites and pesticide-resistant vectors (see Chapters 3 and 7). Changing population demographics, particularly increasing numbers of individuals over age 65, suggest that there will be increased vulnerability to a number of the potential health impacts of climate change (e.g., death due to heat stress as the frequency and intensity of heat waves increase). Changes in land use, such as conversion of forest to rice cultivation, can create favorable conditions for disease vectors.

Socioeconomic conditions can change, resulting in the easing, exacerbation, or creation of new health problems. For example, Weinstein and Woodward in Chapter 4 identify increasing social inequality as a possible cause of greater exposure to water-borne diseases. In contrast, improved social conditions can result in more favorable conditions to implementing actions to reduce a disease risk [9].

A sustained commitment is required

In their analysis of the reemergence of yellow fever, dengue/dengue hemorrhagic fever, and malaria, particularly in the Americas, Gubler and Wilson (Chapter 3) make a strong case for the importance of a sustained commitment. Mosquito control programs in the 1940s and 1950s resulted in effective control of the diseases in the Americas because mosquitoes carrying the diseases were controlled. This success unfortunately bred complacency and control measures were relaxed (and funds were diverted to other activities). In the 1960s and 1970s, these diseases reemerged.

Gubler and Wilson state that solid political will, adequate resources devoted to adaptation, and institutional commitments are needed to control health problems over the long term. One high-priority need is for adequate trained personnel and resources to be made available to countries where diseases are endemic and investments need to be maintained. The level of investment must address not only the burden of disease in that population but also the potential consequences if the disease moves across national boundaries (as was highlighted by the SARS outbreak of 2003).

A related lesson is that many diseases cannot be eliminated, but they can be controlled. Both Gubler and Wilson (Chapter 3) and Githeko and Schiff (Chapter 7) conclude that we should strive to control and limit diseases such as malaria rather than try to eliminate them. Gubler and Wilson state that elimination of *Anopheles* and *Aedes* mosquitoes is technologically impossible. A vaccine for malaria or dengue also does not exist. They point out the need to maintain the political will to continue disease control programs even when it appears that a disease is under control.

A strategy of eternal vigilance is difficult to follow. Once a problem is sufficiently reduced in scope that it fades from public consciousness, other problems emerge to take its place. We can assume that, in addition to having limited resources, there is a public attention budget that focuses on only a limited number of problems at one time. If a problem has insufficient immediate risk to keep it prominent in the public and political eye, other problems are likely to displace it and to attract financial support.

Intervention on a small scale is warranted to test the effectiveness of a "solution"

An intervention, particularly something new, should be tested on a small scale. Testing can identify potential implementation problems or unintended consequences (e.g., for a new technology). Testing of tubewells in Bangladesh might have identified the high arsenic levels in groundwater, in addition to the difficulties in getting people to properly use the wells. The lessons learned during the testing period can then be applied when an intervention is scaled up.

Monitoring and evaluation of interventions is important

Many of the chapters in this book point out the importance of monitoring and evaluation of interventions. We cannot assume that the interventions implemented will work and will do so in perpetuity. The only way to find out if solutions are working is to monitor the health outcomes. In addition, monitoring and evaluation can identify unanticipated problems. As is noted in Chapter 10 on surveillance, we can count on surprises in a rapidly changing world. Monitoring can identify where solutions have not worked and where further interventions are needed. Evaluation can identify why particular solutions did and did not work. Both monitoring and evaluation are needed to ensure interventions continue to have the intended effect.

Collaboration and coordination are required across sectors

Strategies, policies, and measures for adaptation to climate variability and change in multiple sectors, including public health, are increasingly inseparable. If strong collaboration and coordination do not develop across different sectors and communities, then effective and efficient adaptation to climate change will be less likely, with needless human suffering. Most science is currently conducted within narrow disciplines. Multidisciplinary approaches are required to understand the potential risks and benefits of climate change, and the possible response strategies to protect and enhance human health and well-being in the face of what the future will bring. Taking a more systems-based approach allows the risks from climate change to be put into perspective with other drivers of disease, thus facilitating the identification of the most effective and efficient adaptation options.

4. INSIGHTS FOR POLICY

We turn to the question of what modifications to public health policies and systems might be necessary to enhance adaptive capacity to climate variability and change. To successfully modify public health policies and systems, several prerequisites must exist: an awareness that a problem exists; a sense that the problem matters; an understanding of the causes of the problem; an ability to influence the outcome; and political will to influence the problem [10]. These prerequisites for action can be addressed by education, research, and institutional development.

Public health officials and researchers must be educated to understand that climate is a risk factor for a range of diseases and that increasing resilience to current climate variability will be a step toward increasing adaptive capacity to future change. The climate is always changing, but it is projected to change at an increasingly more rapid rate in the future. This change may be accompanied by increasing climate variability. The resulting increase in floods, droughts, heat waves, etc., would strain the ability of public health systems worldwide to cope unless proactive policies are formulated and implemented.

Research is needed to better understand the relationships between climate and health, and how these relationships could be effectively modified or influenced. Such investments will provide the policy-relevant information needed by decision makers to formulate and implement appropriate response options. In addition, to design effective adaptation options, research is needed to understand which public health interventions have and have not worked in the past, and why. Public health can use the lessons learned to inform the design of future interventions and the processes by which they are implemented.

Because climate change will continue for the foreseeable future and because adaptation to these changes will be an ongoing process, institutions need to be developed with the responsibility for consistently and effectively factoring climate variability and change into all policies. Active management of the risks and benefits of climate change needs to be mainstreamed into the design and implementation of disease control strategies and policies across the institutions and agencies responsible for maintaining and improving population health. Some risk management activities will be extensions of current public health approaches that can be designed within existing operational responsibilities. For example, the design and implementation of early warning systems for climate-sensitive diseases is a win-win strategy that would reduce the current burden of disease while increasing adaptive capacity if the incidence of disease increases. Other risk management activities will require modifications to current public health practices or systems. For example, a number of studies have suggested that the changing temperature and precipitation patterns projected to occur with climate change could increase the current range of epidemic malaria in Africa [6]. Development of closer collaboration between public health and meteorological agencies is needed in the placement of long-term monitoring stations to ensure that prevention activities are proactively implemented.

The precise modifications needed to public health policies will depend on factors such as:

- the health outcome;
- the expected severity of the climate impact;
- the location;
- the timing of when the impact is likely to occur;
- the effectiveness of measures in place to cope with the impact;
- other stressors that could increase or decrease resilience to impacts;
- the capacity of the population to cope with the impact; and
- who is expected to take action.

5. CONCLUSION

Some have argued that the existence of scientific uncertainties precludes policy makers from taking action today in anticipation of climate change. This is not true. In fact, policy makers, resource managers, and other stakeholders make decisions every day despite the existence of uncertainties. The outcomes of these decisions may be affected by climate change. Or the decisions may foreclose future opportunities to adapt to climate change.

Public health has a long history of effectively intervening to reduce risks to the health of individuals and communities. The increasing understanding of the ways that weather and climate variability are likely to change over the coming decades offers an opportunity to proactively design and implement adaptation strategies, policies, and measures to reduce projected future burdens of climate sensitive diseases.

REFERENCES

1. Scheraga, J.D. and Smith, A.E.: Environmental Policy Assessment in the 1990s. *Forum Soc. Econ.* 20 (1990), pp.33–39.
2. Smith, J.B. and Lenhart, S.S.: Climate Change Adaptation Policy Options. *Climate Res.* 6 (1996), pp.193–201.
3. NAS: *Policy Implications of Greenhouse Warming: Mitigation, Adaptation, and the Science Base.* National Academy of Sciences. National Academy Press, Washington, DC, 1992.

4. OTA: *Preparing for an Uncertain Climate*. Office of Technology Assessment. U.S. Government Printing Office, Washington, DC, 1993.
5. Scheraga, J.D., Ebi, K.L., Furlow, J., and Moreno, A.R.: From Science to Policy: Developing Responses to Climate Change. In: A.J. McMichael, D. Campbell-Lendrum, C.F. Corvalan, K.L. Ebi, A. Githeko, J.D. Scheraga, and A. Woodward (eds): *Climate Change and Human Health: Risks and Responses*. WHO/WMO/UNEP, 2003, pp.237–266.
6. IPCC: *Climate Change 2001. Impacts, Adaptation, and Vulnerability. Contribution of Working Group II to the Third Assessment Report of the Intergovernmental Panel on Climate Change*. Intergovernmental Panel on Climate Change. Cambridge University Press, Cambridge, UK, 2001.
7. Smit, B., Pilifosova, O., Burton, I., Challenger, B., Huq, S., Klein, R., and Yohe, G.: Adaptation to Climate Change in the Context of Sustainable Development and Equity. In: IPCC: *Climate Change 2001: Impacts, Adaptation, and Vulnerability*. Intergovernmental Panel on Climate Change. Cambridge University Press, Cambridge, UK, 2001, pp.877–912.
8. Burton, I.: The Growth of Adaptation Capacity: Practice and Policy. In: J.B. Smith, N. Bhatti, G. Menzhulin, R. Benioff, M. Budyko, M. Campos, B. Jallow, and F. Rijsberman (eds): *Adapting to Climate Change*. Springer, New York, 1996.
9. Tol, R.S.J. and Dowlatabadi, H.: Vector-Borne Diseases, Development, and Climate Change. *Integr. Environ. Assess.* 2 (2002), pp.73–181.
10. Last, J.M.: *Public Health and Human Ecology, 2nd Edition*. Prentice Hall International, London, 1998.